Probing the Structure of Quantum Mechanics

Nonlinearity

Nonlocality

Computation

Axiomatics

Probing the Structure of Quantum Mechanics

Nonlinearity

Nonlocality

Computation

Axiomatics

Brussels, Belgium June 2000

Editors

Diederik Aerts
Brussels Free University, Belgium

Marek Czachor
Technical University of Gdansk, Poland

Thomas Durt
Brussels Free University, Belgium

World Scientific
New Jersey • London • Singapore • Hong Kong

Published by

World Scientific Publishing Co. Pte. Ltd.

P O Box 128, Farrer Road, Singapore 912805

USA office: Suite 1B, 1060 Main Street, River Edge, NJ 07661

UK office: 57 Shelton Street, Covent Garden, London WC2H 9HE

British Library Cataloguing-in-Publication Data
A catalogue record for this book is available from the British Library.

PROBING THE STRUCTURE OF QUANTUM MECHANICS
Nonlinearity, Nonlocality, Computation and Axiomatics

ISBN 981-02-4847-4

Printed in Singapore by World Scientific Printers (S) Pte Ltd

CONTENTS

Probing the Structure of Quantum Mechanics
D. Aerts, M. Czachor and T. Durt 1

The Linearity of Quantum Mechanics at Stake: The Description of Separated Quantum Entities
D. Aerts and F. Valckenborgh 20

Linearity and Compound Physical Systems: The Case of Two Separated Spin 1/2 Entities
D. Aerts and F. Valckenborgh 47

Being and Change: Foundations of a Realistic Operational Formalism
D. Aerts 71

The Classical Limit of the Lattice-Theoretical Orthocomplementation in the Framework of the Hidden-Measurement Approach
T. Durt and B. D'Hooghe 111

State Property Systems and Closure Spaces: Extracting the Classical en Non-Classical Parts
D. Aerts and D. Deses 130

Hidden Measurements from Contextual Axiomatics
S. Aerts 149

High Energy Approaches to Low Energy Phenomena in Astrophysics
S. M. Austin 73

Memory Effects in Atomic Interferometry: A Negative Result
T. Durt, J. Baudon, R. Mathevet, J. Robert and B. Viaris de Lesegno 165

Reality and Probability: Introducing a New Type of Probability Calculus
D. Aerts 205

Quantum Computation: Towards the Construction of a 'Between Quantum and Classical Computer'
D. Aerts and B. D'Hooghe 230

Buckley-Siler Connectives for Quantum Logics of Fuzzy Sets
J. Pykacz and B. D'Hooghe 248

Some Notes on Aerts' Interpretation of the EPR-Paradox and the Violation of Bell-Inequalities
W. Christiaens 259

Quantum Cryptographic Encryption in Three Complementary Bases Through a Mach-Zehnder Set Up
T. Durt and B. Nagler 287

Quantum Cryptography Without Quantum Uncertainties
T. Durt 296

How to Construct Darboux-Invariant Equations of von Neumann Type
J. L. Cieśliński 324

Darboux-Integrable Equations with Non-Abelian Nonlinearities
N. V. Ustinov and M. Czachor 335

Dressing Chain Equations Associated with Difference Soliton Systems
S. Leble 354

Covariance Approach to the Free Photon Field
M. Kuna and J. Naudts 368

PROBING THE STRUCTURE OF QUANTUM MECHANICS

DIEDERIK AERTS
Center Leo Apostel (CLEA) and Foundations of the Exact Sciences (FUND), Brussels Free University, Krijgskundestraat 33, 1160 Brussels, Belgium
E-mail: diraerts@vub.ac.be

MAREK CZACHOR
Katedra Fizyki Teoretycznej i Metod Matematycznych Politechnika Gdańska, ul. Narutowicza 11/12, 80-952 Gdańsk, Poland
E-mail: mczachor@pg.gda.pl

THOMAS DURT
Foundations of the Exact Sciences (FUND) and Applied Physics and Photonics (TONA), Brussels Free University, Pleinlaan 2, 1050 Brussels, Belgium
E-mail: thomdurt@vub.ac.be

We believe that in the decades to come quantum theory will play an increasingly important role for many different fields. One of the reasons is that technology aims at manipulating and controlling information and energy in ever smaller regions of space and windows of time. As a consequence the behavior of the entities to manipulate and control will become more and more quantum. This means not only spectacular advances of new techniques and outlooks on revolutionary applications, but also a constant stress and attention on the theory of quantum mechanics itself.

It is well known that quantum mechanics has been scrutinized in all kind of ways during the past decades, but that still a lot of conceptual problems remain. The problems of standard quantum mechanics are however not only of a conceptual nature. Also the formal mathematical structure of quantum mechanics has been investigated with the aim to make the theory more operational and to found the basic concepts in direct correspondence with what happens in the laboratory. Such an operationally founded quantum mechanics may soon become of great value because of the technological advances, that will demand a more straightforward connection between the theory and the type of manipulations and control to be executed in the laboratory.

Although operational quantum mechanics is in full development, we must admit that the time has not yet come for it to function as a 'better to apply and more easy to use' theory for experimentation. The reason is that the operational quantum structures that have been elaborated, while carefully but boldly aiming at physical clearness and transparency, stumble upon a lot of problems of purely technical mathematical nature. Quantum mechanics is not only conceptually a very difficult theory, it entails also a very sophisticated mathematical apparatus.

It becomes even more and more clear that both dimensions of difficulty, the conceptual one and the mathematical structural one, are linked in a profound way. It has been shown that some of the deep conceptual problems of quantum mechanics – the so called quantum paradoxes – that make it impossible for standard quantum mechanics to be a straightforward operational theory, are due to structural shortcomings of the mathematical apparatus of standard quantum mechanics. Let us make clear by an analogy what we mean by the last statement. Suppose that history had evolved differently for classical mechanics and that Hamiltonian mechanics had been formulated without Newton mechanics. If physicists then had to apply Hamiltonian mechanics to the whole of the domain of reality where classical physics is used, strange conceptual problems would certainly have arrived due to the too limited mathematical structure of Hamiltonian mechanics to cope with all type of situations encountered in the macroscopic physical world. For example, the simple situation of a sphere rolling on a plane, which entails a nonholonomic constraint, would not have been possible to describe by the classical physicists only being able to make use of Hamiltonian theory, because Hamiltonian physics cannot describe nonholonomic constraints. If then, in a stubborn way, and because no other theory was available, the classical physicist would have persisted in his or her attempt to deliver a Hamiltonian description for the sphere rolling on a plane, conceptual paradoxes would probably have appeared as a consequence. Certainly if we push the analogy a litter further and also suppose that the classical physicist would only have access to the situation of the rolling sphere by means of sophisticated experiments, and not with his or her eyes and human intuition.

Let us sketch briefly the problem of standard quantum mechanics that we refer to in the analogy of the foregoing paragraph – the deficiency of the mathematical apparatus of standard quantum mechanics – and that part of articles of this book focus on, and where the four mentioned concepts of the subtitle of the book, *nonlinearity*, *nonlocality*, *computation* and *axiomatics*, play a role.

The first approach that is put forward in the book, and that reveals aspects of the mentioned deficiency of standard quantum mechanics, comes from a line of research that is active for some decades, and where it has been shown that the limitations of the mathematical structure of standard quantum mechanics are in great part at the origin of the problems related to the situation of compound quantum entities, hence the Einstein Podolsky Rosen (EPR) type of situations.

The research that we refer to in this first approach is undertaken within what is called the axiomatic approach to standard quantum mechanics. That is why we will call it the 'axiomatic approach'. Standard quantum mechanics is retrieved in this approach by a set of five axioms formulated on the very general structure of a lattice. First some general problems that seemed to be mostly of a technical nature, and with no clear physical significance, were discovered in the attempt to retrieve the standard tensor product procedure to describe the compound entity consisting of two sub entities from a coupling procedure on the level of the axiomatic approach [1,2,3,4,5]. A full blow was given to the standard quantum mechanics formalism when it was proven that one of the most simple of all situations, namely the situation of a compound entity that consists of two 'separated' quantum entities, cannot be described by standard axiomatics. And more specifically it was shown that two of the five axioms that lead to standard quantum mechanics are not satisfied for the situation of a compound entity consisting of two separated quantum entities [6,7,8,9]. Moreover one of these failing axioms is equivalent to the linearity of the state space of the physical entity under consideration. This means that if a nonlinear generalization of standard quantum mechanics would be elaborated, a completely different approach for the Einstein Podolsky Rosen paradox like situations could be worked out. Rapidly other results in axiomatic quantum mechanics confirmed this finding. All indicated a fundamental difficulty with the 'linearity axiom' in relation with the description of the compound entity consisting of two quantum entities [10,11,12].

The second approach that is treated in the book, although from a completely different direction, hits upon the same problem as the one we mentioned in the foregoing paragraphs. This approach comes from a direct attempt to built a nonlinear quantum mechanics, and therefore we will call it the 'nonlinearity approach'. Thinking of quantum mechanics as a limiting case of a more fundamental nonlinear theory one encounters difficulties which are both conceptual and technical. The conceptual problems were from the very beginning deeply related to the question of how to treat separated entities and how to discuss nonlinear dynamics of entangled states.

It seems that the link between separability conditions and the possible forms of nonlinearities was noticed for the first time in [13], the same year as [1] where the problem was identified in the axiomatic approach. The assumption that a nonlinear correction to the Schrödinger equation should be additive on product states led the authors of [13] to the conclusion that only the logarithmic term is acceptable. One of the drawbacks of the analysis given in [13] was that the discussion was limited to product states. The question of entangled states appeared in this context for the first time in [14] with the conclusion that difficulties may be fundamental and hard to overcome. The point was further elaborated, in a rather general setting, in [15]. The explicit definition of nonlinear evolution of entangled states proposed in [16] was quickly shown to lead to unphysical influences between separated systems [17,18,19,20] (for a recent discussion cf. [21]).

However, once the difficulty was formulated for concrete and explicit models it became clear that the problem is more subtle and that some implicit assumptions may be of crucial importance. A part of the way out was suggested in the important paper [22]. From the perspective of the past decade we can say that the first step of the solution is to correctly identify the one-particle space of pure states. The difficulty is always present if one insists on representation of pure states in terms of rays or vectors in a Hilbert space. This appears justified if one works at the level of Schrödinger equations. Still, we know that the Schrödinger equation can be replaced by the von Neumann equation for one dimensional projectors. The advantage of such a viewpoint is that the von Neumann equation can describe evolution of entangled subsystems whereas the same cannot be said of the Schrödinger equation. The solution proposed in [22] was to start with nonlinear evolution equations appropriately defined for density matrices and recover Schrödinger-type evolutions by restricting the dynamics to projectors. One can say that reduced density matrices obtained via partial tracing from projectors on entangled states have to be treated as pure states. Such states are 'pure' in the sense of being 'fully quantum', a point of view which is in a striking agreement with the quantum axiomatic results discussed in the first few papers of this book. The opinion that in nonlinear quantum mechanics one has to distinguish between two types of 'mixtures' was expressed already in 1991 in [24]. The density-matrix perspective was further elaborated by Jordan in [23] who explicitly constructed the dynamics in terms of nonlinear von Neumann equations. The originality of [23] was not in the very form of the evolution equations, which were discussed in the context of generalizations of quantum mechanics earlier in [24], but in the link of such equations to the separability problems for entangled states. Further analysis showed that a consistent application of Polchinski-type multi-particle extensions leads to equations which *look* nonlocal in configuration space but remain physically local in the physical space [25]. Explicit solutions of such physically local equations allowed one to understand various subtle interplays between tensor product structures, nonlinearity, and locality on one hand, and complete positivity of nonlinear maps on the other [26].

Finally, quite recently the proposal from [22] was generalized in [27] to multiple-time correlation experiments of the type discussed in [15,19]. It seems that even though many questions in nonlinear quantum mechanics may remain open, the nonlocality — if appropriately treated — is not a true difficulty.

Let us return to the axiomatic approach, and show that the research there evolved in a way that is parallel and at the same time complementary to what happened in the nonlinearity approach.

Different types of products under slightly different coupling conditions were tried out, but always the structure that was found on the more general axiomatic level, where the failing axioms had been dropped, showed out to be very different from the tensor product structure used in the coupling in standard quantum mechanics [12,28,29]. Meanwhile however also some simple situations of coupled spins had been studied, and there it was revealed that the tensor product structure used in the coupling of standard quantum mechanics could be completely regained if a rigid coupling was introduced representing the entanglement [30,31,32]. In these models not only the rays of the considered Hilbert spaces, but also the density operators appeared as pure states, which at first sight was considered to be a weak point of the models. After reflecting more on these models it became clear however that a generalization of standard quantum mechanics could be built in this way, where the rays as well as the density operators represent pure states, and the density operators also, at the same time, represent mixed states [33]. The fact that from an experimental point of view, by limiting oneself to one quantum entity, it is not possible to make a difference between the pure state and the mixed state represented by the same density operator, is due to the linearity of standard quantum mechanics. The linear structure in some way hides the difference between the pure state and the mixture represented by the same density operator. We have called the quantum mechanics where also density operators represent pure states 'completed quantum mechanics' in [33]. That the problem was revealed by studying the situation of the compound entity of two quantum entities is due to the fact that in this situation nonlinearity shows up at the ontological level.

We do not present a solution in this book, *i.e.* the elaboration of a generalized non linear quantum mechanics. We merely present the material needed to see the way that one could eventually go for the development of such a theory. Future research shall have to make clear whether our analysis of the situation is correct, and hence a generalized nonlinear quantum mechanics can be built, resolving the problems with standard quantum mechanics that we have mentioned.

In the first two articles of this book, 'D. Aerts and F. Valckenborgh, *The linearity of quantum mechanics at stake: the description of separated quantum entities*' and 'D. Aerts and F. Valckenborgh, *Linearity and compound physical systems: the case of two spin 1/2 entities*', the deficiency of the mathematical structure of standard quantum mechanics that we mentioned is analyzed in detail within the axiomatic approach. In the first article a clear account of traditional quantum axiomatics is put forward and it is shown how the last two axioms are at the origin of the impossibility to deliver a model for separated quantum entities. It is also shown how one of these axioms is equivalent with the linearity of the state space and hence with the superposition principle. In the second article the description of two separated spins 1/2 is worked out in detail, such that it can be seen, for this simple case, how the mathematical structure that arises is very different from the standard quantum mechanics description of two spins 1/2 in a complex Hilbert space. Here it can be pointed out concretely where linearity fails, for example no superposition states exists for two couples of states, where both states of one of the spins are different from both states of the other spin, while superpositions do exist for couples of states where one of the states of both spins is equal. Translated into standard quantum mechanical language one could say that super selection rules of a new nature show up, between states that are not orthogonal, such that they cannot be treated as traditional super selection rules, by avoiding superpositions between different orthogonal subspaces of the Hilbert space.

If density operators can also represent pure states of a quantum entity, another one of the five axioms of traditional quantum axiomatics has to be abandoned. In the third article of the book 'D. Aerts, *Being and change: foundations of a realistic operational formalism*, an operational axiomatic approach to quantum mechanics is developed in all its generality. Also the axiom that avoids pure states to be described by density operators is omitted in the formalism proposed here. The article refers to some of the earlier developments, but also introduces the newest advances within this approach. It is a continuation of [33,34], but now more attention is paid to the development of the dynamical aspects of operational axiomatics. The change of state under influence of a measurement and the dynamical change of state are integrated into a 'general change of state under influence of a context', such that 'dynamics' and 'measurement influence' appear as two aspects of a more general type of change. The formalism is also prepared explicitly for wider applications than just an application to quantum mechanics in its description of the microworld. For example, the formalism is made sufficiently general to allow also an influence of the context (measurement or dynamical) on the state, which is not the case, neither for classical entities nor for quantum entities, but which is often the case for applications to other fields where contextual influence is present, *e.g.* biology and cognition. Classical and quantum physical situations are retrieved as special cases where the state of the entity under study does not influence the context (dynamical or measurement), and where in both cases dynamical context influences the state, and in the case of a quantum physical situation also measurement context influences the state.

We mentioned already how the structure of standard quantum mechanics falls short when it comes to the description of compound entities. Some years after the discovery of this shortcoming another shortcoming of the structure of standard quantum mechanics of a similar nature was revealed. By studying the classical limit in a simple quantum model for the spin of a spin 1/2 quantum entity – but a quantum model that is defined in the larger structural context of axiomatic quantum mechanics than standard Hilbert space quantum mechanics – it could be proven that again the last two of the traditional axioms of quantum axiomatics are not satisfied in the region 'between quantum and classical' [35,36,37,56]. This means again that it is the linearity of standard quantum mechanics that makes it impossible to describe a continuous transition from quantum to classical, something that can be done within a generalized nonlinear quantum formalism, as the one used in [35,36,37,56]. The 'between quantum and classical situation' was studied more in detail in [39,40,41,42], and meanwhile it had become clear that there is also a problem with one of the other axioms of quantum axiomatics, the axiom related to the existence of an orthogonality relation on the set of states of the physical entity under consideration.

The fourth article of this book, 'T. Durt and B. D'Hooghe, *The classical limit of the lattice-theoretical orthocomplementation in the framework of the hidden-measurement approach*', investigates the classical limit in this perspective. By looking to different types of orthogonality relations it is proven that determinism is not enough for an entity to entail a full classical structure.

Within traditional quantum axiomatics the classical part and the quantum part of a physical entity can be filtered out, such that a general physical entity can have classical properties and quantum properties and also a mixture of both [43]. In the fifth article of this book, 'D. Aerts and D. Deses, *State property systems and closure spaces: extracting the classical and nonclassical parts*', is investigated in which way this classical and quantum parts can still be filtered out, even if the two last axioms and also the axiom that causes problems with the orthogonality are not satisfied. The categorical equivalence between state property systems, the structures that in quantum axiomatics describe a physical entity by means of its states and its properties, and closure spaces, a mathematical generalization of topologies, that was derived in earlier work

[44,45,46,47], is used to derive a decomposition theorem that is a generalization of the original decomposition theorem as presented in [7]. This decomposition theorem translates through the equivalence of categories to a decomposition theorem of closure spaces into their connected components.

The operational axiomatic approaches to quantum mechanics that we have considered have traditionally concentrated on the description of a physical entity by means of its states and properties. In a certain sense one could say that the probabilistic aspects of quantum theory have been neglected in these approaches. In the foregoing sections we concentrated on the advances of a structural nature that have been made in quantum axiomatics, related to the study of the compound entity of separated entities and the investigation of the classical limit within a formalism that is more general than standard quantum mechanics. There also has been an important step ahead on the conceptual level in relation with quantum probability. It was shown that the structure of the quantum probability model could be derived from a hypothesis about the physical origin of quantum probability that is the following: quantum probability is due to the presence of fluctuations on the interaction between measurement apparatus and physical entity under study [48,49]. The approach that introduces the quantum probabilities in this way has meanwhile been called the 'hidden measurement approach', and different aspects of it have been studied [50,51,52]. In the sixth article of this book, 'S. Aerts, *Hidden measurements from contextual axiomatics*, the hidden measurement approach is investigated, and three simple requirements are put forward that make it possible to uniquely recover the structure of hidden measurements.

It is worth noting that the ontology proposed in hidden variables theories differs from the ontology proposed in other interpretations. Therefore, it is possible in principle to conceive crucial experiments during which the validity of a particular interpretation could be tested. This is what occurred for instance in the numerous EPR-Bell experiments that were realized during the last three decades [53]. Such experiments are crucial experiments during which it is possible to discriminate between local-realistic ontologies and the other ontologies (non-local realistic ones à la Bohm [54] or non-realistic ones à la Bohr [55]). Similarly, it is possible in principle to discriminate between the hidden measurement interpretation and the standard interpretation provided the fluctuations of the hidden state of the apparatus are not instantaneous which means that the detector remembers its hidden state for a while. Then non-standard correlations ought to appear between successive outcomes obtained from the same apparatus [56]. The seventh article of this book 'T. Durt, J. Baudon, R. Mathevet, J. Robert and B. Viaris de Lesegno, *Memory effects in atomic interferometry: a negative result*' describes an attempt to detect such correlations. This experiment was negative in the sense that standard predictions were confirmed. An upper bound could be found for the value of hypothetical hidden measurement memory times. If these times are too small however, they cannot be detected, and the hidden measurement approach gives an ad hoc description of quantum phenomena. Similarly, it is possible, by making use of the inefficiency of presently available detectors (this is the so-called efficiency loophole) to simulate the results of present EPR-Bell experiments by ad hoc local realistic models.

Loose of the hidden measurement approach, the structure of probability, whether it is classical probability or quantum probability, poses another problem of a conceptual nature. As we mentioned already, the generalized axiomatic approaches have been developed focusing on the description of the states and the properties of the physical entity under consideration. A property is linked to a test with 'certain' outcome. But 'certainty' is a concept that cannot easily be recovered by a probability theory that is founded on traditional measure theory. The reason is that events with probability equal to 1 are not completely certain events. This problem is investigated in the eighth article of this book, 'D. Aerts, *Reality and probability: introducing a new type of probability calculus*. It is proven that 'certainty' can be recovered from a probabilistic approach if a new type of probability theory is introduced, called 'subset probability', where the probability is evaluated by a subset of the interval $[0, 1]$ instead of an element of $[0, 1]$, as it is the case in traditional probability theory founded on measure theory. The subset probability is a generalization of traditional probability theory that is retrieved when all subsets are singletons of the interval $[0, 1]$. Not only 'certainty' can be modelled in a natural way by a subset probability, but also situations 'near to certainty' can be described in a way that avoids the problems encountered with traditional probability theory. The structure of a state property system, that has been studied extensively in the axiomatic approach [44,45,46,47], is recovered as corresponding to the description of 'certainty' in the subset probabilistic approach.

Quantum computation constitutes another promising and fascinating contemporary field of research. The basic idea is that quantum systems do not behave as deterministic systems, but exhibit a flexibility that has no classical counterpart. For instance, quantum bits (qubits) can be superposed and teleported, and it can be shown that in principle a processor based on qubits works in certain circumstances (when the quantum entanglement and the superposition principle are optimally exploited) exponentially faster than its classical counterpart. The hidden measurement approach [48,49,50,51,52], combined with the results on separated and nonseparated physical entities in quantum axiomatics [6,7,8,9,10,11], invites us to consider the quantum computation process from a more general perspective. Traditional quantum computation considers different steps in its computational process. The data of a problem are encoded into the quantum state of N entangled spin $\frac{1}{2}$. The quantum computation process, described by a unitary evolution, brings then this state into another state of the N entangled spin $\frac{1}{2}$, such that the outcome of the calculation, and the resolution of the problem, can be extracted from this state. Two of the basic quantum aspects that are at play in this quantum calculation process are entanglement and indeterminism. The models that have been developed within the hidden measurement approach [30,48,49,35,36,37,56,39,40,41,42], and more specifically the $\epsilon\rho$-model [57], make it possible to parametrize the amount of indeterminism and entanglement by means of two variables $\epsilon, \rho \in [0,1]$, such that for $\epsilon = \rho = 0$ we get the classical situation of a Turing calculation process with no entanglement and no indeterminism, while for $\epsilon = \rho = 1$ we get a pure quantum calculation process. For intermediate values of ϵ and ρ an intermediate 'between quantum and classical' calculation process can be modelled. This makes it possible to investigate the influence of the two aspects 'entanglement' and 'indeterminism' in the quantum computation process. In the ninth article of the book, 'D. Aerts and B. D'Hooghe, *Quantum computation: towards the construction of a 'between quantum and classical' computer*, this perspective is considered. The nonlinearity problem that we mentioned already also appears here, since it can be shown that the 'between quantum and classical' situations gives rise to a structure that does not satisfy the linearity axioms of traditional quantum axiomatics.

Also connections between quantum axiomatics and fuzzy set theory have been studied. A quantum axiomatic system defined by a set of experimentally verifiable propositions can be represented by a suitably chosen family of fuzzy sets over the set of states such that conjunction and disjunction are given by Giles'[62] fuzzy set intersection and union[63,64,65]. Each proposition is represented by a fuzzy set whose membership function value in a point is given by the probability of the experimental proposition if the system is in the corresponding state. Although Giles' operations satisfy the law of contradiction and excluded middle, they do not satisfy the law of idempotency. Also, there is an infinite number of possible fuzzy set connectives and hence an infinite number of possible definitions for conjunction and disjunction of two fuzzy sets representing experimental propositions. For instance, the fuzzy set connectives introduced by Zadeh in his historic paper[66] are idempotent but violate the laws of contradiction and excluded middle. However, these and other fuzzy set operations considered usually in the literature are defined pointwise, i.e., the membership function value of the conjunction (disjunction) of two fuzzy sets is completely defined by the membership function values of the fuzzy sets in that point only. As a result, the pointwise defined fuzzy set connectives do not make a distinction when genuinely different fuzzy sets have the same membership function value in a certain point. Amongst others, such fuzzy set connectives can not be both idempotent and satisfy the laws of excluded middle and contradiction at the same time. As such, these fuzzy set connectives are not completely satisfactory to define conjunction and disjunction of fuzzy sets representing propositions of a quantum entity, since meet and join are idempotent and do satisfy the law of excluded middle and contradiction. In an attempt to solve this problem for fuzzy sets representing propositions of classical systems, Buckley and Siler [58,59,60] proposed fuzzy set connectives (*i.e.*, conjunction and disjunction) parametrized by a correlation coefficient between the two fuzzy sets such that the law of contradiction, excluded middle and idempotency hold. In the tenth article of the book, 'B. D'Hooghe and J. Pykacz, *Buckley-Siler connectives for quantum logics of fuzzy sets*, the Buckley-Siler approach is generalized to the case of a quantum entity and illustrated on a fuzzy set representation of the spin properties of a spin-1/2 particle [61].

The eleventh article of the book, 'W. Christiaens, *Some notes on Aerts' interpretation of the EPR-paradox and the violation of Bell-inequalities*' studies the Einstein Podolsky Rosen type of paradoxes [67] in the light of the approaches that we have exposed in the foregoing sections. Cartwright's model for the violation of Bell's inequalities [68] is investigated and compared with Aerts's model [69]. It is shown that a causal view can be advanced for a situation of nonlocality in quantum mechanics if one of the basic assumptions about reality, namely that a physical entity is always present inside space, is relaxed. Also the creation discovery view [69], where it is taken for granted that an experiment on a physical entity contains two fundamental aspects, the discovery of an existing part of reality and the creation of new part of reality, is investigated from a philosophical point of view.

This brings us to the next point, quantum cryptography. Quantum cryptography is the most efficient application of the fundamental and experimental current of research centered around the interpretational problems of quantum mechanics that was developed during the last decades. It is a combination of deep physical insight, new technology, and ingenious reflection. It is highly representative of the influence that could have quantum mechanics on tomorrow's technology and ... last but not least, it works [70]!

An essential difference between classical theories and the quantum theory is the fact that in the latter the influence of the observer (of the apparatus, of the whole measurement context) cannot be neglected, and, moreover, can have dramatic consequences on the properties of the system under study, an idea that is central in the hidden measurement approach as well as in the Copenhagen interpretation. The complementary principle is a direct illustration of this non-classical feature: when observables do not commute (such as position and momentum for instance), it is often impossible to measure them simultaneously without dispersion, and, in general, the dispersions of the outcomes obtained during their individual measurements obey uncertainty relations. Complementarity is exploited in quantum cryptography [71] where, instead of considering uncertainty relations as a negative limitation for the users (traditionally called Alice and Bob), they appear to be useful because they allow Alice and Bob to reveal the presence of a third party (Eve) that would eavesdrop the signal exchanged between them and/or to limit her knowledge of this signal. In order to do so, it is necessary that the signal is encoded in complementary variables (non-commuting bases). In the twelfth article of the book, 'T. Durt and B. Nagler, *Quantum cryptographic encryption in three complementary bases through a Mach-Zehnder set up*', a protocol for quantum key distribution is described in which it is shown that the wave-particle complementarity plays a fundamental role; this complementarity is also related to the complementarity between position and momentum if we consider position to be a corpuscular property and momentum (which is related to de Broglie wave-length) to be an undulatory property.

It is worth noting that, although it is possible in principle to prepare and to measure one photon in a given position (or to emit and to detect one photon in a given temporal window), technologically, the problem is not solved yet. This is due to the fact that at our scale, when large amount of photons are present, they often behave as (classical) waves, and that it is not so easy to reveal their corpuscular properties. Detectors based on the photo-electric effect exploit such properties, but they are not very efficient when few photons are present (this is related to the aforementioned efficiency loophole). Similarly, we are not able yet, today, to produce on request a single photon state. These limitations suggested a semi-classical (or semi-quantum) protocol for key distribution in which the information is encoded in corpuscular properties and in which technological limitations play the same role as quantum uncertainties in quantum cryptography. This protocol is described in the thirteenth article of the book, 'T. Durt, *Quantum cryptography without quantum uncertainties*'. It is a direct illustration of the hidden measurement approach in which unavoidable fluctuations characterize the interaction between the observer and the physical world. The stochasticity of these contextual fluctuations can be exploited in order to send a secure cryptographic key. The semi-classical protocol fills the gap between quantum cryptography and classical techniques in which the message is hidden among a huge random noise. Considered so it is a mesoscopic protocol, that belongs to a framework more general than the standard quantum one, that contains also the classical framework as a limiting case. It is even, more than an illustration, an application of the hidden measurement approach.

If linear quantum mechanics is a special case of a nonlinear theory then there must be some freedom in what is nowadays understood as *canonical quantization*. The papers that follow are related to the problem of quantization. One of the problems that remains unsolved in great part is how to classify equations which are physically admissible. Here different criteria may be introduced, based on locality, positivity, or integrability. A class of candidate equations selected on the basis of positivity, locality, and probabilistic requirements was introduced in [72]. All these equations were of the form $i\dot{\rho} = [H, f(\rho)]$ where $[f(\rho), \rho] = 0$. Their physically nontrivial solutions were found in [73] for $f(\rho) = \rho^2$ and recently generalized to other f's in [74]. A link of such general f's with nonextensive statistical mechanics was described in [75]. More general classes of physically interesting von Neumann-type equations derived from multiple Nambu-type brackets were introduced in [76].

The question of integrability is always a difficult one. One of the reasons is that it is not completely clear what should be actually meant by this notion. The definition which is very useful is that integrability means practical integrability by means of soliton techniques. The following articles contain new results on integrability of generalized von Neumann equations. The fourteenth article of the book, 'J. L. Cieśliński, *How to construct Darboux-invariant equations of von Neumann type*', generalizes earlier results of [77] to a large class of Darboux transformations. Essentially the same family of equations is treated in the fifteenth article of the book, 'M. Czachor, N. V. Ustinov, *Darboux-integrable equations with non-Abelian nonlinearities*' along the lines of [77]. The two articles use different mathematical constructions showing two different aspects of Darboux-covariance of the same set of equations. Strictly speaking one has to admit that it is not fully clear to what extent the two approaches are equivalent but all the explicit examples given in the papers can be formulated in either of the two ways.

The sixteenth article of the book, 'S. Leble, *Dressing chain equations associated with difference soliton systems*', employs still another variant of the Darboux transformation: The chain equations. The elements which are common in all the three papers are: The use of Darboux-covariant Lax pairs, representation of evolution equations by compatibility conditions, the presence of non-Abelian nonlinearities, and the well known Nahm system as a particular example. The latter shows also that the notion of a generalized von Neumann-type equation covers here a very large class of nonlinear evolution equations extending far beyond the standard formalism of linear quantum mechanics.

The collection of articles on aspects of generalized quantization is completed by the seventeenth article of the book, 'M. Kuna, J. Naudts, *Covariance approach to the free photon field*'. The authors start with the notion of a generalized covariance system and a generalized GNS construction, an approach which follows their earlier results published in [78,79]. The generality inherently present in their formalism is here purposefully restricted in order to rederive the standard Fock space representation of free electromagnetic fields. However, the results of [78,79] show that the formalism they propose is flexible enough to incorporate also various non-canonical systems such as those based on non-commutative spacetime [80] or non-canonical vacua [81].

References

1. D. Aerts and I. Daubechies, "Physical justification for using the tensor product to describe two quantum systems as one joint system", *Helv. Phys. Acta* **51**, 661-675 (1978).
2. D. Aerts and I. Daubechies, "A characterization of subsystems in physics", *Lett. Math. Phys.*, **3**, 11-17 (1979).
3. D. Aerts and I. Daubechies, "A mathematical condition for a sub-lattice of a propositional system to represent a physical subsystem with a physical interpretation", *Lett. Math. Phys.*, **3**, 19-27 (1979).
4. D. J. Foulis and C. H. Randall, "Empirical logic and tensor products", in *Interpretations and Foundations of Quantum Theory*, ed. H. Neumann, Wissenschaftsverlag, Mannheim (1981).
5. C. H. Randall and D. J. Foulis, "Operational statistics and tensor products", in *Interpretations and Foundations of Quantum Theory*, ed. H. Neumann, Wissenschaftsverlag, Mannheim (1981).
6. D. Aerts, *The One and the Many: Towards a Unification of the Quantum and the Classical Description of One and Many Physical Entities*, Doctoral Dissertation, Brussels Free University (1981).
7. D. Aerts, "Description of many physical entities without the paradoxes encountered in quantum mechanics", *Found. Phys.* **12**, 1131-1170 (1982).
8. D. Aerts, "How do we have to change quantum mechanics in order to describe separated systems", in *The Wave-Particle Dualism*, eds. S. Diner, *et al.*, Kluwer Academic, Dordrecht, 419-431 (1984).

9. D. Aerts, "The physical origin of the Einstein Podolsky Rosen paradox", in *Open Questions in Quantum Physics: Invited Papers on the Foundations of Microphysics*, eds. G. Tarozzi and A. van der Merwe, Kluwer Academic, Dordrecht, 33-50 (1985).
10. S. Pulmannova, "Coupling of quantum logics", *Int. J. Theor. Phys.* **22**, 837-850 (1983).
11. S. Pulmannova, "Tensor products of quantum logics", *J. Math. Phys.* **26**, 1-5 (1984).
12. D. Aerts, "Construction of the tensor product for lattices of properties of physical entities", *J. Math. Phys.* **25**, 1434-1441 (1984).
13. I. Białynicki-Birula and J. Mycielski, "Nonlinear wave mechanics", *Ann. Phys. (NY)*, **100**, 62 (1978).
14. R. Haag and U. Bannier, "Comments on Mielnik's generalized (non linear) quantum mechanics", *Comm. Math. Phys.*, **60**, 1 (1978).
15. N. Gisin, *Helv. Pys. Acta.*, **62**, 363 (1989).
16. S. Weinberg, "Testing quantum mechanics", *Ann. Phys. (NY)*, **194**, 336 (1989).
17. J. Polchinski, unpublished (1989); cf. [16].
18. M. Czachor, "Nonlinearity can make quantum paradoxes malignant", a talk at the conference *Problems in Quantum Physics, Gdańsk'89*, unpublished; cf. [22].
19. N. Gisin, "Weinberg's non-linear quantum mechanics and superluminal communications", *Phys. Lett. A*, **143**, 1 (1990).
20. M. Czachor, "Mobility and nonseparability", *Found. Phys. Lett.*, **4**, 351 (1991).
21. B. Mielnik, "Nonlinear quantum mechanics: a conflict with Ptolomean structures?", *Phys. Lett. A*, **289**, 1 (2001).
22. J. Polchinski, "Weinberg's nonlinear quantum mechanics and the Einstein-Podolsky-Rosen paradox", *Phys. Rev. Lett.*, **66**, 397 (1991).
23. T. F. Jordan, *Ann. Phys. (NY)*, **225**, 83 (1993).
24. P. Bóna, "Quantum mechanics with mean-field backgrounds", Comenius University Report No. Ph10-91 (1991); for an extended review see P. Bóna, "Extended quantum mechanics", *Acta. Phys. Slov.*, **50**, 1 (2000).
25. M. Czachor, "Nonlocal-looking equations can make nonlinear quantum dynamics local", *Phys. Rev. A*, **57**, 4122 (1998).
26. M. Czachor and M. Kuna, "Complete positivity of nonlinear evolution: a case study", *Phys. Rev. A*, **58**, 128 (1998).
27. M. Czachor and H. D. Doebner, "Correlation experiments in nonlinear quantum mechanics", *quant-ph/0110008* (2001).
28. F. Valckenborgh, "Operational axiomatics and compound systems", in

Current Research in Operational Quantum Logic: Algebras, Categories, Languages, eds. B. Coecke, D. J. Moore and A. Wilce, Kluwer Academic Publishers, Dordrecht, 219-244 (2000).
29. F. Valckenborgh, *Compound Systems in Quantum Axiomatics,* Doctoral Thesis, Brussels Free University (2001).
30. D. Aerts, "A mechanistic classical laboratory situation violating the Bell inequalities with $2\sqrt{2}$, exactly 'in the same way' as its violations by the EPR experiments", *Helv. Phys. Acta* **64**, 1-23 (1991).
31. B. Coecke, "Representation of a spin-1 entity as a joint system of two spin-1/2 entities on which we introduce correlation's of the second kind", *Helv. Phys. Acta* **68**, 396 (1995).
32. B. Coecke, "Superposition states through correlation's of the second kind", *Int. J. Theor. Phys.* **35**, 2371 (1996).
33. D. Aerts, "Foundations of quantum physics: a general realistic and operational approach", *Int. J. Theor. Phys.* **38**, 289 - 358 (1999).
34. D. Aerts, "Quantum mechanics: structures, axioms and paradoxes", in *Quantum Mechanics and the Nature of Reality,* eds. Aerts, D. and Pykacz, J., Kluwer Academic, Dordrecht (1999).
35. D. Aerts, T. Durt and B. Van Bogaert, "Quantum probability, the classical limit and nonlocality", in the *Proceedings of the International Symposium on the Foundations of Modern Physics 1992,* Helsinki, Finland, ed. T. Hyvonen, World Scientific, Singapore, 35-56 (1993).
36. D. Aerts and T. Durt, "Quantum. classical and intermediate, an illustrative example", *Found. Phys.* **24**, 1353-1369 (1994).
37. D. Aerts and T. Durt, "Quantum, classical and intermediate: a measurement model", in *Symposium on the Foundations of Modern Physics,* eds. K. V. Laurikainen, C. Montonen and K. Sunnaborg, Editions Frontieres, Gives Sur Yvettes, France (1994).
38. T. Durt, *From Quantum to Classical, A Toy Model,* Doctoral Dissertation, Brussels Free University (1996).
39. D. Aerts, B. Coecke, T. Durt and F. Valckenborgh, "Quantum, classical and intermediate I: a model on the Poincaré sphere", *Tatra Mt. Math. Publ.*, **10**, 225 (1997).
40. D. Aerts, B. Coecke, T. Durt and F. Valckenborgh, "Quantum, classical and intermediate II: the vanishing vector space structure", *Tatra Mt. Math. Publ.*, **10**, 241 (1997).
41. T. Durt, "Orthogonality relations: from classical to quantum", in *Quantum Structures and the Nature of Reality,* eds. D. Aerts and J. Pykacz, Kluwer Academic, Dordrecht, 207 (1999).
42. B. D'Hooghe, *From Quantum to Classical: A study of the Effect of Vary-*

ing Fluctuations in the Measurement Context and State Transitions due to Experiments, Doctoral Dissertation, Brussels Free University (2000).
43. D. Aerts, "Classical theories and non classical theories as special case of a more general theory", *J. Math. Phys.* **24**, 2441-2453 (1983).
44. D. Aerts, E. Colebunders, A. Van der Voorde and B. Van Steirteghem, "State property systems and closure spaces: a study of categorical equivalence", *Int. J. Theor. Phys.* **38**, 359-385 (1999).
45. D. Aerts, D. Deses and A. Van der Voorde, "Connectedness applied to closure spaces and state property systems", *Journal of Electrical Engineering*, **52**, 18-21 (2001).
46. D. Aerts, D. Deses and A. Van der Voorde, "Classicality and connectedness for state property systems and closure spaces", submitted to *International Journal of Theoretical Physics.*
47. A. Van der Voorde, *Separation axioms in extension theory for closure spaces and their relevance to state property systems*, Doctoral Dissertation, Brussels Free University (2001).
48. D. Aerts, "A possible explanation for the probabilities of quantum mechanics", *J. Math. Phys.*, **27**, 203 (1986).
49. D. Aerts, "The origin of the non-classical character of the quantum probability model", in *Information, Complexity, and Control in Quantum Physics*, eds. A. Blanquiere, S. Diner and G. Lochak, Springer-Verlag, Wien-New York, 77-100 (1987).
50. D. Aerts, "Quantum structures due to fluctuations of the measurement situation", *Int. J. Theor. Phys.*, **32**, 2207-2220, (1993).
51. D. Aerts, S. Aerts, B. Coecke, B. D'Hooghe, T. Durt and F. Valckenborgh, "A model with varying fluctuations in the measurement context", in *Fundamental Problems in Quantum Physics II*, 7-9, eds. M. Ferrero and A. van der Merwe, Plenum, New York, (1996).
52. D. Aerts, "The hidden measurement formalism: what can be explained and where paradoxes remain", *Int. J. Theor. Phys.*, **37**, 291, (1998).
53. J. S. Bell, "On the EPR paradox", *Physics*, **1**, 195 (1964).
54. D. Bohm, "A suggested interpretation of quantum theory in terms of hidden variables", *Phys. Rev.*, **85**, 166 (1952).
55. N. Bohr, "Can quantum mechanical description of physical reality be considered complete?" *Phys. Rev.*, **48**, 696 (1935).
56. T. Durt, *From Quantum to Classical: A Toy Model*, Doctoral Dissertation, Brussels Free University (1996).
57. D. Aerts, S. Aerts, J. Broekaert and L. Gabora, "The violation of Bell inequalities in the macroworld", *Found. Phys.* **30**, 1387-1414 (2000), lanl archive ref: *quant-ph/0007044*.

58. J.J. Buckley and W. Siler, "L$_\infty$ fuzzy logic", preprint, available from the authors on the request sent to: buckley@math.nab.edu.
59. J.J. Buckley and W. Siler, "A new t-norm", preprint, available from the authors on the request sent to: buckley@math.nab.edu.
60. J.J. Buckley, W. Siler and Y. Hayashi, "A new fuzzy intersection and union", in *Proceedings of the 7th IFSA World Congress*, Academia, Prague (1997).
61. B. D'Hooghe and J. Pykacz, "Classical limit in fuzzy set models of spin-$\frac{1}{2}$ quantum logics", *Int. J. Theor. Phys.*, **38**, 387 (1999).
62. R. Giles, "Łukasiewicz logic and fuzzy set theory", *International Journal of Man-Machine Studies*, **8**, 313 (1976).
63. J. Pykacz, "Quantum logics as families of fuzzy subsets of the set of physical states", in *Preprints of the Second IFSA World Congress, Tokyo*, **vol. II**, 437 (1987).
64. J. Pykacz, "Fuzzy quantum logics and infinite-valued Łukasiewicz logic", *Int. J. Theor. Phys.*, **33**, 1403 (1994).
65. J. Pykacz, "Triangular norms-based quantum structures of fuzzy sets", in *Proceedings of the 7th IFSA World Congress*, Academia, Prague (1997).
66. L. A. Zadeh, "Fuzzy sets", *Information and Control*, **8**, 338 (1965).
67. A. Einstein, B. Podolsky and N. Rosen "Can quantum-mechanical description of physical reality be considered complete?", *Phys. Rev.*, **47**, 777 (1935).
68. N. Cartwright, *Nature Capacities and Their Measurement*, Clarendon Press, Oxford (1989).
69. D. Aerts, "The entity and modern physics: The creation-discovery view of reality", in *Interpreting Bodies. Classical and Quantum Objects in Modern Physics*, ed. E. Castellani, Princeton University Press, Princeton (1998).
70. N. Gisin, G. Ribordy, W. Tittel and H. Zbinden, "Quantum cryptography", *quant-ph/0101098*, submitted to *Rev. of Mod. Phys.* (2001).
71. C.H. Bennett and G. Brassard, "Quantum cryptography: public key distribution and coin tossing", *IEEE International Conference on Computing, Signal and Processing*, Bangalore, India, 175 (1984).
72. M. Czachor, "Nambu-type generalization of the Dirac equation", *Phys. Lett. A*, **225**, 1 (1997).
73. S. B. Leble and M. Czachor, "Darboux-integrable nonlinear Liouville-von Neumann equation", *Phys. Rev. E*, **58**, 7091 (1998).
74. N. V. Ustinov, M. Czachor, M. Kuna and S. B. Leble, "Darboux-integration of $i\dot{\rho} = [H, f(\rho)]$", *Phys. Lett. A*, **279**, 333 (2001).
75. M. Czachor and J. Naudts, "Microscopic foundation of nonextensive

statistics", *Phys. Rev. E*, **59**, R2497 (1999).
76. M. Czachor, "Lie-Nambu and beyond", *Int. J. Theor. Physics.*, **38**, 475 (1999).
77. N. V. Ustinov and M. Czachor, "New class of Darboux-integrable von Neumann-type equations", *nlin.SI/0011013* (2000).
78. J. Naudts and M. Kuna, "Covariance systems", *J. Phys. A: Math. Gen.*, **34**, 9265 (2001).
79. J. Naudts and M. Kuna, "Model of a quantum particle in spacetime", *J. Phys. A: Math. Gen.*, **34**, 4227 (2001).
80. S. Doplicher, K. Friedenhagen and J. E. Roberts, "The quantum structure of spacetime at the Planck scale and quantum fields", *Comm. Math. Phys.*, **172**, 187 (1995).
81. M. Czachor, "Non-canonical quantum optics", *J. Phys. A: Math. Gen.*, **33**, 8081 (2000).

THE LINEARITY OF QUANTUM MECHANICS AT STAKE: THE DESCRIPTION OF SEPARATED QUANTUM ENTITIES

DIEDERIK AERTS
Center Leo Apostel (CLEA) and Foundations of the Exact Sciences (FUND), Brussels Free University, Krijgskundestraat 33, 1160 Brussels, Belgium
E-mail: diraerts@vub.ac.be

FRANK VALCKENBORGH
Foundations of the Exact Sciences (FUND), Department of Mathematics, Brussels Free University, Pleinlaan 2, 1050 Brussels, Belgium
E-mail: fvalcken@vub.ac.be

We consider the situation of a physical entity that is the compound entity consisting of two 'separated' quantum entities. In earlier work it has been proved by one of the authors that such a physical entity cannot be described by standard quantum mechanics. More precisely, it was shown that two of the axioms of traditional quantum axiomatics are at the origin of the impossibility for standard quantum mechanics to describe this type of compound entity. One of these axioms is equivalent with the superposition principle, which means that separated quantum entities put the linearity of quantum mechanics at stake. We analyze the conceptual steps that are involved in this proof, and expose the necessary material of quantum axiomatics to be able to understand the argument.

1 Introduction

It is often stated that quantum mechanics is basically a linear theory. Let us reflect somewhat about what one usually means when expressing this statement.

The Schrödinger equation that describes the change of the state of a quantum entity under the influence of the external world is a linear equation. This means that if the wave function $\psi_1(x)$ and the wave function $\psi_2(x)$ are both solutions of the Schrödinger equation, then, for $\lambda_1, \lambda_2 \in \mathbb{C}$, the wave function $\lambda_1\psi_1(x)+\lambda_2\psi_2(x)$ is also a solution. Hence the set of solutions of the Schrödinger equation forms a vector space over the field of complex numbers: solutions can be added and multiplied by a complex number and the results remain solutions. This is the way how the linearity of the evolution equation is linked to the linearity, or vector space structure, of the set of states.

There is another type of 'change of state' in quantum mechanics, namely the one provoked by a measurement or an experiment. This type of change, often called the 'collapse of the wave function', is nonlinear. It is described by the action of a projection operator associated with the self-adjoint operator that represents the considered measurement – and hence for this part it is linear, because a projection operator is a linear function – followed by a renormalization of the state. The two effects, projection and renormalization, one after the other, give rise to a nonlinear transformation.

The fundamental nature of the linearity of the vector space used to represent the states of a quantum mechanical entity is expressed by adopting the 'superposition principle' as one of the basic principles of quantum mechanics. Because linearity in general appears very often as an idealized case of the real situation, some suspicion towards the fundamental linear nature of quantum mechanics is at its place. Would there not be a more general theory that as a first order linear approximation gives rise to quantum mechanics?

In relation with this question it is good to know that the situation in quantum mechanics is very different from the situation in classical physics. Nonlinear situations in classical mechanics exist at all places, and it can easily be understood how the linearized version of the theory is an idealization of the general situations (e.g. the linearization used to study small movements around an equilibrium position). Quantum mechanics on the contrary was immediately formulated as a linear theory, and no nonlinear version of quantum mechanics has ever been proposed in a general way.

The fundamentally different way in which linearity presents itself in quantum mechanics as compared to classical mechanics makes that quite a few physicists believe that quantum linearity is a profound property of the world. The way in which classical mechanics works as a theory for the macroworld with at a 'more basic level' quantum mechanics as a description for the microworld, and additionally the hypothesis that this macroworld is built from building blocks that are quantum, makes that some physicists propose that the nonlinearity of macrophenomena should emerge from an underlying linearity of the microworld. This type of reflections is however very speculative. Mostly because nobody has been able to solve in a satisfactory way the problem of the classical limit, and explain how the microworld described by quantum mechanics gives rise to a macroworld described by classical mechanics.

Because of the profound and unsolved nature of this problem it is worth to analyze a result that has been obtained by one of the authors in the eighties. The result is the following:

If we consider the physical entity that consists of two 'separated' quantum entities, then this physical entity cannot be described by standard quantum mechanics [1,2].

The aspect of this result that we want to focus on in this article, is that the origin of the impossibility for standard quantum mechanics to describe the entity consisting of two separated quantum entities is the linearity of the vector space representing the states of a quantum entity.

We analyze the conceptual steps to arrive at this result in the present paper without giving proofs. For the proofs we refer to [1,2].

2 Quantum Axiomatics

As we mentioned in the introduction, there is no straightforward way to conceive of a more general, possibly nonlinear, quantum mechanics if one starts conceptually from the standard quantum mechanical formalism. The reason is that standard quantum mechanics is elaborated completely around the vector space structure of the set of states of a quantum entity and the linear operator algebra on this vector space. If one tries to drop linearity starting from this structure one is left with nothing that remains mathematically relevant to work with.

We also mentioned that there is one transformation in standard quantum mechanics that is nonlinear, namely the transformation of a state under influence of a measurement. The nonlinearity here comes from the fact that also a renormalization procedure is involved, because states of a quantum entity are not represented by vectors, but by normalized vectors of the vector space. This fact gives us a first hint of where to look for possible ways to generalize quantum mechanics and free it from its very strict vector space strait jacket. This is also the way things have happened historically. Physicists and mathematicians noticed that the requirement of normalization and renormalization after projection means that quantum states 'live' in the projective geometry corresponding with the vector space. The standard quantum mechanical representation theory of groups makes full use of this insight: group representations are projective representations and not vector space representations, and experimental results confirm completely that it is the projective representations that are at work in the reality of the microworld and not the vector space representations.

Of course, there is a deep mathematical connection between a projective geometry and a vector space, through what is called the 'fundamental representation theorem of projective geometry' [3]. This theorem states that every projective geometry of dimension greater than two can be represented in a vector space over a division ring, where a ray of the vector space corresponds to a point of the projective geometry, and a plane through two different rays corresponds to a projective line. This means that a projective geometry entails the type of linearity that is encountered in quantum mechanics. Conceptually however a projective geometric structure is quite different from a vector space structure. The aspects of a projective geometry that give rise to linearity can perhaps more easily be generalized than this is the case for the aspects of a vector space related to linearity.

John von Neumann gave the first abstract mathematical formulation of quantum mechanics [4], and proposed an abstract complex Hilbert space as the basic mathematical structure to work with for quantum mechanics. If we refer to standard quantum mechanics we mean quantum mechanics as formulated in the seminal book of von Neumann. Some years later he wrote an important paper, together with Garrett Birkhoff, that initiated the research on quantum axiomatics [5]. In this paper Birkhoff and von Neumann propose to concentrate on the set of closed subspaces of the complex Hilbert space as the basic mathematical structure for the development of a quantum axiomatics. In later years George Mackey wrote an influential book on the mathematical foundations of quantum mechanics where he states explicitly that a physical foundation for the complex Hilbert structure should be looked for [6]. A breakthrough came with the work of Constantin Piron when he proved a fundamental representation theorem [7]. It had been noticed meanwhile that the set of closed subspaces of a complex Hilbert space forms a complete, atomistic, orthocomplemented lattice and Piron proved the converse, namely that a complete, atomistic orthocomplemented lattice, satisfying some extra conditions, could always be represented as the set of closed subspaces of a generalized Hilbert space [7,8]. In his proof Piron first derives a projective geometry and then makes the step to the vector space. Piron's representation theorem is exposed in detail in theorem 2 of the present article.

As we will see, it is exactly the extra conditions, needed to represent the lattice as the lattice of closed subspaces of a generalized Hilbert space, that are not satisfied for the description of the compound entity that consists of two separated quantum entities. Since the aim of this article is to put forward the conceptual steps that are involved in the failure of standard quantum mechanics to describe such an entity, we will start by explaining the general aspects of quantum axiomatics in some detail, omitting all proofs, for the sake of readability. For the reader who is interested in a more detailed exposition, references to the literature are given.

2.1 What Is a Complete Lattice?

A lattice $\mathcal{L}$ is a set that is equipped with a partial order relation $<$. This means a relation such that for $a, b, c \in \mathcal{L}$ we have:

$$a < a \tag{1}$$
$$a < b \text{ and } b < a \Rightarrow a = b \tag{2}$$
$$a < b \text{ and } b < c \Rightarrow a < c \tag{3}$$

(1) is called reflexivity, (2) is called antisymmetry and (3) is called transitivity of the relation $<$. Hence a partial order relation is a reflexive, antisymmetric and transitive relation.

Let us give two examples of partial order relations. First, consider a set Ω and let $\mathcal{P}(\Omega)$ be the set of all subsets of Ω. The inclusion operation $\subset$ on $\mathcal{P}(\Omega)$ is a partial order on $\mathcal{P}(\Omega)$. Second, consider a complex Hilbert space $\mathcal{H}$, and let $\mathcal{P}(\mathcal{H})$ be the set of all closed subspaces of $\mathcal{H}$. The inclusion operation $\subset$ on $\mathcal{P}(\mathcal{H})$ is a partial order on $\mathcal{P}(\mathcal{H})$. The two examples that we propose here are the archetypical examples for quantum axiomatics. The first example represents (part of) the mathematical structure related to a classical physical entity, where Ω corresponds with its state space, and the second example the one of a quantum physical entity, where $\mathcal{H}$ is the complex Hilbert space representing the states of the quantum entity.

A lattice is a mathematical object that has some more structure than just this partial order relation. Suppose we consider two elements $a, b \in \mathcal{L}$ of the lattice, then we demand that there exists an infimum and a supremum for a and b for the partial order relation $<$ in $\mathcal{L}$. The infimum and supremum are denoted respectively $a \wedge b$ and $a \vee b$. An infimum of a and b is a greatest lower bound, this means a maximum of all the elements of $\mathcal{L}$ that are smaller than a and smaller than b. A supremum is a least upper bound, this means a minimum of all the elements of $\mathcal{L}$ that are greater than a and greater than b. Let us repeat this in a formal way. For $a, b \in \mathcal{L}$ there exist $a \wedge b, a \vee b \in \mathcal{L}$ such that for $x, y \in \mathcal{L}$ we have:

$$x < a \text{ and } x < b \Leftrightarrow x < a \wedge b \tag{4}$$
$$a < y \text{ and } b < y \Leftrightarrow a \vee b < y \tag{5}$$

A structure $(\mathcal{L}, <, \wedge, \vee)$, where $\mathcal{L}$ is a set, $<$ is a partial order relation satisfying (1), (2) and (3), and $\wedge$ and $\vee$ are an infimum and a supremum, satisfying (4) and (5) is a **lattice**.

Our two previous examples are both examples of lattices. For the case of $\mathcal{P}(\Omega)$ the infimum and supremum of two subsets $A, B \in \mathcal{P}(\Omega)$ are given respectively by the intersection $A \cap B \in \mathcal{P}(\Omega)$ and the union $A \cup B \in \mathcal{P}(\Omega)$

of subsets. For the case of $\mathcal{P}(\mathcal{H})$ the infimum of two closed subspaces $A, B \in \mathcal{P}(\mathcal{H})$ is given by the intersection $A \cap B \in \mathcal{P}(\mathcal{H})$. On the other hand, the union of two closed subspaces is in general not a closed subspace. This means that the union does not give us the supremum in this case. For two closed subspaces $A, B \in \mathcal{P}(\mathcal{H})$, the smallest closed subspace that contains both, is $\overline{A+B}$, the topological closure of the sum of the two subspaces. Hence this is the supremum of A and B in $\mathcal{P}(\mathcal{H})$.

It is worth remarking that we can see already at this level of the discussion one of the fundamental differences between a classical entity and a quantum entity. The vectors that are contained in the topological closure $\overline{A+B}$ of the sum of A and B are exactly the vectors that are superpositions of vectors in A and vectors in B. Hence for a quantum entity, described by $\mathcal{P}(\mathcal{H})$, there are additional vectors in the supremum of A and B, contained neither in A nor in B, while for the classical entity, described by $\mathcal{P}(\Omega)$, there are no such additional elements, because the supremum of A and B is the union $A \cup B$. This means that by changing our description to the level of the lattice, we can express both cases, the classical case and the quantum case, in the same mathematical language, which is not true if we describe the quantum entity in a Hilbert space and the classical entity in a state (or phase) space.

A **complete lattice** is a lattice that contains an infimum and a supremum for every subset of its elements, hence not only for each pair of elements as it is the case for a lattice. Let us express this requirement in a formal way. For $A \subset \mathcal{L}$ there exists $\wedge A$ and $\vee A$ such that:

$$x < a \ \forall a \in A \Leftrightarrow x < \wedge A \tag{6}$$

$$a < y \ \forall a \in A \Leftrightarrow \vee A < y \tag{7}$$

Note that a complete lattice $\mathcal{L}$ contains always a minimal element, namely the element $\wedge \mathcal{L}$, that we denote by 0, and a maximal element, namely $\vee \mathcal{L}$, that we denote by I.

Our two examples, $(\mathcal{P}(\Omega), \subset, \cap, \cup)$ and $(\mathcal{P}(\mathcal{H}), \subset, \cap, \overline{+})$ are both complete lattices. The minimal elements of $\mathcal{P}(\Omega)$ and $\mathcal{P}(\mathcal{H})$ are respectively $\emptyset$ and $\{0\}$, and the maximal elements are Ω and $\mathcal{H}$.

In the axiomatic approach the elements of the lattice represent the **properties** of the physical entity under study. Let us explain how the states of a physical entity are represented in the axiomatic approach.

2.2 Atoms of a Lattice

An element of a (complete) lattice is called an atom, if it is a smallest element different from 0. Let us define precisely what we mean. We say that $p \in \mathcal{L}$ is an atom of $\mathcal{L}$ if for $x \in \mathcal{L}$ we have:

$$0 < x < p \Rightarrow x = 0 \text{ or } x = p \tag{8}$$

The atoms of $(\mathcal{P}(\Omega), \subset, \cap, \cup)$ are the singletons of the phase space Ω, and the atoms of $(\mathcal{P}(\mathcal{H}), \subset, \cap, \overline{+})$ are the one dimensional subspaces (rays) of the Hilbert space $\mathcal{H}$.

In the traditional versions of the axiomatic approach the states of the physical entity under study were represented by the atoms of the lattice $\mathcal{L}$. Of course, this is largely due to the fact that for the case of standard quantum mechanics, the rays can indeed be identified as atoms of the lattice of closed subspaces of the Hilbert space. On the other hand, from a physical point of view, it is obvious that a state is a completely different concept than a property, and hence for an operational approach states should not be properties. Already in [1,2] it can be seen that when it comes to calculating and proving theorems, the states are treated differently from the properties, although the old tradition of representing both within the same mathematical structure, reminiscent of the 'states are represented by atoms of the lattice' idea, is maintained in [1,2]. Only slowly the insight grew about how to handle this problem in a more profound way. The new way, that is fully operationally founded, is to introduce two sets from the start, the set of **states** of the physical entity, denoted by Σ, and the set of properties, denoted by $\mathcal{L}$. It follows from the operational part of the construction that additionally to these two sets one needs to consider a one-to-one function $\kappa : \mathcal{L} \to \mathcal{P}(\Sigma)$, called the Cartan map, such that $p \in \kappa(a)$ expresses the following fundamental physical situation: "The property $a \in \mathcal{L}$ is 'actual' if the physical entity is in state $p \in \Sigma$". Moreover it follows from the operational aspects of the axiomatic approach that κ satisfies the three following additional requirements:

$$\kappa(\wedge_i a_i) = \cap_i \kappa(a_i) \tag{9}$$

$$\kappa(0) = \emptyset \tag{10}$$

$$\kappa(I) = \Sigma \tag{11}$$

for $(a_i)_i \in \mathcal{L}$ and 0 and I being respectively the minimal and maximal element of $\mathcal{L}$. The triple $(\Sigma, \mathcal{L}, \kappa)$, where Σ is a set, $\mathcal{L}$ a complete lattice, and κ a map satisfying (9), (10), and (11) has been called a state-property system [9,10], and it is this mathematical structure that can be derived from the operational aspects of the axiomatic approach.

Let us consider our two examples and see how this structure appears there. For the quantum case, Σ is the set of rays of the Hilbert space and $\mathcal{L}$ the set of closed subspaces. The Cartan map maps each closed subspace on the set of rays that are contained in this closed subspace, and indeed (9), (10), and (11) are satisfied. For the classical case, Σ is the phase space and $\mathcal{L}$ is the set of subsets of the phase space. The Cartan map is the identity. Of course, in the two examples, quantum and classical, the states correspond with atoms of the lattice of properties. By means of two axioms we regain this property for the general axiomatic situation. Since, as we mentioned already, the structure of a state-property system is derived from the operational aspects of the axiomatic approach, these two axioms should be considered as the first two axioms of the new axiomatic approach where states are not identified *a priori* with the atoms of the property lattice. Let us give these two axioms.

2.3 The First Two Axioms: State Determination and Atomisticity

A physical entity S is described by its state-property system $(\Sigma, \mathcal{L}, \kappa)$, where Σ is a set, its elements representing the states of S, $\mathcal{L}$ is a complete lattice, its elements representing the properties of S, and κ is a map from $\mathcal{L}$ to $\mathcal{P}(\Sigma)$, satisfying (9), (10), and (11), and expressing the physical situation: "The property $a \in \mathcal{L}$ is actual if the entity S is in state $p \in \Sigma$" by $p \in \kappa(a)$. This is the structure that we derive from only operational aspects of the axiomatic approach. The first axiom that we introduce consists in demanding that a state is determined by the set of properties that are actual in this state.

Axiom 1 (State Determination) *For $p, q \in \Sigma$ such that*

$$\bigwedge_{p \in \kappa(a)} a = \bigwedge_{q \in \kappa(b)} b \tag{12}$$

we have $p = q$.

We remark that in [1,2,8,12] this axiom is considered to be satisfied *a priori*. The second axiom consists in demanding that the states can be considered as atoms of the property lattice.

Axiom 2 (Atomisticity) *For $p \in \Sigma$ we have that*

$$\bigwedge_{p \in \kappa(a)} a \tag{13}$$

is an atom of $\mathcal{L}$.

Obviously these two axioms are satisfied for the two examples $\mathcal{P}(\Omega)$ and $\mathcal{P}(\mathcal{H})$ that we considered.

2.4 The Third Axiom: Orthocomplementation

For the third axiom it is already very difficult to give a complete physical interpretation. This third axiom introduces the structure of an orthocomplementation for the lattice of properties. At first sight the orthocomplementation could be seen as a structure that plays a similar role for properties as the negation in logic plays for propositions. But that is not a very careful way of looking at things. We cannot go into the details of the attempts that have been made to interpret the orthocomplementation in a physical way, and refer to [8,12,1,2,11] for those that are interested in this problem. Also in [13,14,15,16] the problem is considered in depth.

Axiom 3 (Orthocomplementation) *The lattice $\mathcal{L}$ of properties of the physical entity under study is orthocomplemented. This means that there exists a function $' : \mathcal{L} \rightarrow \mathcal{L}$ such that for $a, b \in \mathcal{L}$ we have:*

$$(a')' = a \tag{14}$$

$$a < b \Rightarrow b' < a' \tag{15}$$

$$a \wedge a' = 0 \text{ and } a \vee a' = I \tag{16}$$

For $\mathcal{P}(\Omega)$ the orthocomplement of a subset is given by the complement of this subset, and for $\mathcal{P}(\mathcal{H})$ the orthocomplement of a closed subspace is given by the subspace orthogonal to this closed subspace.

2.5 The Fourth and Fifth Axiom: The Covering Law and Weak Modularity

The next two axioms are called the covering law and weak modularity. There is no obvious physical interpretation for them. They have been put forward mainly because they are satisfied in the lattice of closed subspaces of a complex Hilbert space. These two axioms are what we have called the 'extra conditions' when we talked about Piron's representation theorem in the introduction of this section.

Axiom 4 (Covering Law) *The lattice $\mathcal{L}$ of properties of the physical entity under study satisfies the covering law. This means that for $a, x \in \mathcal{L}$ and $p \in \Sigma$ we have:*

$$a < x < a \vee p \Rightarrow x = a \text{ or } x = a \vee p \tag{17}$$

Axiom 5 (Weak Modularity) *The orthocomplemented lattice $\mathcal{L}$ of properties of the physical entity under study is weakly modular. This means that for $a, b \in \mathcal{L}$ we have:*

$$a < b \Rightarrow (b \wedge a') \vee a = b \tag{18}$$

These are the five axioms of standard quantum axiomatics. It can be shown that both axioms, the covering law and weak modularity, are satisfied for the two examples $\mathcal{P}(\Omega)$ and $\mathcal{P}(\mathcal{H})$ [7,8].

The two examples that we have mentioned show that both classical entities and quantum entities can be described by the common structure of a complete atomistic orthocomplemented lattice that satisfies the covering law and is weakly modular. Now we have to consider the converse, namely how this structure leads us to classical physics and to quantum physics.

3 The Representation Theorem

First we show how the classical and nonclassical parts can be extracted from the general structure, and second we show how the nonclassical parts can be represented by so-called generalized Hilbert spaces.

3.1 The Classical and Nonclassical Parts

Since both examples $\mathcal{P}(\Omega)$ and $\mathcal{P}(\mathcal{H})$ satisfy the five axioms, it is clear that a theory where the five axioms are satisfied can give rise to a classical theory, as well as to a quantum theory. It is possible to filter out the classical part by introducing the notions of clássical property and classical state.

Definition 1 (Classical Property) *Suppose that* $(\Sigma, \mathcal{L}, \kappa)$ *is the state property system representing a physical entity, satisfying axioms 1, 2 and 3. We say that a property* $a \in \mathcal{L}$ *is a classical property if for all* $p \in \Sigma$ *we have*

$$p \in \kappa(a) \text{ or } p \in \kappa(a') \tag{19}$$

The set of all classical properties we denote by $\mathcal{C}$.

Again considering our two examples, it is easy to see that for the quantum case, hence for $\mathcal{L} = \mathcal{P}(\mathcal{H})$, we have no nontrivial classical properties. Indeed, for any closed subspace $A \in \mathcal{H}$, different from 0 and $\mathcal{H}$, we have rays of $\mathcal{H}$ that are neither contained in A nor contained in A'. These are exactly the rays that correspond to states that are superposition states of states in A and states in A'. It is the superposition principle in standard quantum mechanics that makes that the only classical properties of a quantum entity are the trivial ones, represented by 0 and $\mathcal{H}$. It can also easily be seen that for the case of a classical entity, described by $\mathcal{P}(\Omega)$, all the properties are classical properties. Indeed, consider an arbitrary property $A \in \mathcal{P}(\Omega)$, then for any singleton $\{p\} \in \Sigma$ representing a state of the classical entity, we have $\{p\} \subset A$ or $\{p\} \subset A'$, since A' is the set theoretical complement of A.

Definition 2 (Classical State) *Suppose that* $(\Sigma, \mathcal{L}, \kappa)$ *is the state property system of a physical entity satisfying axioms 1, 2 and 3. For* $p \in \Sigma$ *we* *introduce*

$$\omega(p) = \bigwedge_{p\in\kappa(a), a\in C} a \tag{20}$$

$$\kappa_c(a) = \{\omega(p) \mid p \in \kappa(a)\} \tag{21}$$

and call $\omega(p)$ the classical state of the physical entity whenever it is in a state $p \in \Sigma$, and κ_c the classical Cartan map. The set of all classical states will be denoted by Ω.

Definition 3 (Classical State Property System) ***Suppose that $(\Sigma, \mathcal{L}, \kappa)$ is the state property system of a physical entity satisfying axioms 1, 2 and 3. The classical state property system corresponding with $(\Sigma, \mathcal{L}, \kappa)$ is $(\Omega, \mathcal{C}, \kappa_c)$.***

Let us look at our two examples. For the quantum case, with $\mathcal{L} = \mathcal{P}(\mathcal{H})$, we have only two classical properties, namely 0 and $\mathcal{H}$. This means that there is only one classical state, namely $\mathcal{H}$. It is the classical state that corresponds to 'considering the quantum entity under study' and the state does not specify anything more than that. For the classical case, every state is a classical state.

It can be proven that $\kappa_c : \mathcal{C} \to \mathcal{P}(\Omega)$ is an isomorphism[1,11]. This means that if we filter out the classical part and limit the description of our general physical entity to its classical properties and classical states, the description becomes a standard classical physical description.

Let us filter out the nonclassical part.

Definition 4 (Nonclassical Part) ***Suppose that $(\Sigma, \mathcal{L}, \kappa)$ is the state property system of a physical entity satisfying axioms 1, 2 and 3. For $\omega \in \Omega$ we introduce***

$$\mathcal{L}_\omega = \{a \mid a < \omega, a \in \mathcal{L}\} \tag{22}$$

$$\Sigma_\omega = \{p \mid p \in \kappa(\omega), p \in \Sigma\} \tag{23}$$

$$\kappa_\omega(a) = \kappa(a) \text{ for } a \in \mathcal{L}_\omega \tag{24}$$

and we call $(\Sigma_\omega, \mathcal{L}_\omega, \kappa_\omega)$ the nonclassical components of $(\Sigma, \mathcal{L}, \kappa)$.

For the quantum case, hence $\mathcal{L} = \mathcal{P}(\mathcal{H})$, we have only one classical state $\mathcal{H}$, and obviously $\mathcal{L}_\mathcal{H} = \mathcal{L}$. Similarly we have $\Sigma_\mathcal{H} = \Sigma$. This means that the only nonclassical component is $(\Sigma, \mathcal{L}, \kappa)$ itself. For the classical case, since all properties are classical properties and all states are classical states, we have $\mathcal{L}_\omega = \{0, \omega\}$, which is the trivial lattice, containing only its minimal and maximal element, and $\Sigma_\omega = \{\omega\}$. This means that the nonclassical components are all trivial.

For the general situation of a physical entity described by $(\Sigma, \mathcal{L}, \kappa)$ it can be shown that $\mathcal{L}_\omega$ contains no classical properties with respect to Σ_ω except 0 and ω, the minimal and maximal element of $\mathcal{L}_\omega$, and that if $(\Sigma, \mathcal{L}, \kappa)$ satisfies axioms 1, 2, 3, 4, and 5, then also $(\Sigma_\omega, \mathcal{L}_\omega, \kappa_\omega)$ $\forall \omega \in \Omega$ satisfy axioms 1, 2, 3, 4 and 5 (see [1,11]).

We remark that if axioms 1, 2 and 3 are satisfied we can identify a state $p \in \Sigma$ with the element of the lattice of properties $\mathcal{L}$ given by:

$$s(p) = \bigwedge_{p \in \kappa(a), a \in \mathcal{L}} a \tag{25}$$

which is an atom of $\mathcal{L}$. More precisely, it is not difficult to verify that, under the assumption of axioms 1 and 2, $s : \Sigma \to \Sigma_{\mathcal{L}}$ is a well-defined mapping that is one-to-one and onto, $\Sigma_{\mathcal{L}}$ being the collection of all atoms in $\mathcal{L}$. Moreover, $p \in \kappa(a)$ iff $s(p) < a$. We can call $s(p)$ the state property corresponding to p and define

$$\Sigma' = \{s(p) \mid p \in \Sigma\} \tag{26}$$

the set of state properties. It is easy to verify that if we introduce

$$\kappa' : \mathcal{L} \to \mathcal{P}(\Sigma') \tag{27}$$

where

$$\kappa'(a) = \{s(p) \mid p \in \kappa(a)\} \tag{28}$$

that

$$(\Sigma', \mathcal{L}, \kappa') \cong (\Sigma, \mathcal{L}, \kappa) \tag{29}$$

when axioms 1, 2 and 3 are satisfied.

To see in more detail in which way the classical and nonclassical parts are structured within the lattice $\mathcal{L}$, we make use of this isomorphism and introduce the direct union of a set of complete, atomistic orthocomplemented lattices, making use of this identification.

Definition 5 (Direct Union) *Consider a set $\{\mathcal{L}_\omega \mid \omega \in \Omega\}$ of complete, atomistic orthocomplemented lattices. The direct union $\textcircled{\vee}_{\omega \in \Omega} \mathcal{L}_\omega$ of these lattices consists of the sequences $a = (a_\omega)_\omega$, such that*

$$(a_\omega)_\omega < (b_\omega)_\omega \Leftrightarrow a_\omega < b_\omega \ \forall \omega \in \Omega \tag{30}$$

$$(a_\omega)_\omega \wedge (b_\omega)_\omega = (a_\omega \wedge b_\omega)_\omega \tag{31}$$

$$(a_\omega)_\omega \vee (b_\omega)_\omega = (a_\omega \vee b_\omega)_\omega \tag{32}$$

$$(a_\omega)'_\omega = (a'_\omega)_\omega \tag{33}$$

The atoms of $\textcircled{\vee}_{\omega \in \Omega} \mathcal{L}_\omega$ are of the form $(a_\omega)_\omega$ where $a_{\omega_1} = p$ for some ω_1 and $p \in \Sigma_{\omega_1}$, and $a_\omega = 0$ for $\omega \neq \omega_1$.

It can be proven that if $\mathcal{L}_\omega$ are complete, atomistic, orthocomplemented lattices, then also $\circledvee_{\omega \in \Omega} \mathcal{L}_\omega$ is a complete, atomistic, orthocomplemented lattice (see [1,11]). The structure of direct union of complete, atomistic, orthocomplemented lattices makes it possible to define the direct union of state property systems in the case axioms 1, 2, and 3 are satisfied.

Definition 6 (Direct Union of State Property Systems) *Consider a set of state property systems $(\Sigma_\omega, \mathcal{L}_\omega, \kappa_\omega)$, where $\mathcal{L}_\omega$ are complete, atomistic, orthocomplemented lattices and for each ω we have that Σ_ω is the set of atoms of $\mathcal{L}_\omega$. The direct union $\circledvee_\omega(\Sigma_\omega, \mathcal{L}_\omega, \kappa_\omega)$ of these state property systems is the state property system $(\cup_\omega \Sigma_\omega, \circledvee_\omega \mathcal{L}_\omega, \circledvee_\omega \kappa_\omega)$, where $\cup_\omega \Sigma_\omega$ is the disjoint union of the sets Σ_ω, $\circledvee_\omega \mathcal{L}_\omega$ is the direct union of the lattices $\mathcal{L}_\omega$, and*

$$\circledvee_\omega \kappa_\omega((a_\omega)_\omega) = \cup_\omega \kappa_\omega(a_\omega) \tag{34}$$

The first part of a fundamental representation theorem can now be stated. For this part it is sufficient that axioms 1, 2 and 3 are satisfied.

Theorem 1 (Representation Theorem: Part 1) *We consider a physical entity described by its state property system $(\Sigma, \mathcal{L}, \kappa)$. Suppose that axioms 1, 2 and 3 are satisfied. Then*

$$(\Sigma, \mathcal{L}, \kappa) \cong \circledvee_{\omega \in \Omega}(\Sigma'_\omega, \mathcal{L}_\omega, \kappa'_\omega) \tag{35}$$

where Ω is the set of classical states of $(\Sigma, \mathcal{L}, \kappa)$ (see definition 2), Σ'_ω is the set of state properties, κ'_ω the corresponding Cartan map, (see (26) and (28)), and $\mathcal{L}_\omega$ the lattice of properties (see definition 4) of the nonclassical component $(\Sigma_\omega, \mathcal{L}_\omega, \kappa_\omega)$. If axioms 4 and 5 are satisfied for $(\Sigma, \mathcal{L}, \kappa)$, then they are also satisfied for $(\Sigma'_\omega, \mathcal{L}_\omega, \kappa'_\omega)$ for all $\omega \in \Omega$.

Proof: see [1,11]

3.2 Further Representation of the Nonclassical Components

From the previous section follows that if axioms 1, 2, 3, 4 and 5 are satisfied we can write the state property system $(\Sigma, \mathcal{L}, \kappa)$ of the physical entity under study as the direct union $\circledvee_{\omega \in \Omega}(\Sigma'_\omega, \mathcal{L}_\omega, \kappa'_\omega)$ over its classical state space Ω of its nonclassical components $(\Sigma'_\omega, \mathcal{L}_\omega, \kappa'_\omega)$, and that each of these nonclassical components also satisfies axiom 1, 2, 3, 4 and 5. Additionally for each one of these nonclassical components $(\Sigma'_\omega, \mathcal{L}_\omega, \kappa'_\omega)$ no classical properties except 0 and ω exist. It is for these nonclassical components that a further representation theorem can be proven such that a vector space structure emerges for each one of the nonclassical components. To do this we rely on the original representation theorem that Piron proved in [7].

Theorem 2 (Representation Theorem: Part 2) *Consider the same situation as in theorem 1, with additionally axiom 4 and 5 satisfied. For each nonclassical component $(\Sigma'_\omega, \mathcal{L}_\omega, \kappa'_\omega)$, of which the lattice $\mathcal{L}_\omega$ has at least four orthogonal states*[a]*, there exists a vector space V_ω, over a division ring K_ω, with an involution of K_ω, which means a function*

$$^* : K_\omega \to K_\omega \tag{36}$$

such that for $k, l \in K_\omega$ we have:

$$(k^*)^* = k \tag{37}$$

$$(k \cdot l)^* = l^* \cdot k^* \tag{38}$$

and an Hermitian product on V_ω, which means a function

$$\langle\ ,\ \rangle : V_\omega \times V_\omega \to K_\omega \tag{39}$$

such that for $x, y, z \in V_\omega$ and $k \in K_\omega$ we have:

$$\langle x + ky, z\rangle = \langle x, z\rangle + k\langle x, y\rangle \tag{40}$$

$$\langle x, y\rangle^* = \langle y, x\rangle \tag{41}$$

$$\langle x, x\rangle = 0 \Leftrightarrow x = 0 \tag{42}$$

and such that for $M \subset V_\omega$ we have:

$$M^\perp + (M^\perp)^\perp = V_\omega \tag{43}$$

where $M^\perp = \{y \mid y \in V_\omega, \langle y, x\rangle = 0, \forall x \in M\}$. Such a vector space is called a generalized Hilbert space or an orthomodular vector space. And we have that:

$$(\Sigma'_\omega, \mathcal{L}_\omega, \kappa'_\omega) \cong (\mathcal{R}(V), \mathcal{L}(V), \nu) \tag{44}$$

where $\mathcal{R}(V)$ is the set of rays of V, $\mathcal{L}(V)$ is the set of biorthogonally closed subspaces (subspaces that are equal to their biorthogonal) of V, and ν makes correspond with each such biorthogonal subspace the set of rays that are contained in it.

Proof: See [7,8].

4 The Two Failing Axioms of Standard Quantum Mechanics

We have introduced all that is necessary to be able to put forward the theorem that has been proved regarding the failure of standard quantum mechanics for the description of the joint entity consisting of two separated quantum entities [1,2]. Let us first explain what is meant by separated physical entities.

[a]Two states $p, q \in \Sigma_\omega$ are orthogonal if there exists $a \in \mathcal{L}_\omega$ such that $p < a$ and $q < a'$.

4.1 *What Are Separated Physical Entities?*

We consider the situation of a physical entity S that consists of two physical entities S_1 and S_2. The definition of 'separated' that has been used in [1,2] is the following. Suppose that we consider two experiments e_1 and e_2 that can be performed respectively on the entity S_1 and on the entity S_2, such that the joint experiments $e_1 \times e_2$ can be performed on the joint entity S consisting of S_1 and S_2. We say that experiments e_1 and e_2 are separated experiments whenever for an arbitrary state p of S we have that (x_1, x_2) is a possible outcome for experiment $e_1 \times e_2$ if and only if x_1 is a possible outcome for e_1 and x_2 is a possible outcome for e_2. We say that S_1 and S_2 are separated entities if and only if all the experiments e_1 on S_1 are separated from the experiments e_2 on S_2.

Let us remark that S_1 and S_2 being separated does not mean that there is no interaction between S_1 and S_2. Most entities in the macroscopic world are separated entities. Let us consider some examples to make this clear.

The earth and the moon, for example, are separated entities. Indeed, consider any experiment e_1 that can be performed on the physical entity earth (for example measuring its position), and any experiment e_2 that can be performed on the physical entity moon (for example measuring its velocity). The joint experiment $e_1 \times e_2$ consists of performing e_1 and e_2 together on the joint entity of earth and moon (measuring the position of the earth and the velocity of the moon at once). Obviously the requirement of separation is satisfied. The pair (x_1, x_2) (position of the earth and velocity of the moon) is a possible outcome for $e_1 \times e_2$ if and only if x_1 (position of the earth) is a possible outcome of e_1 and x_2 (velocity of the moon) is a possible outcome of e_2. This is what we mean when we say that the earth has position x_1 and the moon velocity x_2 at once. Clearly this is independent of whether there is an interaction, the gravitational interaction in this case, between the earth and the moon.

It is not easy to find an example of two physical entities that are not separated in the macroscopic world, because usually nonseparated entities are described as one entity and not as two. In earlier work we have given examples of nonseparated macroscopic entities [17,33,19]. The example of connected vessels of water is a good example to give an intuitive idea of what nonseparation means.

Consider two vessels V_1 and V_2 each containing 10 liters of water. The vessels are connected by a tube, which means that they form a connected set of vessels. Also the tube contains some water, but this does not play any role for what we want to show. Experiment e_1 consists of taking out water of vessel V_1 by a siphon, and measuring the amount of water that comes out. We give the outcome x_1 if the amount of water coming out is greater than 10 liters. Experiment e_2 consists of doing exactly the same on vessel V_2. We give outcome x_2 to e_2 if the amount of water coming out is greater than 10 liters. The joint experiment $e_1 \times e_2$ consists of performing e_1 and e_2 together on the joint entity of the two connected vessels of water. Because of the connection, and the physical principles that govern connected vessels, for e_1 and for e_2 performed alone we find 20 liters of water coming out. This means that x_1 is a possible (even certain) outcome for e_1 and x_2 is a possible (also certain) outcome for e_2. If we perform the joint experiment $e_1 \times e_2$ the following happens. If there is more than 10 liters coming out of vessel V_1 there is less than 10 liters coming out of vessel V_2 and if there is more than 10 liters coming out of vessel V_2 there is less than 10 liters coming out of vessel V_1. This means that (x_1, x_2) is not a possible outcome for the joint experiment $e_1 \times e_2$. Hence e_1 and e_2 are nonseparated experiments and as a consequence V_1 and V_2 are nonseparated entities.

The nonseparated entities that we find in the macroscopic world are entities that are very similar to the connected vessels of water. There must be an ontological connection between the two entities, and that is also the reason that usually the joint entity will be treated as one entity again. A connection through dynamic interaction, as it is the case between the earth and the moon, interacting by gravitation, leaves the entities separated.

For quantum entities it can be shown that only when the joint entity of two quantum entities contains entangled states the entities are nonseparated quantum entities. It can be proven [17,33,19] that experiments are separated if and only if they do not violate Bell's inequalities. All this has been explored and investigated in many ways, and several papers have been published on the matter [17,33,19,20,21]. Interesting consequences for the Einstein Podolsky Rosen paradox and the violation of Bell's inequalities have been investigated [22,23].

4.2 The Separated Quantum Entities Theorem

We are ready now to state the theorem about the impossibility for standard quantum mechanics to describe separated quantum entities [1,2].

Theorem 3 (Separated Quantum Entities Theorem) *Suppose that S is a physical entity consisting of two separated physical entities S_1 and S_2. Let us suppose that axiom 1, 2 and 3 are satisfied and call $(\Sigma, \mathcal{L}, \kappa)$ the state property system describing S, and $(\Sigma_1, \mathcal{L}_1, \kappa_1)$ and $(\Sigma_2, \mathcal{L}_2, \kappa_2)$ the state property systems describing S_1 and S_2.*

If the fourth axiom is satisfied, namely the covering law, then one of the two entities S_1 or S_2 is a classical entity, in the sense that one of the two state property systems $(\Sigma_1, \mathcal{L}_1, \kappa_1)$ or $(\Sigma_2, \mathcal{L}_2, \kappa_2)$ contains only classical states and classical properties.

If the fifth axiom is satisfied, namely weak modularity, then one of the two entities S_1 or S_2 is a classical entity, in the sense that one of the two state property systems $(\Sigma_1, \mathcal{L}_1, \kappa_1)$ or $(\Sigma_2, \mathcal{L}_2, \kappa_2)$ contains only classical states and classical properties.

Proof: see [1,2]

The theorem proves that two separated quantum entities cannot be described by standard quantum mechanics. A classical entity that is separated from a quantum entity and two separated classical entities do not cause any problem, but two separated quantum entities need a structure where neither the covering law nor weak modularity are satisfied.

One of the possible ways out is that there would not exist separated quantum entities in nature. This would mean that all quantum entities are entangled in some way or another. If this is true, perhaps the standard formalism could be saved. Let us remark that even standard quantum mechanics presupposes the existence of separated quantum entities. Indeed, if we describe one quantum entity by means of the standard formalism, we take one Hilbert space to represent the states of this entity. In this sense we suppose the rest of the universe to be separated from this one quantum entity. If not, we would have to modify the description and consider two Hilbert spaces, one for the entity and one for the rest of the universe, and the states would be entangled states of the states of the entity and the states of the rest of the universe. But, this would mean that the one quantum entity that we considered is never in a well-defined state. It would mean that the only possibility that remains is to describe the whole universe at once by using one huge Hilbert space. It goes without saying that such an approach will lead to many other problems. For example, if this one Hilbert space has to describe the whole universe, will it also contain itself, as a description, because as a description, a human activity, it is part of the whole universe. Another, more down to earth problem is, that in this one Hilbert space of the whole universe also all classical macroscopical entities have to be described. But classical entities are not described by a Hilbert space, as we have seen in section 2. If the hypothesis that we can only describe the whole universe at once is correct, it would anyhow be more plausible that the theory that does deliver such a description would be the direct union structure of different Hilbert spaces. But if this is the case, we anyhow are already using a more general theory than standard quantum mechanics. So we can as well use the still slightly more general theory, where axioms 4 and 5 are not satisfied, and make the description of separated quantum entities possible.

All this convinces us that the shortcoming of standard quantum mechanics to be able to describe separated quantum entities is really a shortcoming of the mathematical formalism used by standard quantum mechanics, and more notably of the vector space structure of the Hilbert space used in standard quantum mechanics.

4.3 Operational Foundation of Quantum Axiomatics

To be able to explain the conceptual steps that are made to prove theorem 3 we have to explain how the concept of 'separated' is expressed in the quantum axiomatics that we introduced in section 2. Separated entities are defined by means of separated experiments. In the quantum axiomatics of section 2 we do not talk about experiments, which means that there is still a link that is missing. This link is made by what is called the operational foundation of the quantum axiomatic lattice formalism. Within this operational foundation a property of the entity under study is defined by the equivalence class of all experiments that test this property. We will not explain the details of this operational foundation, because some subtle matters are involved, and refer to [8,1,2] for these details. What we need to close the circle in this article is the fact that, making use of the operational foundations, it is possible to introduce 'separated properties' as properties that are defined by equivalence classes of separated experiments.

4.4 The Separated Quantum Entities Theorem Bis

Theorem 3 can then be reformulated completely in the language of the axiomatic quantum formalism that we introduced in section 2 in the following way:

Theorem 4 (Separated Quantum Entities Theorem Bis) *Suppose that we consider the compound entity S that consists of two physical entities S_1 and S_2. Suppose that axioms 1, 2 and 3 are satisfied for S, S_1 and S_2. Suppose that each property of S_1 is 'separated' from each property of S_2. If axiom 4 is satisfied for S, then one of the two entities S_1 or S_2 contains only classical properties and classical states, and hence S_1 or S_2 is a classical entity. If axiom 5 is satisfied for S, then one of the two entities S_1 or S_2 contains only classical properties and classical states, and hence S_1 or S_2 is a classical entity.*

4.5 Linearity at Stake

If the covering law is not satisfied for the lattice $\mathcal{L}$ that describes the compound entity consisting of two separated quantum entities, then this lattice cannot be represented into a vector space. This means that the superposition principle will not be valid for S. In standard quantum mechanics, situations have been encountered where the superposition principle is not valid, and one refers to these situations as 'the presence of superselection rules'. For example the property 'charge' for a microparticle entails such a superselection rule. There are no superpositions of states with different charge. It has always been possible to incorporate superselection rules into the standard formalism by demanding that there should be no superpositions between different sectors of the Hilbert space. The reason that this could be done for superselection rules as the ones that arise from the property charge, is because the states that correspond to different values of a physical quantity are always orthogonal. This has made it possible to circumvent the problem by considering different orthogonal sectors of a common Hilbert space, and not allowing superpositions between states of different sectors. It can be shown that the superselection rules that arise from the situation of separated quantum entities correspond to states that are not orthogonal, which means that the traditional way of avoiding the problem cannot work. In [1,2] explicit examples of states that are separated by a super selection rule are given. Also in [25] we give examples of such states.

4.6 Some Subtle Aspects of the Separated Quantum Entities Theorem

The 'Separated Quantum Entities Theorem' that was proved in [1,2] was correctly criticized by Cattaneo and Nisticó [26]. As we mentioned already, the proof in [1,2] is made by introducing separated experiments, where separated is defined as explained in section 4.1. Then separated properties are defined as properties that are tested by separated experiments, and once the property lattice of the joint entity is constructed in this way, the theorem can be proven. The whole construction in [1,2] is built by starting with only yes/no-experiments, hence experiments that have only two possible outcomes. The reason that the construction in [1,2] is made by means of yes/no-experiments has a purely historical origin. The version of operational quantum axiomatics elaborated in Geneva, where one of the authors was working when proving the separated quantum entities theorem, was a version where only yes/no-experiments are considered as basic operational concepts. There did exist at that time versions of operational quantum axiomatics that incorporated right from the start experiments with any number of possible outcomes as basic operational concepts, as for example the approach elaborated by Randall and Foulis [28,29,30]. Cattaneo and Nisticó proved in [26] that, by considering only yes/no-experiments as an operational basis for the construction of the property lattice of the compound entity consisting of separated entities, some of the possible experiments that can be performed on this compound entity are overlooked. It could well be that the experiments that had been overlooked in the construction of [1,2] were exactly the ones that, once added, would give rise to additional properties and make the lattice of properties satisfy again axiom 4 and 5. That is the reason that Cattaneo and Nisticó state explicitly in their article [26] that they do not question the mathematical argument of the proof, but rather its operational basis. This was indeed a serious critique that had been pondered carefully. Although the author involved in this matter remembers clearly that he was convinced then that the lattice of properties would not change by means of the addition of the lacking experiments indicated by Cattaneo and Nisticó, and that his theorem remained valid, there did not seem an easy way to prove this. The only way out was to redo the construction but now starting with experiments with any number of outcomes as basic operational concepts. This is done in [27], and indeed, the separated quantum entities theorem can also be proved with this operational basis. This means that in [27] the critique of Cattaneo and Nistico has been answered, and the result is that the theorem remains valid. The construction presented in [27] is however much less transparent than the original one to be found in [1,2]. That is the reason why it is interesting to analyze the most simple of all situations of such a compound entity, the one consisting of two separated spin 1/2 objects. This is exactly what we will do in [25]. On this simple example it is easy to go through the full construction of the lattice $\mathcal{L}$ and its set of states Σ, such that we can see how fundamentally different it is from a structure that would entail a vector space type of linearity. Note that since the separated quantum entities theorem is a no-go theorem, also the simple example of [25] contains a proof of the no-go aspect of the original theorem.

5 Attempts and Perspectives for Solutions

In this section we mention briefly what are the attempts that have meanwhile taken place to find a solution to the problem that we have considered in this paper.

If we consider the aspect of the Separated Quantum Entities Theorem where an explicit construction of the lattice of properties and set of states of the compound entity consisting of separated subentities is made, then the theorem proves that this construction cannot be made within standard quantum mechanics, from which follows that standard quantum mechanics cannot describe separated quantum entities. Of course, in its profound logical form the Separated Quantum Entities Theorem is a no-go theorem, which means that also some of the other hypotheses that are used to prove the theorem can be false and hence also at the origin of the problem. Research, which partially took place even before the Separated Quantum Entities Theorem, and partially afterwards, gives us some valuable extra information about what are the possible directions that could be explored to 'solve' the problem connected with the Separated Quantum Entities Theorem.

5.1 Earlier Research on the Compound Entity Problem

At the end of the seventies, one of the authors studied the problem of the description of compound entities in quantum axiomatics, but this time staying within the quantum axiomatic framework where each considered entity is described by a complex Hilbert space, as in standard quantum mechanics [31,32]. This means that the quantum axiomatic framework was only used to give an alternative but equivalent description of standard quantum mechanics, because even then the quantum axiomatic framework makes it possible to translate physical requirements in relation with the situation of a compound physical entity consisting of two quantum mechanical subentities. The main aim of this research on the problem was to find back the tensor product procedure of standard quantum mechanics for the description of the compound entity, but this time not as an *ad hoc* procedure, which it is in standard quantum mechanics, but from physically interpretable requirements. For these requirements, some so-called 'coupling conditions' were put forward.

Theorem 5 *We describe quantum entities S_1, S_2 and S, respectively by their Hilbert space lattices (sets of closed subspaces of the Hilbert space), $\mathcal{L}(\mathcal{H}_1)$, $\mathcal{L}(\mathcal{H}_2)$ and $\mathcal{L}(\mathcal{H})$, and by their Hilbert space state spaces (sets of rays of the Hilbert spaces) $\Sigma(\mathcal{H}_1)$, $\Sigma(\mathcal{H}_2)$ and $\Sigma(\mathcal{H})$. Suppose that dim $\mathcal{H}_1 > 2$ and dim $\mathcal{H}_2 > 2$. Suppose that h_1, h_2 are functions:*

$$h_1 : \mathcal{L}(\mathcal{H}_1) \to \mathcal{L}(\mathcal{H}) \tag{45}$$

$$h_2 : \mathcal{L}(\mathcal{H}_2) \to \mathcal{L}(\mathcal{H}) \tag{46}$$

such that for all $A_1, B_1, C_1, (A_1^i)_i \in \mathcal{L}(\mathcal{H}_1)$, $A_2, B_2, C_2, (A_2^j)_j \in \mathcal{L}(\mathcal{H}_2)$, $p_1 \in \Sigma(\mathcal{H}_1)$ and $p_2 \in \Sigma(\mathcal{H}_2)$ we have

$$A_1 \subset B_1 \Rightarrow h_1(A_1) \subset h_1(B_1) \tag{47}$$

$$A_2 \subset B_2 \Rightarrow h_2(A_2) \subset h_2(B_2) \tag{48}$$

$$h_1(\vee_i A_1^i)) = \vee_i h_1(A_1^i) \tag{49}$$

$$h_2(\vee_i A_2^i)) = \vee_i h_2(A_2^i) \tag{50}$$
$$h_1(\mathcal{H}_1) = h_2(\mathcal{H}_2) = \mathcal{H} \tag{51}$$
$$h_1(C_1) \leftrightarrow h_2(C_2) \tag{52}$$
$$h_1(p_1) \wedge h_2(p_2) \in \Sigma(\mathcal{H}) \tag{53}$$

where $\leftrightarrow$ *is the symbol for 'compatible', then* $\mathcal{P}(\mathcal{H})$ *is canonically isomorphic to* $\mathcal{P}(\mathcal{H}_1 \otimes \mathcal{H}_2)$ *or to* $\mathcal{P}(\mathcal{H}_1 \otimes \mathcal{H}_2^*)$.
Proof: see [31,32]

The conditions (47), (48),(49),(50), (51),(52) and (53) are called the 'coupling conditions' in [31,32]. The physical interpretation for the different conditions is quite straightforward. Conditions (47), (48),(49),(50) and (51) mean that the functions h_1 and h_2 are morphisms of the lattice structure. Hence they express that S_1 and S_2 can be recognized as subentities of S. Condition (52) expresses that properties of S_1 are compatible with properties of S_2, and condition (53) expresses that when S_1 and S_2 are in certain states, then S is in a state uniquely determined by these states of S_1 and S_2.

When the article [32] was written, the authors aimed at giving a physical justification for using the tensor product in standard quantum mechanics for the description of the compound entity consisting of two quantum entities. The theorem succeeds well in doing so. There is however one remarkable aspect of the theorem. Two possible solutions appear and they are not canonically isomorphic. This means that for the category of Hilbert space lattices and their morphisms none of the two solutions can be a categorical product, because than the two solutions should be canonically isomorphic. This is amazing, because one would expect that if one moves to the mathematical level that corresponds well with the physics, which should be the Hilbert space lattice rather than the Hilbert space itself, one would find one of the categorical products to correspond to what is needed for the description of the compound entity. Let us remark that the theorem shows that if the Hilbert spaces would all be real Hilbert spaces instead of complex, there is only one solution, in which case it could be a categorical product. Of course, it is well known that the complex numbers play an essential role in quantum mechanics, such that the two solutions do represent different entities.

5.2 Investigating Further a Categorical Approach

Becoming aware of the fact that no categorical solution can be inferred from theorem 5, it became interesting to look straightforwardly for a categorical construction. This is what was done in [33]. A categorical product, more specifically a co-product, can be constructed, but it gives a structure that is very different from what one gets in standard quantum mechanics (the tensor product of Hilbert spaces), and from what one gets from theorem 5. A theorem that is very similar to the Separated Quantum Entities Theorem can be proven for the co-product. Again two of the axioms of traditional quantum axiomatics are never satisfied for the compound entity of two quantum entities if we would describe this compound entity by means of the co-product, except when one of the subentities is a trivial entity, with a lattice of properties containing only 0 and I [33]. One of the failing axioms is again the covering law, which means that also here, if we choose to use the co-product instead of the separated product to describe the compound entity, linearity is gone. We cannot go more into detail on this matter in this paper, but refer to [13,14] where the situation of the three products, the separated product, the co-product and the Hilbert space tensor product, is studied in detail by one of the authors.

5.3 *The Problem of Pure States and Mixed States*

From what we have explained in the foregoing, the situation is such that (1) the compound entity consisting of two separated quantum entities cannot be described by standard quantum mechanics, and (2) there remains an unsolved problem in relation with the description of the compound entity of two (not necessarily separated) quantum entities, in the sense that traditional quantum mechanics only knows the tensor product of Hilbert spaces procedure, but this procedure cannot be fitted into an operational scheme at the axiomatic level. These results seem to indicate strongly that standard quantum mechanics must be generalized in the sense that a mathematical formalism should be worked out where the covering law is dropped, and hence linearity is lost. More recently however another possibility has come to the surface. Since also this possibility is relevant for the whole of the book where this article appears, we want to explain it briefly.

The Separated Quantum Entities Theorem of [2] and the Co-Product Theorem of [33] are in essence no-go theorems. And although both theorems give a very strong argument in favor of the view that standard quantum mechanics should be generalized by dropping the covering law and hence loosing linearity, we have to be careful. The most profound conclusion that has to be drawn from any no-go theorem is that at least one of the hypotheses that is used to prove the theorem is false. This means that the situation may even be worse, namely that not only the covering law (and weak modularity) should be dropped, but that there is even another, perhaps more important, hypothesis false in standard quantum mechanics. Of course, normally one would start to elaborate a generalization by dropping the least possible number of hypotheses necessary. But our research shows that even if we drop the covering law (and weak modularity), and construct then the co-product, we still do not find a satisfactory way to describe the compound entity of two quantum entities. Moreover, as we mentioned, the co-product structure is so different from the tensor product of Hilbert spaces structure used in standard quantum mechanics in a more or less fruitful way, that it might well be that we are not looking at the right category. This type of reflections and other ones have led one of the authors to consider the following possibility: *perhaps we should reconsider the way in which pure states are described in quantum mechanics by means of rays of the Hilbert space. And more concretely, perhaps also density operators, that normally are interpreted as only describing mixed states, represent pure states as well as mixed states.* This idea has been considered and introduced in [9] and the physical and philosophical situation connected with it has been analyzed in [34]. The quantum formalism where one allows density operators to represent pure states has been called 'extended quantum mechanics' in [9].

It is clear that this conceptual change will not solve all of the problems, for example, the fact that separated quantum entities cannot be described by extended standard quantum mechanics is still true. This means that also extended quantum mechanics shall have to be formulated by means of a structure that is different from the complex Hilbert space that is used in standard quantum mechanics. But something is also gained that might make that the mathematical change that is needed under extended quantum mechanics is less drastic than the one that is needed under standard quantum mechanics. We can see this by noting that for extended quantum mechanics the state of a compound entity, whether it is a ray state or a density operator state, is always a product state of states of the subentities. This comes from the 'mathematical' fact that any density operator in the tensor product Hilbert space is a product of density operators in the component Hilbert spaces. This means that there is more hope that a categorical construction for the lattice of properties and set of states of an extended quantum mechanics would give rise to a co-product that is closer to the tensor product of Hilbert spaces structure that is now used in standard quantum mechanics. We cannot say much more about this now, because we did not have the time to investigate sufficiently the operational and categorical structures that go along with extended quantum mechanics. We plan to engage in this investigation in the future. What we can see immediately is that if density operators also represent pure states, the axiom of atomicity will not be fulfilled for the pure states that arise from density operators.

References

1. D. Aerts, *The One and the Many: Towards a Unification of the Quantum and the Classical Description of One and Many Physical Entities,* Doctoral Thesis, Brussels Free University (1981).
2. D. Aerts, "Description of many physical entities without the paradoxes encountered in quantum mechanics", *Found. Phys.* **12**, 1131-1170 (1982).
3. E. Artin, *Geometric Algebra,* Interscience Publishers, Inc., New York (1957).
4. J. von Neumann, *Grundlagen der Quantenmechanik,* Springer-Verlag, Berlin, Heidelberg, New York (1932).
5. G. Birkhoff and J. von Neumann, "The logic of quantum mechanics", *Ann. Math.* **37**, 823-843 (1936).
6. G. Mackey, *The Mathematical Foundations of Quantum Mechanics,* Benjamin, New York (1963).
7. C. Piron, "Axiomatique quantique", *Helv. Phys. Acta* **37**, 439 (1964).
8. C. Piron, *Foundations of Quantum Physics,* Reading, Mass., W. A. Benjamin (1976).
9. D. Aerts, "Foundations of quantum physics: a general realistic and operational approach", *Int. J. Theor. Phys.* **38**, 289-358 (1999), lanl archive ref: *quant-ph/0105109.*
10. D. Aerts, E. Colebunders, A. Van der Voorde and B. Van Steirteghem, "State property systems and closure spaces: a study of categorical equivalence", *Int. J. Theor. Phys.* **38**, 359-385 (1999), lanl archive ref: *quant-ph/0105108.*
11. D. Aerts, "Classical theories and nonclassical theories as a special case of a more general theory", *J. Math. Phys.* **24**, 2441-2453 (1983).
12. C. Piron, *Mècanique Quantique: Bases et Applications,,* Press Polytechnique de Lausanne (1990).
13. F. Valckenborgh, "Operational axiomatics and compound systems", in *Current Research in Operational Quantum Logic: Algebras, Categories, Languages,* eds. Coecke, B., Moore, D. J. and Wilce, A., Kluwer Academic Publishers, Dordrecht, 219-244 (2000).
14. F. Valckenborgh, *Compound Systems in Quantum Axiomatics,* Doctoral Thesis, Brussels Free University (2001).
15. T. Durt and B. D'Hooghe, "The classical limit of the lattice-theoretical orthocomplementation in the framework of the hidden-measurement approach", this volume.
16. D. Aerts, "Reality and probability: introducing a new type of probability calculus", this volume.

17. D. Aerts, "Example of a macroscopical situation that violates Bell inequalities", *Lett. Nuovo Cimento*, **34**, 107-111 (1982).
18. D. Aerts, "How do we have to change quantum mechanics in order to describe separated systems", in *The Wave-Particle Dualism*, eds. Diner, S., et al., Kluwer Academic, Dordrecht, 419-431 (1984).
19. D. Aerts, "The description of separated systems and quantum mechanics and a possible explanation for the probabilities of quantum mechanics", in *Micro-physical Reality and Quantum Formalism*, eds. van der Merwe, A., et al., Kluwer Academic Publishers, 97-115 (1988).
20. D. Aerts, "An attempt to imagine parts of the reality of the microworld", in the *Proceedings of the Conference 'Problems in Quantum Physics; Gdansk '89'*, World Scientific Publishing Company, Singapore, 3-25 (1990).
21. D. Aerts, S. Aerts, J. Broekaert and L. Gabora, "The violation of Bell inequalities in the macroworld", *Found. Phys.* **30**, 1387-1414 (2000).
22. D. Aerts, "The physical origin of the Einstein Podolsky Rosen paradox", in *Open Questions in Quantum Physics: Invited Papers on the Foundations of Microphysics*, eds. Tarozzi, G. and van der Merwe, A., Kluwer Academic, Dordrecht, 33-50 (1985).
23. W. Christiaens, "Some notes on Aerts' interpretation of the EPR-paradox and the violation of Bell-inequalities", this volume.
24. D. Aerts, "Being and change: foundations of a realistic operational formalism", this volume.
25. D. Aerts and F. Valckenborgh, "Linearity and compound physical systems: the case of two separated spin 1/2 entities", this volume.
26. G. Cattaneo and G. Nisticó, "A note on Aerts' description of separated entities", *Found. Phys.* **20**, 119 (1990).
27. D. Aerts, "Quantum structures, separated physical entities and probability", *Found. Phys.* **24**, 1227-1259 (1994).
28. C. Randall and D. Foulis, "A mathematical setting for inductive reasoning", in *Foundations of Probability Theory, Statistical Inference, and Statistical Theories of Science III*, ed. C. Hooker, Kluwer Academic, Dordrecht, 169 (1976).
29. C. Randall and D. Foulis, "The operational approach to quantum mechanics", in *Physical Theories as Logico-Operational Structures*, ed. C. A. Hooker, Kluwer Academic, Dordrecht, 167 (1978).
30. D. Foulis and C. Randall, "What are quantum logics and what ought they to be?" in *Current Issues in Quantum Logic*, eds. E. Beltrametti and B. van Fraassen, Kluwer Academic, **35** (1981).
31. D. Aerts and I. Daubechies, "About the structure preserving maps of a

quantum mechanical propositional system", *Helv. Phys. Acta* **51**, 637-660 (1978).

32. D. Aerts and I. Daubechies, "Physical justification for using the tensor product to describe two quantum systems as one joint system", *Helv. Phys. Acta* **51**, 661-675 (1978).
33. D. Aerts, "Construction of the tensor product for lattices of properties of physical entities", *J. Math. Phys.* **25**, 1434-1441 (1984).
34. D. Aerts, "The description of joint quantum entities and the formulation of a paradox", *Int. J. Theor. Phys.* **39**, 485-496 (2000).

LINEARITY AND COMPOUND PHYSICAL SYSTEMS: THE CASE OF TWO SEPARATED SPIN 1/2 ENTITIES

DIEDERIK AERTS

Center Leo Apostel (CLEA) and Foundations of the Exact Sciences (FUND), Brussels Free University, Krijgskundestraat 33, 1160 Brussels, Belgium
E-mail: diraerts@vub.ac.be

FRANK VALCKENBORGH

Foundations of the Exact Sciences (FUND), Department of Mathematics, Brussels Free University, Pleinlaan 2, 1050 Brussels, Belgium
E-mail: fvalcken@vub.ac.be

We illustrate some problems that are related to the existence of an underlying linear structure at the level of the property lattice associated with a physical system, for the particular case of two explicitly separated spin 1/2 objects that are considered, and mathematically described, as one compound system. It is shown that the separated product of the property lattices corresponding with the two spin 1/2 objects does not have an underlying linear structure, although the property lattices associated with the subobjects in isolation manifestly do. This is related at a fundamental level to the fact that separated products do not behave well with respect to the covering law (and orthomodularity) of elementary lattice theory. In addition, we discuss the orthogonality relation associated with the separated product in general and consider the related problem of the behavior of the corresponding Sasaki projections as partial state space mappings.

1 Introduction

In another contribution in this volume [1], we have given an overview of a general mathematical framework, known under several names, that can be used for the description of physical systems in general and compound physical systems in particular. This framework was developed in its most important aspects in Geneva and Brussels [2,3,4,5,6,7,8,9,10,11]. One of the characteristics of this approach is the fact that the basic, primitive elements of the formalism have a sound realistic and operational interpretation. Indeed, a physical entity is described by means of its states, and the experimental projects which can be performed on samples of this system. Additional structure is gradually introduced as a series of physical postulates or mathematical axioms, ranging from the physically very plausible to axioms of an admittedly more technical nature, the latter introduced with the aim of bringing the structure closer to standard classical and quantum physics. We want to emphasize the generality of such an axiomatic approach and the fact that the results are valid in general, independently of the particularities of the formalism.

It has been shown that two of the more technical of these axioms — that are definitely satisfied for standard quantum systems — are not valid in the mathematical model that results from these general prescriptions for a compound physical system that consists of two operationally separated quantum objects [6,7,10]. One of the two failing axioms is equivalent with the linearity of the set of states for a quantum entity, hence with the superposition principle.

One of the themes of this book is to investigate how the failure of this "linearity" axiom is related to other perspectives on the problem of a "non-linear" quantum mechanics. In this paper we want to apply our axiomatic approach to the particular case of two separated spin 1/2 objects that are described as a whole. According to standard quantum physics, an isolated spin 1/2 system can be mathematically represented by the complex Hilbert space $\mathbb{C}^2$. More precisely, its set of possible states corresponds with the collection of all one-dimensional subspaces (rays) in this space, and observables with (some of the) self-adjoint operators on $\mathbb{C}^2$. The advantage is that for this relatively simple situation we can not only explicitly construct a mathematical model, but also keep an eye on the physical meaning of the mathematical objects and understand why the linearity axiom of standard quantum mechanics fails, at least in this case.

Let us give a brief overview of the basic ideas of the approach. In the next section, these ideas will become more clear, when we apply them to a particular example, the spin part of a single spin 1/2 object, *in extenso*. According to the prescriptions of the axiomatic approach, one should first construct the property lattice $\mathcal{L}$ and set of (pure) states Σ associated with the physical system under investigation, reflecting an underlying program of realism that is pursued [4]. In general, the state space is an orthogonality space,[a] while the property lattice, which is constructed from a class of *yes/no-experiments*, is always a complete atomistic lattice, usually taken to be orthocomplemented as well [8]. The connection between both structures is given by the Cartan map

$$\kappa : \mathcal{L} \to \mathcal{P}(\Sigma) : a \mapsto \{p \in \Sigma \mid p \lhd a\} \tag{1}$$

where $\lhd$ implements the physical idea of actuality of a, if the physical system is in a state p. The Cartan map is always a meet-preserving unital injection, hence $\mathcal{L} \cong \kappa[\mathcal{L}] \subseteq \mathcal{P}(\Sigma)$, leading to a state space representation of the property

[a] An orthogonality space consists of a set Σ and an orthogonality relation $\perp$, that is, a relation that is anti-reflexive and symmetric. One writes $A^{\perp} = \{q \in \Sigma \mid q \perp p \text{ for all } p \in A\}$, for $A \subseteq \Sigma$.

lattice. In addition, denoting the collection of all atoms in $\mathcal{L}$ by $\Sigma_{\mathcal{L}}$, we have $\kappa[\Sigma_{\mathcal{L}}] = \{\{p\} \mid p \in \Sigma\} \cong \Sigma$, hence we can identify these two sets, which we will often do. From a physical perspective, this relation reflects the fact that a physical state should embody a maximal amount of information at the level of the property lattice $\mathcal{L}$, even for individual samples of the physical system. In the axiomatic approach, a prominent role is played by the collection of biorthogonally closed subsets $\mathcal{F}(\Sigma) = \{A \subseteq \Sigma \mid A = A^{\perp\perp}\}$ of Σ. Indeed, the orthocomplementation can be introduced under the form of two axioms, which imply that $\kappa[\mathcal{L}] \subseteq \mathcal{F}(\Sigma)$ and $\kappa[\mathcal{L}] \supseteq \mathcal{F}(\Sigma)$, respectively. This *state-property duality* lies at the heart of the axiomatic approach [10,11].

Using this general framework, one of the basic aims is to establish a set of additional specific axioms, free from any probabilistic notions at its most basic level, to recover the formalism of standard quantum physics. Therefore, this approach is a theory of individual physical systems, rather than statistical ensembles. In doing so, a general theory is developed not only for quantal systems, but that also incorporates classical physical systems. The classical parts of a physical system are mathematically reflected in a decomposition of the property lattice in irreducible components [5,6,7,12]. For a genuine quantum system then, that satisfies all the requirements put forward in [5] and [6,7], the celebrated representation theorem of Piron states that these property lattices can be represented in a suitable generalized Hilbert (or orthomodular) space. More precisely, he showed that every irreducible complete atomistic orthocomplemented lattice $\mathcal{L}$ of length ≥ 4 that is orthomodular and satisfies the covering law (sometimes called a Piron lattice), can be represented as the collection of all closed subspaces $\mathcal{L}(\mathcal{H})$ of an appropriate orthomodular space $\mathcal{H}$ [2]. Mathematically speaking, there then exists a c-isomorphism $\mathcal{L} \cong \mathcal{L}(\mathcal{H})$.[b] The physical motivation for this particular lattice structure comes mainly from realistic and operational considerations. At first sight, the mathematical demands of orthomodularity and covering law look rather technical. They are usually justified by taking a more active (and ideal) point of view with respect to the physical meaning of the elements in the property lattice (for an overview, see [13]).

2 A Single Spin 1/2 System

To illustrate the physical meaning of these mathematical considerations, we shall treat some relatively simple particular cases *in extenso*. First, we il-

[b]A unital c-morphism between two complete ortholattices is a mapping that preserves arbitrary joins and orthocomplements.

lustrate the construction of the property lattice and state space for the spin part of a single spin 1/2 physical system. Denote the collection of possible states or, alternatively, preparations, for such a physical system by Σ. As we have seen, empirical access to the physical system is formalized by a set of yes/no-experiments $\mathcal{Q}$, and we proceed with an investigation of $\mathcal{Q}$, which will correspond with Stern-Gerlach experiments.

More precisely, for each spatial direction, a non-trivial definite experimental project is associated with a Stern-Gerlach experiment in that direction, relative to some reference direction; $\alpha_{\theta,\phi}$ denotes the experimental project associated with such an experiment in the direction given by (θ, ϕ), with the following prescription for the attribution of results, if the experiment is properly conducted on a particular sample of the physical system:

> Attribute the positive result (outcome "yes") if the spin 1/2 object is detected at the upper position; otherwise, attribute a negative result (outcome "no").

The collection of all yes/no-experiments will be denoted by Q. Consequently, at this point

$$Q \supseteq \{\alpha_{\theta,\phi} \mid 0 \leq \theta < \pi,\ 0 \leq \phi < 2\pi\} \tag{2}$$

The states of the spin 1/2 particle are the spin states $p(\theta, \phi)$ in the different spatial directions:

$$\Sigma = \{p(\theta, \phi) \mid 0 \leq \theta < \pi,\ 0 \leq \phi < 2\pi\} \tag{3}$$

One of the fundamental ingredients of any physical theory is linked with the following somewhat imprecise statement:

> The yes/no-experiment α gives with certainty the outcome "yes" whenever the sample object happens to be in a state p.

This statement will be expressed symbolically by a binary relation between the set of states and the class of yes/no-experiments. More precisely, the connection between the experimental access to the physical system and physical reality itself can be formalized by a binary relation $\triangleleft \subseteq \Sigma \times Q$. This relation symbolizes the following idea: $p \triangleleft \alpha$ means that if the physical system is (prepared) in a state p, the positive result for α would be obtained, should one execute the yes/no-experiment. In this case, the yes/no-experiment is said to be *true* for the object, if it is in the state p. It is conceptually important to note the counterfactual locution. Indeed, this formulation will allow us to attribute many properties to a particular sample of a physical system. The binary relation induces in a natural way a map, which is intimately related to the Cartan map:

$$S_T : Q \rightarrow \mathcal{P}(\Sigma) : \alpha \mapsto \{p \in \Sigma \mid p \triangleleft \alpha\} \tag{4}$$

For the spin 1/2 particle it is an experimental fact that $p(\theta, \phi) \triangleleft \alpha_{\theta',\phi'}$ iff $(\theta, \phi) = (\theta', \phi')$. There is no relation $\triangleleft$ between $p(\theta, \phi)$ and $\alpha_{\theta',\phi'}$ when $(\theta, \phi) \neq (\theta', \phi')$.

Q is naturally equipped with an inversion relation

$$\tilde{} : Q \rightarrow Q : \alpha \mapsto \tilde{\alpha} \tag{5}$$

the yes/no-experiment $\tilde{\alpha}$ has by definition the same experimental set-up as α, but the positive and negative alternatives are interchanged. This means that $p \triangleleft \tilde{\alpha}$ if the yes/no-experiment α gives with certainty the outcome "no" whenever the state of the physical entity is p. One then has the induction of a natural, physically motivated pre-order structure on Q:

$$\alpha < \beta \text{ iff } S_T(\alpha) \subseteq S_T(\beta) \tag{6}$$

which is used to generate the property lattice. Indeed, it is natural to call two yes/no-experiments equivalent if they cannot be distinguished experimentally, that is, $\alpha \approx \beta$ iff $S_T(\alpha) = S_T(\beta)$ iff $p \triangleleft \alpha \Leftrightarrow p \triangleleft \beta$. For a quantum spin 1/2 particle, it is well known that, according to experiment, one has

$$\tilde{\alpha}_{\theta,\phi} \approx \alpha_{\pi-\theta,\phi+\pi} \tag{7}$$

At this moment, we have made Q into a pre-ordered class, with some sort of an inversion relation. There is a fundamental operation which associates with any collection of yes/no-questions a new yes/no-experiment. Thus, the set of yes/no-experiments should be closed under products. More formally, we have an operation

$$\Pi : \mathcal{P}(Q) \rightarrow Q : \{\alpha_j \mid j \in J\} \mapsto \Pi\{\alpha_j \mid j \in J\} \tag{8}$$

The experimental procedure for this yes/no-experiment consists in choosing randomly one of the α_i and executing the associated experiment. With this specification, we obviously have

$$\tilde{\Pi}\{\alpha_j \mid j \in J\} = \Pi\{\tilde{\alpha}_j \mid j \in J\} \tag{9}$$

which appears somewhat strange at first, and its misunderstanding has been a point of some dispute in the past. In fact, this clever definition of product experiments allows us to attribute unambiguously various different properties to (some preparation of) a particular physical system, without having to explicitly test for all properties on the same object [6]. According to our prescriptions the binary relation $\lhd$ should satisfy $p \lhd \Pi\{\alpha_j \mid j \in J\} \Leftrightarrow p \lhd \alpha_j$ for all $j \in J$, or equivalently

$$S_T(\Pi\{\alpha_j \mid j \in J\}) = \bigcap\{S_T(\alpha_j) \mid j \in J\} \tag{10}$$

For example, for a spin 1/2 particle it is experimentally known that $\Pi(\{\alpha_{\theta,\phi}, \alpha_{\theta',\phi'}\}) \approx \widetilde{\tau}$ unless (θ, ϕ) and (θ', ϕ') represent the same spatial directions.

Finally, there exist *trivial* yes/no-experiments τ and $\widetilde{\tau}$. A possible experimental procedure for τ would consist in doing nothing with the physical system under consideration and always give the positive result. Both yes/no-experiments are in some sense ideal elements, and can be viewed as being added for technical reasons.

The equivalence relation $\approx$ on Q partitions Q in the collection of equivalence classes, according to a standard argument. Moreover, the pre-ordered structure on Q collapses into a partial order on $\mathcal{L} := Q/\approx = \{[\alpha] \mid \alpha \in Q\}$, with $[\alpha]$ denoting the equivalence class of α; S_T lifts to the Cartan map κ, and $\mathcal{L}$ can be mentally put in a one-to-one correspondence with the set of properties or elements of reality of the physical system, in a sense derived from that of Einstein, Podolsky and Rosen [14]. Moreover, it is not very difficult to show that $\mathcal{L}$ becomes a complete lattice [5], with

$$\bigwedge\{[\alpha_j] \mid j \in J\} = [\Pi\{\alpha_j \mid j \in J\}] \tag{11}$$

Given the physical meaning of the equivalence relation, one can unambiguously state that a property is *actual* if one of its corresponding yes/no-experiments is true. It is the lattice $\mathcal{L}$ that we use to describe the properties of a physical system. Note that $\mathcal{L}$, like any complete lattice, always contains a maximal element $I = [\tau]$ and a minimal element $0 = [\widetilde{\tau}]$.

From now on, we will denote equivalence classes $[\alpha]$ by a. For our particular example, we put $a(\theta, \phi) := [\alpha_{\theta,\phi}]$, to make the distinction very clear.

Also the binary relation $\lhd$ lifts to the level of the property lattice $\mathcal{L}$. With some abuse of notation, we then have $p \in \kappa(a) \Leftrightarrow p \lhd a$. The physical interpretation is the following: $p \in \kappa(a)$ stands for "The property a is actual if the physical system is in a state p". For our example, we have $\kappa(a(\theta, \phi)) = \{p(\theta, \phi)\}$.

In a complete lattice, any subcollection of elements has a join or supremum. For a set of properties $([\alpha_i])_i \in \mathcal{L}$ the join can be defined in a purely mathematical way as follows:

$$\bigvee\{[\alpha_j] \mid j \in J\} = \bigwedge\{[\beta] \mid \alpha_j < \beta \text{ for all } j \in J\} \tag{12}$$

It is for this reason that the join of a collection of properties has no obvious physical interpretation.[c]

Let us reconsider our example. Identifying, with some abuse of terminology, the properties with their corresponding equivalence classes $a(\theta,\phi) = [\alpha_{\theta,\phi}]$, it is true that

$$a(\theta,\phi) \wedge a(\theta',\phi') = 0 \tag{13}$$

if (θ,ϕ) and (θ',ϕ') represent different spatial directions, since $\Pi\{\alpha_{\theta,\phi}, \alpha_{\theta',\phi'}\} \approx \tilde{\tau}$. Indeed, there are no preparations for a spin 1/2 system in which both properties can be actual at the same time.

For our example, it is an experimental fact that if we consider two yes/no-experiments $\alpha_{\theta,\phi}$ and $\alpha_{\theta',\phi'}$ where (θ,ϕ) and (θ',ϕ') represent different spatial directions, there is no third yes/no-experiment of the type $\alpha_{\theta'',\phi''}$ such that $\alpha_{\theta,\phi} < \alpha_{\theta'',\phi''}$ and $\alpha_{\theta',\phi'} < \alpha_{\theta'',\phi''}$. This proves that the only yes/no-experiment β such that $\alpha_{\theta,\phi} < \beta$ and $\alpha_{\theta',\phi'} < \beta$ is τ. Hence for $(\theta,\phi) \neq (\theta',\phi')$ we have:

$$a(\theta,\phi) \vee a(\theta',\phi') = I \tag{14}$$

At this moment, we have found from operational considerations all the structural ingredients to define the basic mathematical structure attributed to the compound system that consists of two (operationally) separated spin 1/2 particles. This structure consists in a triple $(\Sigma, \mathcal{L}, \kappa)$ or $(\Sigma, \mathcal{L}, \triangleleft)$, that we have called a state-property system elsewhere [15,16]. The elements of Σ are the states attributed to the physical system under investigation, the elements of $\mathcal{L}$ correspond with its possible properties, and the connection between both sets is given by a Cartan map or, equivalently, a suitable binary relation, as we have seen.

$$\Sigma = \{p(\theta,\phi) \mid 0 \leq \theta < \pi,\ 0 \leq \phi < 2\pi\} \tag{15}$$

$$\mathcal{L} = \{a(\theta,\phi) \mid 0 \leq \theta < \pi,\ 0 \leq \phi < 2\pi\} \cup \{0\} \cup \{I\} \tag{16}$$

$$\kappa(0) = \emptyset, \quad \kappa(a(\theta,\phi)) = \{p(\theta,\phi)\}, \text{ and } \kappa(I) = \Sigma \tag{17}$$

In particular, note that κ maps atoms in $\mathcal{L}$ to singletons in Σ, and that κ gives a state space representation of $\mathcal{L}$.

[c] We remark that the meet operation is equivalent to logical conjunction. The "join" operation is however in general not equivalent to the "or" operation of logic. For classical physical entities the "join" operation is equivalent to logical disjunction, but this is not the case for quantum entities. This fact is at the origin of the common use of the word "quantum logic" for the lattice structure that arises in this way.

The (infinite) property lattice $\mathcal{L}$ for a single spin 1/2 object can be visually displayed by giving its Hasse diagram:

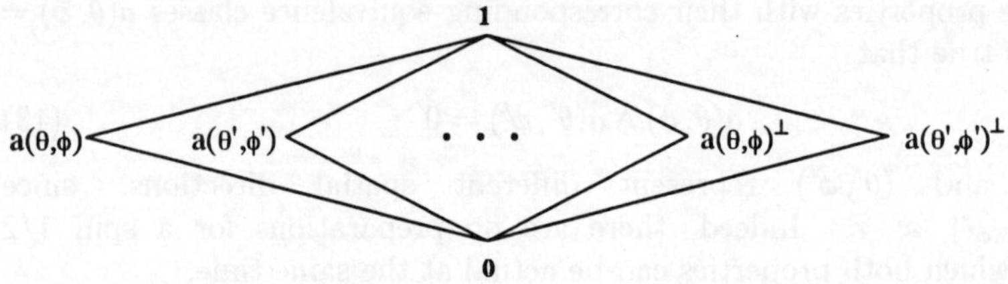

In the axiomatic approach, two states are defined to be *orthogonal* if there exists a yes/no-experiment $\alpha \in Q$ such that $p \triangleleft \alpha$ and $q \triangleleft \widetilde{\alpha}$. For our example, only one state will be orthogonal to a given state $p(\theta, \phi)$, being the state $p(\pi - \theta, \phi + \pi)$. In this way, Σ becomes an orthogonality space.

It is an experimental fact that $a(\theta, \phi)$ is never a *classical* property.[d] Indeed, if we prepare a spin 1/2 object in a state that corresponds to a direction orthogonal to (θ, ϕ), then neither $\alpha_{\theta,\phi}$ nor $\widetilde{\alpha}_{\theta,\phi}$ is true.

For the sake of illustration, let us also consider a second measurement scheme that would be deemed equivalent with $\alpha_{\theta,\phi}$ according to the axiomatic approach. Let $\beta_{\theta,\phi}$ have the same experimental arrangement, except for the fact that the bottom channel is blocked by a suitable absorbing device, in order to prevent an object to be localized below. It is experimentally known that $S_T(\alpha_{\theta,\phi}) = S_T(\beta_{\theta,\phi})$, hence $\alpha_{\theta,\phi} \approx \beta_{\theta,\phi}$. Observe also that $\Pi\{\alpha_{\theta,\phi}, I\} \approx \alpha_{\theta,\phi}$, but $\widetilde{\Pi}\{\alpha_{\theta,\phi}, I\} \not\approx \widetilde{\alpha}_{\theta,\phi}$.

The connection between the axiomatic approach and the standard quantum mechanical description of the spin part of a single spin 1/2 particle, is given by the well known (and easy to see) fact that this lattice can be represented as the collection of all closed subspaces of $\mathbb{C}^2$, that is, $\mathcal{L}(\mathbb{C}^2)$, with

$$a(\theta, \phi) \mapsto [(\exp(-i\frac{\phi}{2}) \cos\frac{\theta}{2}, \exp(i\frac{\phi}{2}) \sin\frac{\theta}{2})] \tag{18}$$

Note that this mapping indeed preserves the orthogonality relation.

For a single spin 1/2 object, we thus have a relatively simple property lattice, in which all non-trivial elements are also representatives of (pure) states. Denoting the collection of one-dimensional subspaces of the Hilbert space $\mathcal{H} = \mathbb{C}^2$ by $\Sigma_{\mathcal{H}}$, we can also put $\Sigma \cong \Sigma_{\mathcal{H}} \cong \mathbb{C}P^1$, this last set being complex projective 1-space.

---

[d] A "property" $a = [\alpha]$ is said to be classical, if for any state of the physical system either α or $\widetilde{\alpha}$ is true.

Once one has arrived at the basic structure of a state-property system, the axiomatic approach proceeds by introducing further axioms on this structure, with the aim of bringing the structure closer to standard quantum mechanics. It is an easy task to verify that all the axioms, as stated in [1], are satisfied for the property lattice displayed above. In the next section, however, we will give an explicit example in which the axioms of orthomodularity and the covering law both fail.

3 The Separated Product of Two Spin 1/2 Systems

One of the easiest compound physical systems that intuitively and conceptually presents itself, is the case of two separated spin 1/2 objects that are described as one whole. Consider two such systems, respectively represented by property lattices $\mathcal{L}_i(\mathbb{C}^2)$, for $i = 1, 2$, and suppose that we want to give a mathematical description for this situation. In this section, we will explicitly construct the property lattice and state space that corresponds to this physical situation.

In general, the separated product — the mathematical description of this situation — $\mathcal{L}_1 \textcircled{\scriptsize$\wedge$} \mathcal{L}_2$ of $\mathcal{L}_1$ and $\mathcal{L}_2$ can be constructed in two different ways. First, one can give an explicit construction from the bottom up, starting from the collection of yes/no-experiments for this system. This construction has the advantage that every property corresponds to an equivalence class of experimental projects, so in principle one has at one's disposal an experimental procedure that tests for any property. Second, the separated product can be mathematically generated through a biorthocomplementation procedure, starting from the orthogonality space $(\Sigma_1 \times \Sigma_2, \perp)$, with the orthogonality relation given by

$$(p_1, p_2) \perp (q_1, q_2) \text{ iff } (p_1 \perp_1 q_1 \text{ or } p_2 \perp_2 q_2) \tag{19}$$

This construction is more convenient from a mathematical point of view, but has the drawback that it is a purely formal construction, which needs an *a posteriori* physical interpretation. Here, we will give an overview of the first approach, at least for the particular case that is the main subject of this paper. For a more detailed exposition of the general case, we refer to [6,7,10].

First, we should be slightly more specific about what we mean with two objects being separated. Intuitively speaking, a necessary operational condition should be the following: it should be possible to devise an experimental procedure, say $e_1 \times e_2$, with outcome set $O_{e_1} \times O_{e_2}$, on the compound system as a whole for every pair of experiments (e_1, e_2), with e_1 an experiment with outcome set O_{e_1} on the first object and similarly for e_2. Moreover, whatever experiment we decide to perform on one of the objects, should yield a result that is independent of the state of the other object and *vice versa*. That is, if the compound system is in a state such that (x_1, x_2) is a possible outcome for the experiment $e_1 \times e_2$, then the first object is in a state such that x_1 is a possible result for the experiment e_1, and similarly for the second object. In addition, any experiment corresponding to one of the subobjects, can be executed independent of the presence or absence of the other subobject. Moreover, if an outcome is possible for an experiment e_1 to be performed on the first object, then this outcome can be obtained irrespective of the presence or absence of the other object. Note that this operational idea of separation is closely related to the notion presented by Einstein, Podolsky and Rosen [14]. Also, note that there is a big conceptual difference between the physical notions of *separation* and *interaction*, the latter notion being related to the causal structure of physical reality.

As before, we will mainly restrict ourselves to spin measurements on a spin 1/2 object, because in this case any experiment on one of the subobjects has only two possible results. On the other hand, an arbitrary experiment of the form $\alpha_1(\theta_1, \phi_1) \times \alpha_2(\theta_2, \phi_2)$ on the compound physical system has 4 possible outcomes: $(y, y), (y, n), (n, y)$ and (n, n), where for notational reasons we have slightly adapted our notation. According to the prescriptions of the axiomatic approach, we have to construct the collection of yes/no-experiments associated with all these product experiments. First, observe that product experiments which have at least one component equivalent with a trivial experiment on the corresponding subobject, are equivalent with either a trivial experiment on the compound system, or an experiment which only involves one of the subobjects. For example, $\alpha_1(\theta_1, \phi_1) \times \tau_2$ is true iff $\alpha_1(\theta_1, \phi_1)$ is true, with respect to the first subobject. To make the distinction, we will put $C\alpha_1$ to be the experimental project on the compound system that consists in performing the experiment corresponding to α_1 on the first subobject, and a similar convention for the second subobject.

Thus, let us consider an arbitrary product experiment of the form $\alpha_1(\theta_1, \phi_1) \times \alpha_2(\theta_2, \phi_2)$, which has 4 possible outcomes. With this product experiment, one can *a priori* associate 2^4 different yes/no-experiments, corresponding to all subsets of the outcome set. The two trivial subsets are equivalent with trivial yes/no-experiments on the compound system, hence will be left out of the rest of the discussion. Temporarily abbreviating $\alpha_i(\theta_i, \phi_i)$ by α_i and $\alpha_i(\pi - \theta_i, \phi_i + \pi)$ by $\widetilde{\alpha}_i$ for a particular direction (θ_i, ϕ_i), we will use the following conventional notations for yes/no-experiments associated with subsets of a particular form, displayed in a well-organized table form below. At the same time, we indicate the corresponding inverse yes/no-experiments:

Notation	Outcome set	Inverse yes/no-experiment
$C\alpha_1$	(y,y), (y,n)	$C\widetilde{\alpha}_1$
$C\alpha_2$	(y,y), (n,y)	$C\widetilde{\alpha}_2$
$\alpha_1\Delta\alpha_2$	(y,y)	$\widetilde{\alpha}_1\nabla\widetilde{\alpha}_2$
$\alpha_1\nabla\alpha_2$	(y,y), (y,n), (n,y)	$\widetilde{\alpha}_1\Delta\widetilde{\alpha}_2$
$\alpha_1\Theta\alpha_2$	(y,y),(n,n)	$\widetilde{\alpha}_1\Theta\alpha_2 \approx \alpha_1\Theta\widetilde{\alpha}_2$

Observe that the notation $C\widetilde{\alpha}_j$ is unambiguous, in the sense that we have $(C\widetilde{\alpha}_j) \approx C(\widetilde{\alpha}_j)$. Considering yes/no-experiments of this general form, we can generate all yes/no-experiments associated with the product experiment $\alpha_1(\theta_1, \phi_1) \times \alpha_2(\theta_2, \phi_2)$. For example, a yes/no-experiment that tests for the result (n, n) could be constructed as $\alpha_1(\pi - \theta_1, \pi + \phi_1)\Delta\alpha_2(\pi - \theta_2, \pi + \phi_2)$. Indeed, this yes/no-experiment would be true if the compound system is in a state such that $\alpha_1(\pi - \theta_1, \pi + \phi_1)$ is true with respect to the first subobject, and $\alpha_2(\pi - \theta_2, \pi + \phi_2)$ is true with respect to the second subobject.

In this way, we can obtain all properties for the compound system that consists of two separated spin 1/2 objects, and we can direct our attention towards the construction of the property lattice. Because of the demand that both subobjects are separated, we have

$$p \triangleleft C\alpha_1(\theta_1, \phi_1) \text{ iff } p_1 \triangleleft_1 \alpha_1(\theta_1, \phi_1) \tag{20}$$

$$p \triangleleft C\alpha_2(\theta_2, \phi_2) \text{ iff } p_2 \triangleleft_2 \alpha_2(\theta_2, \phi_2) \tag{21}$$

$$p \triangleleft \alpha_1(\theta_1, \phi_1)\Delta\alpha_2(\theta_2, \phi_2) \text{ iff } p_1 \triangleleft_1 \alpha_1(\theta_1, \phi_1), \text{ and } p_2 \triangleleft_2 \alpha_2(\theta_2, \phi_2) \tag{22}$$

$$p \triangleleft \alpha_1(\theta_1, \phi_1)\nabla\alpha_2(\theta_2, \phi_2) \text{ iff } p_1 \triangleleft_1 \alpha_1(\theta_1, \phi_1) \text{ or } p_2 \triangleleft_2 \alpha_2(\theta_2, \phi_2) \tag{23}$$

$$\begin{aligned} p \triangleleft \alpha_1(\theta_1, \phi_1)\Theta\alpha_2(\theta_2, \phi_2) \text{ iff either } & p_1 \triangleleft_1 \alpha_1(\theta_1, \phi_1) \text{ and } p_2 \triangleleft_2 \alpha_2(\theta_2, \phi_2), \\ \text{or } & p_1 \triangleleft_1 \alpha_1(\pi - \theta_1, \pi + \phi_1) \text{ and } p_2 \triangleleft_2 \alpha_2(\pi - \theta_2, \pi + \phi_2) \end{aligned} \tag{24}$$

It is not very difficult to see from these prescriptions that the state of the global system is completely known whenever one knows the states of the two separated spin 1/2 objects that make up the compound system. Consequently, the set of states can be taken as $\Sigma_1 \times \Sigma_2$, with

$$\Sigma_1 = \{p_1(\theta_1, \phi_1) \mid 0 \le \theta_1 < \pi,\ 0 \le \phi_1 < 2\pi\} \tag{25}$$

$$\Sigma_2 = \{p_2(\theta_2, \phi_2) \mid 0 \leq \theta_2 < \pi,\ 0 \leq \phi_2 < 2\pi\} \tag{26}$$

However, for notational reasons we shall often use abbreviations of the form p_j for a general element of Σ_j.

Consequently, we can represent the property lattice corresponding to this situation as a subcollection of $\mathcal{P}(\Sigma_1 \times \Sigma_2)$. Properties of the first kind would be represented by singletons $(p_1(\theta_1, \phi_1), p_2(\theta_2, \phi_2))$; properties of the second kind consist of all sets of the general form $\{p_1(\theta_1, \phi_1)\} \times \Sigma_2 \cup \Sigma_1 \times \{p_2(\theta_2, \phi_2)\}$; finally, properties of the third kind are two-element sets of the general form $\{(p_1(\theta_1, \phi_1), p_2(\theta_2, \phi_2)), (p_1(\pi - \theta_1, \pi + \phi_1), p_2(\pi - \theta_2, \pi + \phi_2))\}$. The full property lattice is then generated by taking arbitrary products of corresponding yes/no-experiments, which amounts to taking intersections of the corresponding images of the Cartan map, as we have seen before, because the Cartan map is one-to-one and preserves intersections, due to the corresponding property of the map S_T. Denoting $S^2 := [0, \pi) \times [0, 2\pi)$, we then obtain for the full property lattice corresponding to this physical situation the intersection system $\mathcal{I}(\Omega)$ generated by the collection of all these sets Ω, and it is geometrically clear, by considering the set $\Sigma_1 \times \Sigma_2$, that the second equation also holds:

$$\mathcal{L}_1 \textcircled{\scriptsize$\wedge$} \mathcal{L}_2 \cong \mathcal{I}(\mathcal{T} \cup \mathcal{A}_1 \cup \mathcal{A}_2 \cup \mathcal{S} \cup \mathcal{U} \cup \mathcal{B}) \tag{27}$$

$$\cong \mathcal{T} \cup \mathcal{A}_1 \cup \mathcal{A}_2 \cup \mathcal{S} \cup \mathcal{U} \cup \mathcal{P} \tag{28}$$

with

$$\mathcal{T} = \{\emptyset, \Sigma_1 \times \Sigma_2\}$$

$$\mathcal{A}_1 = \{\{p_1(\theta_1, \phi_1)\} \times \Sigma_2 \mid (\theta_1, \phi_1) \in S^2\}$$

$$\mathcal{A}_2 = \{\Sigma_1 \times \{p_2(\theta_2, \phi_2)\} \mid (\theta_2, \phi_2) \in S^2\}$$

$$\mathcal{S} = \{(p_1(\theta_1, \phi_1), p_2(\theta_2, \phi_2)) \mid (\theta_i, \phi_i) \in S^2\}$$

$$\mathcal{U} = \{\{p_1(\theta_1, \phi_1)\} \times \Sigma_2 \cup \Sigma_1 \times \{p_2(\theta_2, \phi_2)\} \mid (\theta_i, \phi_i) \in S^2\}$$

$$\mathcal{B} = \{\{(p_1(\theta_1, \phi_1), p_2(\theta_2, \phi_2)), (p_1(\pi - \theta_1, \pi + \phi_1), p_2(\pi - \theta_2, \pi + \phi_2))\} \mid (\theta_i, \phi_i) \in S^2\}$$

$$\mathcal{P} = \{(p_1(\theta_1, \phi_1), p_2(\theta_2, \phi_2)), (p_1(\theta_1', \phi_1'), p_2(\theta_2', \phi_2')) \mid \theta_i \neq \theta_i', \phi_j \neq \phi_j'\}$$

Note that $\mathcal{B} \subset \mathcal{P}$. In this way, we have obtained a collection of properties that together make up the property lattice that describes a compound system consisting of two separated spin 1/2 objects, the infimum of a collection of properties being their intersection.

Some of these properties appear familiar, given the fact that the global system consists of two subsystems of which the mathematical description is known. On the other hand, there are also some properties, notably in $\mathcal{P}$, which have a more classical appearance, in the sense that they consist of a set theoretical union of two states, without any new superpositions that do arise in this particular way. It follows from geometrical considerations that these elements arise from intersections of elements in $\mathcal{A}_1$ and $\mathcal{A}_2$. On the other hand, taking two different elements of $\Sigma_1 \times \Sigma_2$ such that one of the coordinates coincides, the property a generated by these two elements has $\pi_j(a) = \Sigma_j$, with π_j the canonical projection on the other coordinate, hence contains plenty of other elements of $\Sigma_1 \times \Sigma_2$ than the two generating states. This situation is reminiscent to the notion of a superselection rule in standard quantum mechanics, to which we come back later. Observe that all these properties, even the more enigmatic ones, have a clear operational meaning, in the sense that there exists a corresponding experimental procedure that can test for this property.

What about the orthogonality relation? Suppose that $(p_1, p_2), (q_1, q_2) \in \Sigma_1 \times \Sigma_2$. If $p_1 \perp_1 q_1$, we have seen that there is some direction (θ_1, ϕ_1) such that p_1 is represented by $a_1(\theta_1, \phi_1)$ and q_1 by $a_1(\pi - \theta_1, \phi_1 + \pi)$. Because the corresponding experiments $C\alpha_1(\theta_1, \phi_1)$ and $C\alpha_1(\pi - \theta_1, \phi_1 + \pi)$ can also be performed on the compound system, it follows that $(p_1, p_2) \perp (q_1, q_2)$. Conversely, if $(p_1, p_2) \perp (q_1, q_2)$, there exists a yes/no-experiment $\alpha \in Q$ such that $(p_1, p_2) \triangleleft \alpha$ and $(q_1, q_2) \triangleleft \widetilde{\alpha}$. Recall that $(p_1, p_2) \triangleleft \alpha$ formalizes the physical idea that the execution of the experiment α should yield a positive result with certainty, would the experiment be properly performed on a particular sample of the compound physical system that happens to be in a state given by (p_1, p_2).

There are several basic forms to be considered for α. Suppose first that $\alpha = C\alpha_1$ for some $\alpha_1 \in Q_1$; we have seen that $\widetilde{\alpha} = C\widetilde{\alpha}_1$, hence $p_1 \perp_1 q_1$. The same type of argument works if $\alpha = C\alpha_2$ for some $\alpha_2 \in Q_2$ to conclude that $p_2 \perp_2 q_2$. Next, suppose that $\alpha = \alpha_1 \Delta \alpha_2$, then $\widetilde{\alpha} = \widetilde{\alpha}_1 \nabla \widetilde{\alpha}_2$. By the prescription of the experimental procedure and by the separation of the two objects, formally encoded in (20) - (24), we have on the one hand, $p_1 \triangleleft_1 \alpha_1$ and $p_2 \triangleleft_2 \alpha_2$, and on the other hand $q_1 \triangleleft_1 \widetilde{\alpha}_1$ or $q_2 \triangleleft_2 \widetilde{\alpha}_2$, which implies that $p_1 \perp_1 q_1$ or $p_2 \perp_2 q_2$. All other cases being similar, we conclude that

$$(p_1, p_2) \perp (q_1, q_2) \text{ iff } p_1 \perp_1 q_1 \text{ or } p_2 \perp_2 q_2 \tag{29}$$

The same type of argument also shows that the property lattice is orthocomplemented. The reader can verify that orthocomplements for the elements of various forms in $\mathcal{L}_1(\mathbb{C}^2) \circledwedge \mathcal{L}_2(\mathbb{C}^2)$ are those that are given in the following convenient table. In this table, the p_i and q_j stand for points in the state spaces that correspond to the two different subobjects. In addition, we demand that $p_1 \neq q_1$ and $p_2 \neq q_2$.

Element	Orthocomplement
$\{(p_1,p_2)\}$	$\{p_1\}^\perp \times \Sigma_2 \cup \Sigma_1 \times \{p_2\}^\perp$
$\{(p_1,p_2),(q_1,q_2)\}$	$\{(p_1^\perp,q_2^\perp),(q_1^\perp,q_2^\perp)\}$
$\{p_1\} \times \Sigma_2$	$\{p_1\}^\perp \times \Sigma_2$
$\Sigma_1 \times \{p_2\}$	$\Sigma_1 \times \{p_2\}^\perp$
$\{p_1\} \times \Sigma_2 \cup \Sigma_1 \times \{p_2\}$	$\{(p_1^\perp,p_2^\perp)\}$

Next, we will show that both the orthomodularity and the covering law fail, taking the mathematical representation for our compound physical system to be $\mathcal{L}_1(\mathbb{C}^2) \textcircled{\wedge} \mathcal{L}_2(\mathbb{C}^2)$, which can be taken, as we have seen, as the property lattice for our compound system. We start with orthomodularity. Denoting by $[x]$ the one-dimensional subspace spanned by an element $x \in \mathbb{C}^2$, let $a = \{([\psi_1],[\psi_2])\}$ and $b = \{([\psi_1],[\psi_2]),([\phi_1],[\phi_2])\}$, with $[\psi_1] \not\perp_1 [\phi_1]$ and $[\psi_2] \not\perp [\phi_2]$. Then $a^\perp = \{[\psi_1]\}^\perp \times \Sigma_2 \cup \Sigma_1 \times \{[\psi_2]\}^\perp$. Consequently, meets corresponding to intersections, we have $b \wedge a^\perp = \emptyset$, hence $a = a \vee (b \wedge a^\perp) < b$ (a strict inequality!). If orthomodularity were valid, we would have obtained $a \vee (b \wedge a^\perp) = b$, which proves our assertion.

It is also easy to show that the covering law cannot be valid for this particular example, too. Indeed, take a lattice element of $\mathcal{P}$, say $\{([\psi_1],[\psi_2]),([\phi_1],[\phi_2])\}$. It is always possible to choose a third element $([\xi_1],[\xi_2])$, such that ξ_1 and ψ_1 are two linearly independent elements, and also ξ_2 and ϕ_2. Then

$$\{([\psi_1],[\psi_2]),([\phi_1],[\phi_2])\} \wedge \{([\xi_1],[\xi_2])\} = \emptyset \tag{30}$$

$$\{([\psi_1],[\psi_2]),([\phi_1],[\phi_2])\} \vee \{([\xi_1],[\xi_2])\} = \Sigma_1 \times \Sigma_2 \tag{31}$$

If the covering law were valid, this element should cover $\{([\psi_1],[\psi_2]),([\phi_1],[\phi_2])\}$. However, the element $\{[\psi_1]\} \times \Sigma_2 \cup \Sigma_1 \times \{[\phi_2]\}$ belongs to $\mathcal{L}_1(\mathbb{C}^2) \textcircled{\wedge} \mathcal{L}_2(\mathbb{C}^2)$ and

$$\{([\psi_1],[\psi_2]),([\phi_1],[\phi_2])\} \subset \{[\psi_1]\} \times \Sigma_2 \cup \Sigma_1 \times \{[\phi_2]\} \subset \Sigma_1 \times \Sigma_2 \tag{32}$$

which is a contradiction, because these are strict inclusions.

We can then safely conclude that the property lattice $\mathcal{L}_1(\mathbb{C}^2) \textcircled{\wedge} \mathcal{L}_2(\mathbb{C}^2)$ is *not* isomorphic to a Piron lattice (associated with an orthomodular space), due to the fact that orthomodularity and the covering law fail. Consequently, an underlying linear structure such that $\mathcal{L}_1(\mathbb{C}^2) \textcircled{\wedge} \mathcal{L}_2(\mathbb{C}^2)$ would correspond to the complete lattice of all closed subspaces is out of the question: one cannot construct an underlying Hilbert space for which the collection of all closed subspaces would correspond with the property lattice associated with this physical situation.

4 The Orthogonality Relation

In this section, we want to take a closer look at the orthogonality relation on a general $\Sigma_1 \times \Sigma_2$ that generates the property lattice corresponding to the separated product. It will be convenient to demonstrate some general results, the first for a general orthogonality space, the second valid for the particular orthogonality relation given by (19).

Lemma 1 *In an arbitrary orthogonality space* $(\Sigma, \perp)$*, we have*

$$\Big(\bigcup_{j\in J} M_j \Big)^{\perp} = \bigcap_{j\in J} M_j^{\perp} \tag{33}$$

Proof: Recall that an orthogonality relation is by definition irreflexive and symmetric. If $A \subseteq B$, then $B^{\perp} \subseteq A^{\perp}$, hence $\left(\cup_{j\in J} M_j \right)^{\perp} \subseteq M_k^{\perp}$, for each $k \in J$. Observe also that $A \subseteq A^{\perp\perp}$ for any $A \subseteq \Sigma$, by symmetry. Consequently, if F is any subset of Σ, we obtain $F \subseteq \cap_{j\in J} M_j^{\perp}$ iff $F \subseteq M_j^{\perp}$ for each $j \in J$ iff $M_j \subseteq F^{\perp}$ for each $j \in J$ iff $\cup_{j\in J} M_j \subseteq F^{\perp}$ iff $F \subseteq \left(\cup_{j\in J} M_j \right)^{\perp}$, which proves the other inclusion. □

Proposition 1 *Suppose that* $\Sigma_1 \times \Sigma_2$ *is an orthogonality space, equipped with the orthogonality relation (19). Let* $M_j \subseteq \Sigma_j$, $j = 1, 2$ *and* $(p_1, p_2) \in \Sigma_1 \times \Sigma_2$. *Then*

$$\{(p_1,p_2)\}^{\perp} = (\{p_1\}^{\perp} \times \Sigma_2) \cup (\Sigma_1 \times \{p_2\}^{\perp}) \tag{34}$$

$$(\{p_1\} \times M_2)^{\perp} = (\{p_1\}^{\perp} \times \Sigma_2) \cup (\Sigma_1 \times M_2^{\perp}) \tag{35}$$

$$(M_1 \times M_2)^{\perp} = (M_1^{\perp} \times \Sigma_2) \cup (\Sigma_1 \times M_2^{\perp}) \tag{36}$$

Proof: First, $(r_1, r_2) \perp (p_1, p_2)$ iff $r_1 \perp_1 p_1$ or $r_2 \perp_2 p_2$ iff $(r_1, r_2) \in \{p_1\}^{\perp} \times \Sigma_2$ or $(r_1, r_2) \in \Sigma_1 \times \{p_2\}^{\perp}$. Second, if $r_1 \in \{p_1\}^{\perp}$ or $r_2 \in M_2^{\perp}$, then $(r_1, r_2) \in (\{p_1\} \times M_2)^{\perp}$; conversely, let $(r_1, r_2) \in (\{p_1\} \times M_2)^{\perp}$; if $r_1 \in \{p_1\}^{\perp}$, there is nothing to prove; if not, take an arbitrary $m_2 \in M_2$; because $(r_1, r_2) \perp (p_1, m_2)$ and $r_1 \not\perp_1 p_1$, it follows that $r_2 \perp_2 m_2$, hence $r_2 \in M_2^{\perp}$. The final equation follows from the next calculation, using some of the previous results:

$$(M_1 \times M_2)^{\perp} = \Big(\bigcup_{r_1\in M_1} (\{r_1\} \times M_2) \Big)^{\perp}$$

$$
\begin{aligned}
&= \bigcap_{r_1\in M_1} \big((\{r_1\}\times M_2)^{\perp} \big) \\
&= \bigcap_{r_1\in M_1} \big((\{r_1\}^{\perp}\times \Sigma_2) \cup (\Sigma_1 \times M_2^{\perp}) \big) \\
&= \big(\bigcap_{r_1\in M_1} (\{r_1\}^{\perp}\times \Sigma_2) \big) \cup (\Sigma_1 \times M_2^{\perp}) \\
&= \big(\bigcup_{r_1\in M_1} \{r_1\} \big)^{\perp}\times \Sigma_2 \cup (\Sigma_1 \times M_2^{\perp}) \\
&= (M_1^{\perp}\times \Sigma_2) \cup (\Sigma_1 \times M_2^{\perp})
\end{aligned}
$$

what was to be proved. □

Let $(\Sigma_j, \perp_j)$, $j = 1, 2$, be two T_1 orthogonality spaces, that is, we additionally demand that $\forall p_j \in \Sigma_j : \{p_j\}^{\perp_j\perp_j} = \{p_j\}$. Suppose that there exist $p_1, q_1 \in \Sigma_1$ such that $p_1 \neq q_1$, and similarly for Σ_2. The following straightforward calculation shows that the two-element set $\{(p_1,p_2),(q_1,q_2)\}$ is always a closed subspace of the orthogonality space $(\Sigma_1 \times \Sigma_2, \perp)$, with the orthogonality given by (19):

$$
\begin{aligned}
\{(p_1,p_2),(q_1,q_2)\}^{\perp\perp} &= (\{(p_1,p_2)\}^{\perp} \cap \{(q_1,q_2)\}^{\perp})^{\perp} \\
&= ((\{p_1\}^{\perp}\times\Sigma_2)\cup(\Sigma_1\times\{p_2\}^{\perp})\cap \\
&\qquad \cap(\{q_1\}_1^{\perp}\times\Sigma_2)\cup(\Sigma_1\times\{q_2\}^{\perp}))^{\perp} \\
&= (((\{p_1\}^{\perp}\cap\{q_1\}^{\perp})\times\Sigma_2)\cup(\{p_1\}^{\perp}\times\{q_2\}^{\perp})\cup \\
&\qquad \cup(\Sigma_1\times(\{p_2\}^{\perp}\cap\{q_2\}^{\perp}))\cup(\{q_1\}^{\perp}\times\{p_2\}^{\perp}))^{\perp} \\
&= ((\{p_1\}^{\perp}\cap\{q_1\}^{\perp})\times\Sigma_2)^{\perp}\cap(\{p_1\}^{\perp}\times\{q_2\}^{\perp})^{\perp}\cap \\
&\qquad \cap(\Sigma_1\times(\{p_2\}^{\perp}\cap\{q_2\}^{\perp}))^{\perp}\cap(\{q_1\}^{\perp}\times\{p_2\}^{\perp})^{\perp} \\
&= (\{p_1,q_1\}^{\perp\perp}\times\Sigma_2)\cap(\{p_1\}\times\Sigma_2\cup\Sigma_1\times\{q_2\})\cap \\
&\qquad \cap(\Sigma_1\times\{p_2,q_2\}^{\perp\perp})\cap(\{q_1\}\times\Sigma_2\cup\Sigma_1\times\{p_2\}) \\
&= (\{p_1\}\times\Sigma_2\cup\{p_1,q_1\}^{\perp\perp}\times\{q_2\})\cap \\
&\qquad \cap(\{q_1\}\times\{p_2,q_2\}^{\perp\perp}\cup\Sigma_1\times\{p_2\}) \\
&= \{(p_1,p_2)\}\cup\{(q_1,q_2)\} \\
&= \{(p_1,p_2),(q_1,q_2)\}
\end{aligned}
$$

Consequently, these two elements do not generate an irreducible projective plane. So in general there exist in the property lattice corresponding to the separated product, a host of two-element sets that form closed subspaces, relative to this orthogonality relation, a situation that is unheard off in standard quantum physics.

In a usage derived from that of standard quantum physics, one can say that two properties a and b in a property lattice are separated by a *superselection rule* whenever $p \lhd a \vee b$ implies either $p \lhd a$ or $p \lhd b$. In standard quantum physics, all known superselection rules can be accommodated for by restricting some global Hilbert space, attributed to the physical system under investigation, to a suitable collection of mutually orthogonal subspaces, not allowing states that do not belong to one of these orthogonal components. However, observe that for the separated product of two spin 1/2 objects there do even exist *non-orthogonal* states that are separated by a superselection rule, in particular pairs of states that constitute many of the properties in $\mathcal{P}$.

5 Sasaki Regularity

Yet another characterization of the covering law can be formulated for (complete) atomistic orthomodular lattices, using the projections in a suitable involution semigroup of mappings associated with the property lattice. Let $\mathcal{L}$ be a complete atomistic orthomodular lattice, with orthocomplementation $a \mapsto a^{\perp}$, then $\mathcal{L}$ satisfies the covering law iff each so-called Sasaki projection

$$\phi_a : \mathcal{L} \to \mathcal{L} : x \mapsto (x \vee a^{\perp}) \wedge a \tag{37}$$

maps any atom not smaller than $a^{\perp}$ to an atom, that is, for any $a \in \mathcal{L}$ the restriction and corestriction

$$\phi_a : \Sigma \setminus \{p \in \Sigma \mid p < a^{\perp}\} \to \Sigma : p \mapsto (p \vee a^{\perp}) \wedge a \tag{38}$$

is well-defined [5]. In general, it is convenient to give a special name to all Sasaki projections that satisfy this last condition. We will call them *regular* Sasaki projections.

Because $\mathcal{L}$ is isomorphic with the orthomodular lattice of all Sasaki projections under some suitable conditions [13], and the Sasaki projections can be interpreted as representing state transitions corresponding to a positive response for idealized measurement procedures associated with the properties, this procedure refers to a more active point of view on physical systems. Indeed, one assumes the existence of an ideal class of measurement procedures, such that the state before such a measurement becomes a well-defined state after the experiment, whenever one has obtained the positive result. In view of this interpretation, it seems indeed more natural to consider Sasaki projections as partially defined state space mappings. Given the fact that $\kappa[\mathcal{L}] = \mathcal{F}(\Sigma)$ under the usual orthocomplementation axioms of the axiomatic approach, we then have to consider a family of mappings

$$\phi_M : \mathcal{D}(\phi_M) \to \Sigma : p \mapsto (\{p\} \cup M^{\perp})^{\perp\perp} \cap M \tag{39}$$

with $\mathcal{D}(\phi_M) \subseteq \Sigma$, and $M = \kappa(a)$ for some $a \in \mathcal{L}$. The latter condition arises because it is exactly subsets of this form that represent properties attributed to the physical system. As we have seen, ϕ_M is regular iff $\mathcal{D}(\phi_M) = \{p \in \Sigma \mid p \notin M^{\perp}\}$.

Given the role of the covering law in the representation theorems and its interpretation, it is then of considerable interest to investigate the presence of any aberrant Sasaki projections for operationally separated objects that are described as one compound physical system, both in general and with respect to our example. Because of their putative interpretation as state transitions, we consider the Sasaki projections as partial state space mappings, and investigate them at the level of the state space description $(\Sigma_1 \times \Sigma_2, \perp)$.

Consequently, let $(\Sigma_1, \perp_1)$ and $(\Sigma_2, \perp_2)$ be two Sasaki regular T_1 orthogonality spaces, in the sense that Sasaki projections associated with biorthogonally closed sets are regular, and take $(p_1, p_2) \in \Sigma_1 \times \Sigma_2$ such that $(p_1, p_2) \not\perp M_1 \times M_2$, with $M_1 = M_1^{\perp\perp}$ and $M_2 = M_2^{\perp\perp}$. According to our previous results, this implies that $p_1 \not\perp_1 M_1$ and $p_2 \not\perp_2 M_2$. After some calculation efforts, one obtains

$$\begin{aligned}
\phi_{M_1 \times M_2}(p_1, p_2) &= \left(\{(p_1, p_2)\} \cup (M_1 \times M_2)^{\perp}\right)^{\perp\perp} \cap (M_1 \times M_2) \\
&= \left(\{(p_1, p_2)\} \cup (M_1^{\perp} \times \Sigma_2) \cup (\Sigma_1 \times M_2^{\perp})\right)^{\perp\perp} \cap (M_1 \times M_2) \\
&= \left((\{p_1\}^{\perp} \times \Sigma_2 \cup \Sigma_1 \times \{p_2\}^{\perp}) \cap (M_1 \times M_2)\right)^{\perp} \cap (M_1 \times M_2) \\
&= \left((\{p_1\}^{\perp} \cap M_1) \times M_2 \cup M_1 \times (\{p_2\}^{\perp} \cap M_2)\right)^{\perp} \cap (M_1 \times M_2) \\
&= \left((\{p_1\}^{\perp} \cap M_1)^{\perp} \times \Sigma_2 \ \cup \ \Sigma_1 \times M_2^{\perp}\right) \cap \\
&\qquad \cap \left(M_1^{\perp} \times \Sigma_2 \ \cup \ \Sigma_1 \times (\{p_2\}^{\perp} \cap M_2)^{\perp}\right) \ \cap \ (M_1 \times M_2) \\
&= \left((\{p_1\}^{\perp} \cap M_1)^{\perp} \times \Sigma_2 \ \cup \ \Sigma_1 \times M_2^{\perp}\right) \cap \\
&\qquad \cap M_1 \times \left((\{p_2\} \cup M_2^{\perp})^{\perp\perp} \cap M_2\right) \\
&= \left((\{p_1\} \cup M_1^{\perp})^{\perp\perp} \cap M_1\right) \times \left((\{p_2\} \cup M_2^{\perp})^{\perp\perp} \cap M_2\right)
\end{aligned}$$

and the right hand side belongs to $\Sigma_1 \times \Sigma_2$, by assumption. In particular, with some abuse of notation

$$\phi_{(q_1, q_2)}(p_1, p_2) = (q_1, q_2) \tag{40}$$

$$\phi_{\{q_1\} \times \Sigma_2}(p_1, p_2) = (q_1, p_2) \tag{41}$$

Consequently, regularity is preserved for all Sasaki projections corresponding to biorthogonally closed subsets of the general form $M_1 \times M_2$. Therefore, we have to screen for other candidates that could violate our regularity condition. Luckily, we don't have to look too far. Indeed, consider one of the peculiar biorthogonally closed sets of the form $M = \{(q_1, q_2), (r_1, r_2)\}$, which, as we have seen, can be found in any property lattice corresponding to a separated product. We will show that, for $(p_1, p_2) \notin M^\perp$:

$$\phi_M(p_1, p_2) = M = \{(q_1, q_2), (r_1, r_2)\} \tag{42}$$

whenever $p_1 \not\perp_1 q_1, p_1 \not\perp_1 r_1, p_2 \not\perp_2 q_2, p_2 \not\perp_2 r_2$. Indeed, if $(p_1, p_2) \notin M^\perp$, then either (1) $p_1 \not\perp_1 q_1$ and $p_2 \not\perp_2 q_2$, or (2) $p_1 \not\perp_1 r_1$ and $p_2 \not\perp_2 r_2$, or (3) both. Consequently,

$$\begin{aligned}
(\{(p_1, p_2)\} \cup M^\perp)^{\perp\perp} &= (\{(p_1, p_2)\}^\perp \cap M)^\perp \\
&= (\{(q_1, q_2), (r_1, r_2)\} \cap (\{p_1\}^\perp \times \Sigma_2 \cup \Sigma_1 \times \{p_2\}^\perp))^\perp \\
&= \begin{cases} \Sigma_1 \times \Sigma_2 & \text{if (1) and (2) are valid} \\ \{q_1\}^\perp \times \Sigma_2 \cup \Sigma_1 \times \{q_2\}^\perp & \text{if only (2) is valid} \\ \{r_1\}^\perp \times \Sigma_2 \cup \Sigma_1 \times \{r_2\}^\perp & \text{if only (1) is valid} \end{cases}
\end{aligned}$$

from which (42) easily follows. Because M is not an atom, ϕ_M is not regular, and ϕ_M can no longer be interpreted as a state transition resulting from a positive response for the yes/no-experiment that corresponds with M. This is apparently due to the construction of the yes/no-experiments and the properties associated with product experiments, and is a manifestation of the symmetry with respect to the possible state transitions the two separated constituents can undergo for one and the same positive outcome, attributed to the corresponding yes/no-experiment.

In summary, if both Σ_1 and Σ_2 contain at least two different states and if in both state spaces there exists a state that is not orthogonal to both these states, that is, for all non-trivial orthogonality spaces, the previous argument is valid and we have demonstrated the following

Theorem 1 *If $(\Sigma_1, \perp_1)$ and $(\Sigma_2, \perp_2)$ are two Sasaki regular T_1 orthogonality spaces, the orthogonality space $(\Sigma_1 \times \Sigma_2, \perp)$, with the orthogonality given by (19), is not Sasaki regular, whenever $\perp_1$ and $\perp_2$ are non-trivial.*

Because the property lattice attributed to a classical physical system usually corresponds with the collection of all subsets of some set Σ, one easily sees that the orthogonality relation in this case becomes trivial: every pair of distinct states is orthogonal. Consequently, the theorem is not valid whenever at least one of the two orthogonality spaces represents a classical physical system.

" f course, it then also follows for the same reason that in our particular example Sasaki regularity is not preserved.

6 Discussion

Several combinatorial mathematical constructions have been proposed for the description of compound physical systems, starting from the representations of the hypothetical subobjects in those compound physical systems. Two of them were studied by one of the authors: the so-called separated product, that constructs the property lattice for the compound system that consists of two explicitly separated physical objects [6,7]; the coproduct, that generates the property lattice of two separated physical systems, for which only experimental projects on one of the subobjects at a time, chosen at random, are taken into account [17,18,19]. The name of the latter comes from the fact that one can show that it corresponds with (the underlying object of) the mathematical coproduct in an appropriate categorical sense. In addition, in [20], the property lattice associated with the more traditional Hilbert tensor product space representation for compound physical systems has been studied.

In this paper, we have examined in some detail the problem of the mathematical description of the conceptually important physical situation that consists in two separated spin 1/2 objects that are considered as one compound physical system. In particular, we have shown that a representation as a collection of closed subspaces of a linear space is impossible in general, due to the fact that the covering law fails. Thus, there seems to be some relation between the notion of a *compound* physical system and the mathematical property of *linearity*. We have also spent some time on studying another perspective which is intimately related to the covering law: the regularity of the corresponding collection of Sasaki projections. It is tempting to speculate how the possible development of a generalized "non-linear" quantum physics could eventually put in a different light the problems that quantum mechanics experiences to describe (operationally) separated quantum objects.

For atomistic lattices, one can show that the covering law is equivalent with the so-called *exchange property*, which states that for each $x \in \mathcal{L}$ and for any pair of atoms $p, q \in \Sigma_{\mathcal{L}}$, the following condition is valid: $p \wedge x = 0$ and $p < q \vee x$ together imply $q < p \vee x$ [21,22]. This condition is reminiscent of the superposition principle of quantum mechanics (and is trivially satisfied for the property lattice of a classical physical system, which is usually taken to be the collection of all subsets of state space). Consequently, the covering law seems deeply related to the possibility of attributing a linear structure to the state space of an arbitrary physical system.

The fact that it is mainly the covering law that is responsible for a linear representation of the property lattice, also follows from the following theorem [21]: For any irreducible complete atomistic orthocomplemented lattice of length ≥ 4 that satisfies the covering law, there exists a division ring K with an involution $\lambda \mapsto \lambda^*$, and a vector space V over K with a hermitian form $f : V \times V \to K$, such that $\mathcal{L}$ is ortho-isomorphic to the lattice of all closed subspaces of V, relative to f. Consequently, the stronger condition of orthomodularity is not necessary to obtain a linear structure.[e]

In our opinion, the mathematical description of this physical situation has at least some relevance with respect to the enigmatic classical limit. At the very least, this approach yields another perspective on the problematic associated with the one and the many, as it was aptly called by one of the authors [6]. Indeed, the construction in this particular case seems to be empirically and operationally adequate in that it incorporates all experiments one can possibly perform on two separated spin 1/2 objects separately and as a whole, and therefore seems to evade the critique of Cattaneo and Nisticó [23].

Of course, if two physical objects are separated in the operational sense that was used in this paper, one usually does not bother about representing the properties that explicitly account for the separation. From this point of view, the so-called coproduct property lattice arises if one considers the collection of properties $\mathcal{L}_1 \cup \mathcal{L}_2$, together with all possible products (in the sense that we have explained), as empirically adequate for the description of the compound physical system. In other words, the fact that one describes two physical systems as a whole does not lead one to consider global experiments on both objects at once. One can object that a description of a compound physical system that takes only into account the possible properties of the subobjects and not of the compound system as a whole is necessarily incomplete.

The underlying set of the coproduct for our example would be isomorphic with the collection of all ordered pairs of non-zero (closed) subspaces of $\mathbb{C}^2$, with an additional global least element pasted at the bottom:

$$\mathcal{L}_1(\mathbb{C}^2) \coprod \mathcal{L}_2(\mathbb{C}^2) = \{(M_1, M_2) \mid M_i \in \mathcal{L}_i^0(\mathbb{C}^2),\ i = 1, 2\} \uplus \{0\} \qquad (43)$$

For a more profound study of the properties of this structure, we refer to [17,18,19]. In general, also in this case the covering law fails, although all corresponding Sasaki projections seem to behave regularly, which is possible because orthomodularity is in general not valid.

---

[e] Actually, there exists an even weaker representation theorem, which states that a linear representation holds for irreducible complete lattices $\mathcal{L}$ of length ≥ 4, if both $\mathcal{L}$ and its opposite lattice $\mathcal{L}^*$ are atomistic and satisfy the covering law [21].

The standard quantum physical prescriptions for the construction of a mathematical representation of a physical system that is conceived as being made up of several components, require one to construct the Hilbert tensor product of the Hilbert spaces corresponding to the putative subobjects, and the possible selection of an appropriate closed subspace, to account for the fermionic or bosonic nature of these constituents. It is clear from our analysis that this procedure is not possible for the case of two separated spin 1/2 particles. On the other hand, the tensor product procedure can be justified in the axiomatic approach, given the fact that the putative compound system satisfies the standard prescriptions of the axiomatic approach [18,20,24].

Last but not least, we think that the standard notion of so-called "identity of elementary physical objects in a compound system", which is so problematic at a fundamental conceptual level, is not particularly problematic in our approach. Indeed, there may not be such thing as a physical system consisting of two identical subobjects. Indeed, such a system may have to be considered as one global physical system, that may even manifest itself at spatially separated regions, a problem that would more properly be related to our *a priori*, possibly macroscopically biased, ideas on localization in space. Indeed, experimental evidence suggests that the property of being localized in space, is in general not a classical property (see [8] and references therein). In that case, the putative compoundness would be a mental construction that is ascribed in retrospect to the physical system before it actually interacted with a suitable measuring device.

References

1. D. Aerts and F. Valckenborgh, "The linearity of quantum mechanics at stake: the description of separated quantum entities", this volume.
2. C. Piron, "Axiomatique quantique", *Helv. Phys. Acta* **37**, 439 (1964).
3. J. Jauch, *Foundations of Quantum Mechanics,* Addison-Wesley, Reading, Massachusetts (1968).
4. J. Jauch and C. Piron, "On the structure of quantal proposition systems", *Helv. Phys. Acta* **42**, 842-848 (1969).
5. C. Piron, *Foundations of Quantum Physics,* Reading, Mass., W. A. Benjamin (1976).
6. D. Aerts, *The One and the Many: Towards a Unification of the Quantum and the Classical Description of One and Many Physical Entities*, Doctoral Thesis, Vrije Universiteit Brussel (1981).
7. D. Aerts, "Description of many physical entities without the paradoxes encountered in quantum mechanics", *Found. Phys.* **12**, 1131-1170 (1982).

8. C. Piron, *Mècanique Quantique: Bases et Applications,* Presse Polytechnique de Lausanne (1990).
9. G. Cattaneo and G. Nisticó, "Axiomatic foundations of quantum physics: critiques and misunderstandings. Piron's Question-Proposition System", *Int. J. Theor. Phys.* **30**, 1293-1336 (1991).
10. D. Aerts, "Quantum structures, separated physical entities and probability", *Found. Phys.* **24**, 1227-1259 (1994).
11. D. J. Moore, "On state spaces and property lattices", *Stud. Hist. Phil. Mod. Phys.* **30**, 61-83 (1999).
12. F. Valckenborgh, "Closure structures and the theorem of decomposition in classical components", *Tatra Mt. Math. Publ.* **10**, 75-86 (1997).
13. E. G. Beltrametti and G. Cassinelli, *The Logic of Quantum Mechanics,* Addison-Wesley, Reading, Massachusetts (1981).
14. A. Einstein, B. Podolsky and N. Rosen, "Can quantum mechanical description of physical reality be considered complete?" *Phys. Rev.* **47**, 777-780 (1935).
15. D. Aerts, "Foundations of quantum physics: a general realistic and operational approach", *Int. J. Theor. Phys.* **38**, 289-358 (1999), lanl archive ref: *quant-ph/0105109.*
16. D. Aerts, E. Colebunders, A. Van der Voorde and B. Van Steirteghem, "State property systems and closure spaces: a study of categorical equivalence", *Int. J. Theor. Phys.* **38**, 359-385 (1999), lanl archive ref: *quant-ph/0105108.*
17. D. Aerts, "Construction of the tensor product for the lattices of properties of physical entities", *J. Math. Phys.* **25**, 1434-1441 (1984).
18. F. Valckenborgh, "Operational axiomatics and compound systems", in *Current Research in Operational Quantum Logic: Algebras, Categories, Languages,* eds. Coecke, B., Moore, D. J. and Wilce, A., Kluwer Academic Publishers, Dordrecht, 219-244 (2000).
19. F. Valckenborgh, *Compound Systems in Quantum Axiomatics,* Doctoral Thesis, Vrije Universiteit Brussel (2001).
20. D. Aerts and I. Daubechies, "Physical justification for using the tensor product to describe two quantum systems as one joint system" *Helv. Phys. Acta* **51**, 661-675 (1978).
21. F. Maeda and S. Maeda, *Theory of Symmetric Lattices,,* Springer-Verlag, Berlin (1970).

22. Cl.-A. Faure and A. Frölicher, *Modern Projective Geometry,'* Kluwer Academic Publishers, Dordrecht (2000).
23. G. Cattaneo and G. Nisticó, "A note on Aerts' description of separated entities", *Found. Phys.* **20**, 119-132 (1990).
24. B. Coecke, "Structural characterization of compoundness", *Int. J. Theor. Phys.* **39**, 585-594 (2000).

BEING AND CHANGE: FOUNDATIONS OF A REALISTIC OPERATIONAL FORMALISM

DIEDERIK AERTS
Center Leo Apostel (CLEA) and
Foundation of the Exact Sciences (FUND),
Brussels Free University, Krijgskundestraat 33,
1160 Brussels, Belgium
E-mail: diraerts@vub.ac.be

The aim of this article is to represent the general description of an entity by means of its states, contexts and properties. The entity that we want to describe does not necessarily have to be a physical entity, but can also be an entity of a more abstract nature, for example a concept, or a cultural artifact, or the mind of a person, etc..., which means that we aim at very general description. The effect that a context has on the state of the entity plays a fundamental role, which means that our approach is intrinsically contextual. The approach is inspired by the mathematical formalisms that have been developed in axiomatic quantum mechanics, where a specific type of quantum contextuality is modelled. However, because in general states also influence context – which is not the case in quantum mechanics – we need a more general setting than the one used there. Our focus on context as a fundamental concept makes it possible to unify 'dynamical change' and 'change under influence of measurement', which makes our approach also more general and more powerful than the traditional quantum axiomatic approaches. For this reason an experiment (or measurement) is introduced as a specific kind of context. Mathematically we introduce a state context property system as the structure to describe an entity by means of its states, contexts and properties. We also strive from the start to a categorical setting and derive the morphisms between state context property systems from a merological covariance principle. We introduce the category **SCOP** with as elements the state context property systems and as morphisms the ones that we derived from this merological covariance principle. We introduce property completeness and state completeness and study the operational foundation of the formalism.

1 Introduction

We put forward a formalism that aims at a general description of an entity under influence of a context. The approach followed in this article is a continuation of what has been started in [1,2]. Meanwhile some new elements have come up, and as a consequence what we present here deviates in some ways from what was elaborated in [1,2]. In [1] the formalism is founded on the basic concepts of states, experiments and outcomes of experiments. One of the aims in [1] is to generalize older approaches [3,4,5,6,7,8,9] by incorporating experiments with more than two outcomes from the start. The older approaches are indeed founded on the basic concept of yes/no-experiment, also called test, question or experimental project. An experiment with more than two outcomes is described by the set of yes/no-experiments that can be formed from this experiment by identifying outcomes in such a way that, after identification, only two outcomes remain. The shift away from the old approach with yes/no-experiments was inspired by the extra power, at first sight at least, that formalisms that start from experiments with more than two outcomes as basis posses [10,11,12,13,14,15]. Meanwhile, after [1,2] had been published, we have identified a fundamental problem that exists in approaches that start with experiments with more than two outcomes.

1.1 *The Subexperiment Problem in Quantum Mechanics*

The problem that we have identified is the following. Suppose that in quantum mechanics one considers a subexperiment of an experiment. Then it is always the case that the subexperiment changes the state of the entity under study in a different way than the experiment does. If however a subexperiment is defined as the experiment where some of the outcomes are identified, the subexperiment must change the state in the same way as the original experiment. This means that in quantum mechanics subexperiments do not arise through an identification process on the outcomes of an experiment. This also means that the general scheme inspired from probability theory, as for example in [10,11,12,13,14,15], where outcomes are taken as basic concepts and events and experiments are defined by means of their identifying sets of outcomes, does not work for quantum mechanics. The subexperiments that can be fabricated in these type of approaches are not the subexperiments that one encounters in quantum mechanics.

When we were writing [1], we started to get aware of this fundamental problem. We were very amazed that to the best of our knowledge nobody else working in quantum axiomatics, ourselves included, seemed to have noticed this before. It puts a fundamental limitation on the approaches where one starts with experiments with more outcomes and deduces the subexperiments, for example the yes/no-experiments, by a procedure of identification of outcomes, as in the approach presented in [1]. In [2] we have avoided the problem by again working with yes/no-experiments as basic concepts. In this article we will introduce experiments, and hence also yes/no-experiments, in another way, not avoiding the problem, but tackling it head on. In [16] we analyze in detail the subexperiment problem in quantum mechanics.

1.2 Applying the Formalism to Situations Outside the Microworld

The development of the formalism has always been grounded by applying it to specific examples of physical entities in specific situations. Already in the early years it became clear that the formalism, that originally was developed with the aim of providing a realistic operational axiomatic foundation for quantum mechanics, can be applied to describe entities that are not part of the micro-physical region of reality. Since the formalism delivers a general description of an entity, it is a priori not mysterious that it can be applied to entities that are part of other regions of reality than the microworld. The study of such entities is interesting on its own and also can shed light on the nature of quantum entities. In this sense we presented a first macroscopic mechanical entity entailing a quantum mechanical structure in [17,18,19]. More specifically this entity represents a model for the spin of a spin$\frac{1}{2}$ quantum entity.

The model puts forward an explanation of the quantum probabilities as due to the presence of fluctuations on the interaction between the measuring apparatus and the physical entity under consideration. We have called this aspect of the formalism the 'hidden measurement approach' [21,22,23,24,25,26,27,28,29]. The name refers to the fact that the presence of these fluctuations can be interpreted as the presence of hidden variables in the measuring apparatus (hidden measurements) instead of the presence of hidden variables in the state of the physical entity, which is what traditional hidden variable theories suppose. Concretely it means that, when a measurement is performed, the actual measuring process that makes occur one of the outcomes is deterministic once it starts. But each time a measurement is repeated, which is necessary to obtain the probability as limit of the relative frequency, a new deterministic process starts. And the presence of fluctuations on the interaction between measurement apparatus and physical entity are such that this new repeated measurement can give rise deterministically to another outcome than the first one. The quantum probability arises from the statistics of how each time again a new repeated measurement can give rise to another outcome, although once started it evolves deterministically.

As mentioned already, concrete realizable mechanical models that expose this situation were built, where it is possible to 'see' how the quantum structure arises as a consequence of the presence of fluctuations on the interaction between measurement and entity [17,18,19,20,24,25,27,30,31,32]. Specifically these macroscopic mechanical models that give rise to quantum structure inspired us to try out whether the formalism could be applied to entities in other fields of reality than the micro-physical world.

A first application was elaborated with the aim of describing the situation of decision making. More concretely we made a model [33,34] for an opinion pole situation, where also persons without opinion can be described, and the effect of the questioning during the opinion pole itself can be taken into account.

We also constructed a cognitive situation where Bell inequalities are violated because of the presence of a nonKolmogorovian probability structure in the cognitive situation [35]. We worked out in detail a description of the liar paradox by means of the Hilbert space structure of standard quantum mechanics [36,37].

More recently our attention has gone to studies of cultural evolution [38,39], cognitive science, more specifically the problem of the representation of concepts [38,39], and biology with the aim of elaborating a global evolution theory [39,40]. One of the important insights that has grown during the work on [38], and the work found in Liane Gabora's doctoral thesis [39], is that experiments (or measurements) can be considered to be contexts that influence the state of the entity under consideration in an indeterministic way, due to fluctuations that are present on the interaction between the context and the entity. As we knew already from our work on the hidden measurement approach, such indeterministic effects give rise to a generalized quantum structure for the description of the states and properties of the entity under study. This makes it possible to classify dynamical evolution for classical and quantum entities and change due to measurements on the same level: dynamical change being deterministic change due to dynamical context and measurement change being indeterministic change due to measurement context, the indeterminism finding its origin on the presence of fluctuations on the interaction between context and entity.

In the present article it is the first time that we present the formalism taking into account this new insight that unifies dynamical change with measurement change. As will be seen in the next sections, this makes it necessary to introduce some fundamental changes in the approach.

1.3 States Can Change Context

Our investigations in biological and cultural evolution [38,39,40] have made it clear that in general change will happen not only by means of an influence of the context on the state of the entity under study, but also by means of the influence of the state of the entity on the context itself. Interaction between the entity and the context gives rise to a change of the state of the entity, but also to a change of the context itself. This possibility was not incorporated in earlier versions of the formalism. The version that we propose in this paper takes it into account from the start, and introduces an essential reformulation of the basic setting. In this paper, although we introduce the possibility of change of the context under influence of the state from the start, we will not focus on this aspect. We refer to [38,39] for a more detailed investigation of this effect, and more specifically to [40] for a full elaboration of it.

1.4 Filtering Out the Mathematical Structure

Already in the last versions of the formalism [1,2] the power of making a good distinction between the mathematical aspects of the formalism and its physical foundations had been identified. Let us explain more concretely what we mean. In the older founding papers [3,4,5,6,7,8,9], although the physical foundation of the formalism is defined in a clear way, and the resulting mathematical structures are treated rigorously, it is not always clear what are the 'purely mathematical' properties of the structures that are at the origin of the results. That is the reason that in more recent work on the formalism we have made an attempt to divide up the physical foundation and the resulting mathematical structure as much as possible. We first explain in which way certain aspects of the mathematical structure arise from the physical foundation, but then, in a second step, define these aspects in a strictly mathematical way, such that propositions and theorems can be proven, 'only' using the mathematical structure without physical interpretation. Afterwards, the results of these propositions and theorems can then be interpreted in a physical way again. This not only opens the way for mathematicians to start working on the structures, but also lends a greater axiomatic strength to the whole approach on the fundamental level. More concretely, the mathematical structure of a state property system is the the structure to be used to describe a physical entity by means of its states and properties[1,2,41]. This step turned out to be fruitful from the start, since we could prove that a state property system as a mathematical structure is isomorphic to a closure space [1,2,41]. This means that the mathematics of closure spaces can be translated to the mathematics of state property systems, and in this sense becomes relevant for the foundations of quantum mechanics. The step of dividing up the mathematics from the physics in a systematic way also led to a scheme to derive the morphisms for the structures that we consider from a covariance principle rooted in the relation of a subentity to the entity of which it is a subentity[1,41]. This paved the way to a categorical study of the mathematical structures involved, which is the next new element of the recent advances that we want to mention.

1.5 *Identifying the Categorical Structure*

Not only was it possible to connect with a state property system a closure space in an isomorphic way, but, after we had introduced the morphisms starting from a merological covariance principle, it was possible to prove that the category of state property systems and their morphisms, that we have named **SP**, is equivalent with the category of closure spaces and continuous functions, denoted by **Cls** [1,41]. More specifically we could prove that **SP** is the amnestic modification of **Cls** [42]. Meanwhile these new element in the approach have lead to strong results.

It could be proven that some of the axioms of axiomatic quantum mechanics [43,4,5] correspond to separation properties of the corresponding closure spaces [44]. More concretely, the axiom of state determination in a state property system [1] is equivalent with the T_0 separation axiom of the corresponding closure space [44,45], and the axiom of atomicity in a state property system [1] is equivalent with the T_1 separation axiom of the corresponding closure space [46,47]. More recently it has been shown that 'classical properties' [4,6,8,9] of the state property system correspond to clopen (open and closed) sets of the closure space [48,49,50], and, explicitly making use of the categorical equivalence, a decomposition theorem for a state property system into its nonclassical components can be proved that corresponds to the decomposition of the corresponding closure space into its connected components [48,49,50].

1.6 *The Introduction of Probability*

In [1,2] we introduce for the first time probability on the same level as the other concepts, such as states, properties and experiments. Indeed, in the older approaches, probability was not introduced on an equally fundamental level as it is the case for the concepts of state, property and experiment..

What probability fundamentally tries to do is to provide a 'measure' for the uncertainty that is present in the situation of entity and context. We had introduced probability in a standard way in [1,2] by means of a measure on the interval $[0,1]$ of the set $\mathbb{R}$ of real numbers. Meanwhile however it has become clear that probability should better be introduced in a way that is quite different from the standard way. We have analyzed this problem in detail in [51]. We come to the conclusion that it is necessary to define a probability not as an object that is evaluated by a number in the interval $[0,1]$, as it is the case in standard probability theory, but as an object that is evaluated by a subset of the interval $[0,1]$. We have called this type of generalized probability – standard probability theory is retrieved when the considered subsets of the interval $[0,1]$ are the singletons – a 'subset probability'. Although a lot of work is still needed to make the subset probability into a full grown probability theory, we will take it in account for the elaboration of the formalism that we propose in this paper.

1.7 How We Will Proceed

We will take into account all the aspects mentioned in sections 1.1, 1.2, 1.3, 1.4, 1.5 and 1.6 for the foundations of the formalism that we introduce in this paper. We also try to make the paper as self contained as possible, such that it is not necessary for the reader to go through all the preceding material to be able to understand it.

For sake of completeness we mention other work that has contributed to advances in the approach that is less directly of importance for what we do in this paper, but shows how the approach is developing into other directions as well [52,53,54,55,56,57,58,59,60,61,62,63,64,65,66,67,68,69,70,71,72,73,74,75,76,77,78,79,80,81].

2 Foundations of the Formalism

In this section we introduce the basic ingredients of the formalism. Since we want to be able to apply the formalism to many different types of situations the basic ingredients must be sufficiently general. The strategy that we follow consists of describing more specific situations as special cases of the general situation. The primary concept that we consider is that of an **entity**, that we will denote by S. Such an entity S can for example be a cat, or a genome, or cultural artifact, such as a building, or an abstract idea, or a mind of a person, or a stone, or a quantum particle, or a fluid etc ...

2.1 States, Contexts and Change

At a specific moment, an entity, *is* in a specific **state**. The state represents *what the entity is and how it reacts to different contexts* at that moment. For example, a cat can be awake or asleep, this are two possible states of the cat. The second basic concept that we consider is that of a **context**. A context is a part of the outside reality of the entity that influences the entity in such a way that its state is changed. If the cat is asleep, and confronted with a context of heavy noise, it is probable that it will wake up. The context 'heavy noise' changes then the state 'cat is asleep' into the state 'cat is awake'.

For a specific entity S we denote the states of this entity by $p, q, r, s, \ldots$ and the contexts that the entity can be confronted with by $e, f, g, h, \ldots$. The set of all relevant states of an entity S we denote by Σ and the set of all relevant contexts by $\mathcal{M}$.

Let us express the basic situation that we consider. An entity S, in a state p, interacts with a context e. In general, the interaction between the entity and the context causes a change of the state of the entity and also a change of the context. Sometimes there will be no change of state or no change of context: this situation we consider as a special case of the general situation; we can for example call it a situation of zero change. Sometimes the entity will be destroyed by the context with which it interacts. A context that destroys the entity provokes a change of the state that the entity is in before it interacts with this context, but this change is so strong that most of the characteristic properties of the entity are destroyed, and hence the remaining part of reality will not be identified any longer as the entity. To be able to express the destruction within our formalism we introduce a special state $0 \in \Sigma$. If a context changes a specific state $p \in \Sigma$ to the state $0 \in \Sigma$ it means that the entity in question is destroyed by the context[a]. We denote by Σ_0 the set of all relevant states of the entity S without the state of destruction 0. Let us introduce the basic notions in a formal way.

Basic Notion 1 (State, Context and Change) *For an entity S we introduce its set of relevant states Σ and its set of relevant contexts $\mathcal{M}$. States are denoted by $p, q, r, s \in \Sigma$ and contexts are denoted by $e, f, g, h \in \mathcal{M}$. The state $0 \in \Sigma$ is the state that expresses that the entity is destroyed. By Σ_0 we denote the set of states without the state 0 that represent the destroyed entity. The situation where the entity S is in a state $p \in \Sigma$ and under influence of a context $e \in \mathcal{M}$ will in general lead to a new situation where the entity is in a state $q \in \Sigma$ and the context is $f \in \mathcal{M}$. The transition from the couple (e, p) to the couple (f, q) we call the change of the entity in state p under influence of the context e.*

2.2 Probability

Probability theory has been developed to get a grip on indeterminism. Usually a probability is defined by means of a measure on the interval $[0, 1]$ of the set of real numbers $\mathbb{R}$. As we mentioned already in section 1.6, we are not convinced that the standard way to introduce probability is the good way for our formalism. We have analyzed this problem in detail in [51] and use the results obtained there.

Definition 1 (Probability of Change) *Consider an entity S with a set of*

[a]The state 0 is not really a state in the proper sense, it is introduced specifically to take into account the possibility for a context to destroy the entity that we consider.

states Σ and a set of contexts $\mathcal{M}$. We introduce the function

$$\mu : \mathcal{M} \times \Sigma \times \mathcal{M} \times \Sigma \to \mathcal{P}([0,1]) \quad (1)$$

$$(f, q, e, p) \mapsto \mu(f, q, e, p) \quad (2)$$

where $\mu(f,q,e,p)$ is the probability for the couple (e,p) to change to the couple (f,q). $\mathcal{P}([0,1])$ is the set of all subsets of the interval $[0,1]$. This means that we evaluate the probability by a subset of $[0,1]$ and not a number of $[0,1]$ as in standard probability theory.

This is a generalization of standard probability theory that we retrieve when all the considered subsets are singletons. It will become clear in the following what are the advantages of this generalization as is also explained in detail in [51]. We remark that if the probability that we introduce would be a traditional probability, where all the $\mu(f,q,e,p)$ are singletons, we would demand the following type of property to be satisfied:

$$\sum_{f \in \mathcal{M}, q \in \Sigma} \mu(f, q, e, p) = 1 \quad (3)$$

expressing the fact that the couple (e,p) is changed always to one of the couples (f,q). This rule is usually referred to as the sum rule for probability. For a subset probability the sum rule is much more complicated. We refer to [51] for a more complete reflection on this matter, but admit that the matter has not been solved yet. In the course of this article we will make use of the sum rule for subset probabilities only to express that indeed a couple (e,p) is always changed to a couples (f,q).

2.3 States and Properties

A property is something that the entity 'has' independent of the type of context that the entity is confronted with. That is the reason why we consider properties to be independent basic notions of the formalism. We denote properties by $a, b, c, \ldots$ and the set of all relevant properties of an entity by $\mathcal{L}$.

For a state we require that an entity is in a state at each moment . The state represents the reality of the entity at that moment. Properties are elements of this reality. This means that an entity S in a specific state $p \in \Sigma$ has different properties that are actual. Properties that are actual for the entity being in a specific state can be potential for this entity in another state.

Basic Notion 2 (Property) *For an entity S, with set of states Σ and set of contexts $\mathcal{M}$, we introduce its set of properties. A property can be actual for* *an entity in a specific state and potential for this entity in another state. We denote properties by $a, b, c, \ldots$, and the set of properties of S by $\mathcal{L}$.*

We introduce some additional concepts to be able to express the basic situation that we want to consider in relation with the state and the properties of an entity. Suppose that the entity S is in a specific state $p \in \Sigma$. Then some of the properties of S are actual and some are not and hence are potential. This means that with each state $p \in \Sigma$ corresponds a set of actual properties, subset of $\mathcal{L}$. This defines a function $\xi : \Sigma \to \mathcal{P}(\mathcal{L})$, which maps each state $p \in \Sigma$ to the set $\xi(p)$ of properties that are actual in this state. Introducing this function makes it possible to replace the statement 'property $a \in \mathcal{L}$ is actual for the entity S in state $p \in \Sigma$' by '$a \in \xi(p)$'.

Suppose now that for the entity S a specific property $a \in \mathcal{L}$ is actual. Then this entity is in a certain state $p \in \Sigma$ that makes a actual. With each property $a \in \mathcal{L}$ we can associate the set of states that make this property actual, i.e. a subset of Σ. This defines a function $\kappa : \mathcal{L} \to \mathcal{P}(\Sigma)$, which makes each property $a \in \mathcal{L}$ correspond to the set of states $\kappa(a)$ that make this property actual. Again we can replace the statement 'property $a \in \mathcal{L}$ is actual if the entity S is in state $p \in \Sigma$' by the set theoretical expression '$p \in \kappa(a)$'. Let us introduce these notions in a formal way.

Definition 2 (Aristotle Map) *Consider an entity S, with set of states Σ, set of contexts $\mathcal{M}$ and set of properties $\mathcal{L}$. We define a function $\xi : \Sigma \to \mathcal{P}(\mathcal{L})$ such that $\xi(p)$ is the set of all properties that are actual if the entity is in state p. We call ξ the Aristotle map* [b] *of the entity S.*

Definition 3 (Cartan Map) *Consider an entity S, with set of states Σ, set of contexts $\mathcal{M}$ and set of properties $\mathcal{L}$. We define a function $\kappa : \mathcal{L} \to \mathcal{P}(\Sigma)$ such that $\kappa(a)$ is the set of all states that make the property a actual. We call κ the Cartan map*[c].

By introducing the Aristotle map and the Cartan map we can express the basic situation we want to consider for states and properties as follows:

$$a \in \xi(p) \Leftrightarrow p \in \kappa(a) \Leftrightarrow a \text{ is actual for } S \text{ in state } p \tag{4}$$

If the state $p \in \Sigma$ of an entity is changed under influence of a context $e \in \mathcal{M}$ into a state $q \in \Sigma$, then the set $\xi(p)$ of actual properties in state p is changed into the set $\xi(q)$ of actual properties in state q. Contrary to the state of an

[b] We introduce the name Aristotle map for this function as a homage to Aristotle, because he was the first to consider the set of actual properties as characteristic for the state of the considered entity.

[c] The name Cartan map for this function was introduced in earlier formulations of the formalism [4,5] as an homage to Eli Cartan, who for the first time considered the state space as the fundamental structure in the case of classical mechanics

entity, a property is not changed under influence of the context. What can be changed is its status of actual or potential. This change of status under influence of the context is monitored completely by the change of state by this context.

2.4 Covariance and Morphisms

We derive the morphism of our structure by making use of a merological covariance principle. What we mean is that we will express for the situation of an entity and subentity the covariance of the descriptions, and in this way derive the morphisms of our structure.

Consider two entities S and S' such that S is a subentity of S'. In that case, the following three natural requirements should be satisfied:

i) If the entity S' is in a state p' then the state $m(p')$ of S is determined. This defines a function m from the set of states of S' to the set of states of S.

ii) If we consider a context e that influences the entity S, then this context also influences the entity S'. This defines a function l from the set of contexts of S to the set of contexts of S'.

iii) When we consider a property a of S, then this corresponds to a property $n(a)$ of S', which is the same property, but now considered as a property of the big entity S'. This defines a function n of $\mathcal{L}$ to $\mathcal{L}'$.

iv) We want e and $l(e)$ to be two descriptions of the 'same' context, once considered as a context that influences S and once considered as a context that influences S'. This means that when we consider how e influences $m(p')$, this is the same physical process as when we consider how $l(e)$ influences p', with the only difference that the first time it is considered within the description of the subentity S and the second time within the description of the big entity S'. In a similar way we want property a to behave in the same way towards the states of S as property $n(a)$ behaves towards the states of S'. As a consequence the following covariance principles hold:

$$\mu(f, m(q'), e, m(p')) = \mu'(l(f), q', l(e), p') \tag{5}$$

$$a \in \xi(m(p')) \Leftrightarrow n(a) \in \xi'(p') \tag{6}$$

We have now everything at hand to define a morphism of our structure.

Definition 4 (Morphisms) *Consider two entities S and S', with sets of states Σ, Σ', sets of contexts $\mathcal{M}$, $\mathcal{M}'$ and sets of properties $\mathcal{L}$ and $\mathcal{L}'$, such that probability is defined as in definition 1. We say that the triple $(m,\ l,\ n)$*

is a morphism if m is a function:

$$m : \Sigma' \rightarrow \Sigma \tag{7}$$

l is a function:

$$l : \mathcal{M} \rightarrow \mathcal{M}' \tag{8}$$

and n is a function:

$$n : \mathcal{L} \rightarrow \mathcal{L}' \tag{9}$$

such that for $p' \in \Sigma'$, $e \in \mathcal{M}$ and $a \in \mathcal{L}$ the following holds:

$$\mu(f, m(q'), e, m(p')) = \mu'(l(f), q', l(e), p') \tag{10}$$
$$a \in \xi(m(p')) \Leftrightarrow n(a) \in \xi'(p') \tag{11}$$

Requirements (10) and (11) are the covariance formula's that characterize the morphisms of our structure.

2.5 The Category of State Context Property Systems and Their Morphisms

We define now the mathematical structure that we need to describe an entity by means of its states, context and properties, in a purely mathematical way, so that the structure can be studied mathematically.

Definition 5 (The Category SCOP) *A state context property system $(\Sigma, \mathcal{M}, \mathcal{L}, \mu, \xi)$, consists of three sets Σ, $\mathcal{M}$, $\mathcal{L}$ and two functions μ and ξ, such that*

$$\mu : \mathcal{M} \times \Sigma \times \mathcal{M} \times \Sigma \rightarrow \mathcal{P}([0,1]) \tag{12}$$
$$\xi : \Sigma \rightarrow \mathcal{P}(\mathcal{L}) \tag{13}$$

The sets Σ, $\mathcal{M}$ and $\mathcal{L}$, play the role of the set of states, the set of contexts, and the set of properties of an entity S. The function μ describes transition probabilities between couples of contexts and states, while the function ξ describes the sets of actual properties for the entity S being in different states.

Consider two state context property systems $(\Sigma, \mathcal{M}, \mathcal{L}, \mu, \xi)$ and $(\Sigma', \mathcal{M}', \mathcal{L}', \mu', \xi')$. A morphism is a triple of functions (m, l, n) such that:

$$m : \Sigma' \rightarrow \Sigma \tag{14}$$
$$l : \mathcal{M} \rightarrow \mathcal{M}' \tag{15}$$
$$n : \mathcal{L} \rightarrow \mathcal{L} \tag{16}$$

and the following formula's are satisfied:

$$\mu(f, m(q'), e, m(p')) = \mu'(l(f), q', l(e), p') \tag{17}$$

$$a \in \xi(m(p')) \Leftrightarrow n(a) \in \xi'(p') \tag{18}$$

We denote the category consisting of state context property systems and their morphisms by **SCOP**.

3 Classical and Quantum Entities

As we mentioned, our formalism must be capable to describe different forms of change that are encountered in nature for different types of entities. Let us consider some examples to show how this happens.

3.1 Classical and Quantum Dynamical Change

In this section we outline in which way the deterministic evolution of classical physical entities described by the dynamical laws of classical physics and the deterministic evolution of quantum entities described by the Schrödinger equation are described in the formalism.

For a classical mechanical entity in a certain state p under a context e change is deterministic. A specific couple (e, p) changes deterministically into another couple (f, q). Quantum entities undergo two types of change. One of them, referred to as the dynamical change and described by the Schrödinger equation, is also deterministic. The other one, referred to as collapse and described by von Neumann's projection postulate, is indeterministic. We consider in this section only the deterministic Schrödinger type of change for a quantum entity.

Let us consider first a concrete example of the classical mechanical change. Place a charged iron ball in in a magnetic field. The ball will be influenced by the context which is the magnetic field and start to move according to the classical laws of electromagnetism. Knowing the context at t_0 and the state of the entity at t_0 we can predict with certainty the state of the entity at t_1, $t_2 \ldots$. If we consider the influence of the magnetic field from time t_0 to time t_1 as one context $e(t_0 \mapsto t_1)$ and the influence of the magnetic field from time t_1 till time t_2 as another context $e(t_1 \mapsto t_2)$, then the whole dynamical evolution can be seen as the change monitored continuously by the set of contexts $e(t_1 \mapsto t)$. All these contexts change the state of the entity in a deterministic way.

The deterministic evolution of a quantum entity from time t_1 to time t, described in the standard quantum formalism by means of the Schrödinger equation, can be represented in our formalism in the same way, as the change monitored continuously by the set of contexts $e(t_1 \mapsto t)$.

Classical and quantum dynamical evolution are both characterized as a continuous change by an infinite set of contexts that all change the state of the entity in a deterministic way. We can specify this situation in the general formalism. To be able to express in a nice way the specific characteristics of quantum and classical entities, we introduce the concept of range.

Definition 6 (Range of a Context for a State) *Consider a state context property system $(\Sigma, \mathcal{M}, \mathcal{L}, \mu, \xi)$ describing an entity S. For $p \in \Sigma$ and $e \in \mathcal{M}$ we introduce:*

$$R(e,p) = \{q \mid q \in \Sigma, \ \exists \ f \in \mathcal{M} \text{ such that } \mu(f,q,e,p) \neq \{0\}\} \tag{19}$$

and call $R(e,p)$ the range of e for p.

The range of a context for a state is the set of states that this state can be changed to under influence of this context. We can now easily define what is a deterministic context to a state.

Definition 7 (Deterministic Context to a State) *Consider a state context property system $(\Sigma, \mathcal{M}, \mathcal{L}, \mu, \xi)$ describing an entity S. We call a context $e \in \mathcal{M}$ a deterministic context to a state $p \in \Sigma$ if $R(e,p) = \{q\}$. We call q the image of the state p under the context e.*

If e is a deterministic context to the state p, then the state p is changed deterministically to its image state q under influence of the context e.

Definition 8 (Range of a Context) *Consider a state context property system $(\Sigma, \mathcal{M}, \mathcal{L}, \mu, \xi)$ describing an entity S. For $e \in \mathcal{M}$ we introduce:*

$$R(e) = \cup_{p \in \Sigma} R(e,p) \tag{20}$$

and call $R(e)$ the range of the context e.

The range of a context e is the set of states that is reached by changes provoked by e on any state of the entity. We can of course also define the range of a state for a context. It is the set of contexts that this context can change to under influence of this state.

Definition 9 (Range of a State for a Context) *Consider a state context property system $(\Sigma, \mathcal{M}, \mathcal{L}, \mu, \xi)$ describing an entity S. For $e \in \mathcal{M}$ and $p \in \Sigma$ we introduce:*

$$R(p,e) = \{f \mid f \in \mathcal{M}, \ \exists \ q \in \Sigma \text{ such that } \mu(f,q,e,p) \neq \{0\}\} \tag{21}$$

and call $R(p,e)$ the range of p for e.

This makes it easy to define what is a deterministic state to a context.

Definition 10 (Deterministic State to a Context) *Consider a state context property system* $(\Sigma, \mathcal{M}, \mathcal{L}, \mu, \xi)$ *describing an entity* S. *We call a state* $p \in \Sigma$ *a deterministic state to a context* $e \in \mathcal{M}$ *if* $R(p, e) = \{f\}$. *We call* f *the image of the context* e *under the state* p.

If p is a deterministic state to the context e then the context e is changed deterministically to its imagine context f under influence of the state p.

Definition 11 (Range of a State) *Consider a state context property system* $(\Sigma, \mathcal{M}, \mathcal{L}, \mu, \xi)$ *describing an entity* S. *For* $p \in \Sigma$ *we introduce:*

$$R(p) = \cup_{e \in \mathcal{M}} R(p, e) \tag{22}$$

and call $R(p)$ *the range of the state* p.

The range of a state p is the set of contexts that is reached by changes provoked by p on any context of the entity.

Definition 12 (Deterministic Context State Couple) *Consider a state context property system* $(\Sigma, \mathcal{M}, \mathcal{L}, \mu, \xi)$ *describing an entity* S. *We call the context* $e \in \mathcal{M}$ *and the state* $p \in \Sigma$ *a deterministic context state couple, if* e *is a deterministic context to* p, *and* p *is a deterministic state to* e. *Then* (e, p) *changes deterministically to* (f, q). *We call* (f, q) *the image of* (e, p).

Definition 13 (Deterministic Context) *Consider a state context property system* $(\Sigma, \mathcal{M}, \mathcal{L}, \mu, \xi)$ *describing an entity* S. *We say that a context* $e \in \mathcal{M}$ *is a deterministic context if it is a deterministic context to each state* $p \in \Sigma$.

For quantum entities only the contexts that generate the dynamical change of state described by the Schrödinger equation are deterministic contexts. For classical entities all contexts are deterministic.

Definition 14 (Deterministic State) *Consider a state context property system* $(\Sigma, \mathcal{M}, \mathcal{L}, \mu, \xi)$ *describing an entity* S. *We say that a state* $p \in \Sigma$ *is a deterministic state if it is a deterministic state to any context* $e \in \mathcal{M}$.

For quantum mechanics as well as for classical physics all states are deterministic states. In physics we indeed do not in general consider an influence of the state on the context.

We have now all the material to define what is in our general formalism a d-classical entity.

Definition 15 (D-Classical Entity) *Consider a state context property system* $(\Sigma, \mathcal{M}, \mathcal{L}, \mu, \xi)$ *describing an entity* S. *We say that the entity* S *is a d-classical entity if all its states and contexts are deterministic.*

The reason why we call such an entity a d-classical entity and not just a classical entity is that our definition only demands the classicality of the entity towards the change of state that can be provoked by a context. There exist other possible forms of classicality, for example towards the type of properties that an entity can have. This is the reason the we prefer to call the type of classicality that we introduce here d-classicality. In [50] the structure related to d-classical entities is analyzed in detail.

We note that in our formalism it are the deterministic contexts that produce an entity to behave like the physical entities behave under dynamical evolution, whether this is the evolution of classical physical entities under any kind of context, or the evolution of quantum physical entities described by the Schrödinger equation.

3.2 Quantum Measurement Contexts

In the foregoing section we have seen that the context that gives rise to a quantum evolution is a deterministic context. For quantum entities there are also contexts that originate in the measurement. These contexts are not deterministic. Let us see how their actions fit into the formalism. Consider a quantum entity in a state p represented by a unit vector $u(p)$ of a complex Hilbert space $\mathcal{H}$. A measurement context e in quantum mechanics is described by a self-adjoint operator $A(e)$ on this Hilbert space. In general, a self-adjoint operator has a spectrum that consists of a point-like part and a continuous part. In the point-like part of the spectrum, the state p is transformed into one of the eigenstates represented by the eigenvectors corresponding to the points of the point-like part of the spectrum of $A(e)$. For the continuous part of the spectrum the situation is somewhat more complicated. There are no points as outcomes here, but only intervals. To such an interval corresponds a unique projection operator of the spectral resolution of $A(e)$, and the state p is then projected by this projection onto a vector of the Hilbert space, which represents that state after the measurement.

3.3 Cultural Change and the Human Mind

Cultural change with the human mind as generating entity is very complex. Here states as well as context will in general not be deterministic. How the formalism can be applied there has been studied in [38,39]. More specifically a situation describing 'the invention of the torch' has been modelled in [39]. We will not consider this type of change in more detail in the present article and refer to [40] for a formal approach concentrated on this case.

3.4 Eigen States and Eigen Contexts

There is a type of determinism of a context towards a state and of a state towards a context that we want to consider specifically, namely when there is no change at all.

Definition 16 (Eigenstate of a Context) *Consider a state context property system $(\Sigma, \mathcal{M}, \mathcal{L}, \mu, \xi)$ describing an entity S. A state $p \in \Sigma$ is called an eigenstate of the context $e \in \mathcal{M}$ if the context e is deterministic to the state p and the image of p under e is p itself.*

Proposition 1 *Consider a state context property system $(\Sigma, \mathcal{M}, \mathcal{L}, \mu, \xi)$ describing an entity S. A state $p \in \Sigma$ is an eigenstate of the context $e \in \mathcal{M}$ iff*

$$R(e,p) = \{p\} \tag{23}$$

Definition 17 (Eigencontext of a State) *Consider a state context property system $(\Sigma, \mathcal{M}, \mathcal{L}, \mu, \xi)$ describing an entity S. A context $e \in \mathcal{M}$ is called an eigencontext of the state $p \in \Sigma$ if the state p is deterministic to the context e and the image of e under p is e itself.*

Proposition 2 *Consider a state context property system $(\Sigma, \mathcal{M}, \mathcal{L}, \mu, \xi)$ describing an entity S. A context $e \in \mathcal{M}$ is an eigencontext of the state $p \in \Sigma$ iff*

$$R(p,e) = \{e\} \tag{24}$$

The name eigenstate has been taken from quantum mechanics, because indeed if a quantum entity is in an eigenstate of the operator that represents the considered measurement, then this state is not changed by the context of this measurement. Experiments in classical physics are observations and hence do not change the state of the physical entity.

4 Pre-Order Structures

Step by step we introduce additional structure in our formalism. As much as possible we introduce this structure in an operational way, meaning that we analyze carefully what is the meaning of the structure that we introduce and how it is connected with reality. In this section we limit ourselves to the identification of a natural pre-order structure on the set of states and on the set of properties.

4.1 State and Property Implication and Equivalence

Before introducing the state and property implications that will form pre-order relations on Σ and $\mathcal{L}$, let us first define what is a pre-order relation on a set.

Definition 18 (Pre-order Relation and Equivalence) *Suppose that we have a set Z. We say that $<$ is a pre-order relation on Z iff for $x, y, z \in Z$ we have:*

$$\begin{aligned} &x < x \\ &x < y \text{ and } y < z \Rightarrow x < z \end{aligned} \tag{25}$$

For two elements $x, y \in Z$ such that $x < y$ and $y < x$ we denote $x \approx y$ and we say that x is equivalent to y.

There exist natural 'implication relations' on Σ and on $\mathcal{L}$. If the situation is such that if '$a \in \mathcal{L}$ is actual for S in state $p \in \Sigma$' implies that '$b \in \mathcal{L}$ is actual for S in state $p \in \Sigma$' we say that property a 'implies' property b. If the situation is such that '$a \in \mathcal{L}$ is actual for S in state $q \in \Sigma$' implies that '$a \in \mathcal{L}$ is actual for S in state $p \in \Sigma$' we say that the state p implies the state q. Let us introduce these two implications in a formal way.

Definition 19 (State Implication and Property Implication) *Consider a state context property system $(\Sigma, \mathcal{M}, \mathcal{L}, \mu, \xi)$ describing an entity S. For $a, b \in \mathcal{L}$ we introduce:*

$$a < b \Leftrightarrow \kappa(a) \subset \kappa(b) \tag{26}$$

and we say that a 'implies' b. For $p, q \in \Sigma$ we introduce:

$$p < q \Leftrightarrow \xi(q) \subset \xi(p) \tag{27}$$

and we say that p 'implies' q [d].

It is easy to verify that the implication relations that we have introduced are pre-order relations.

Proposition 3 *Consider a state context property system $(\Sigma, \mathcal{M}, \mathcal{L}, \mu, \xi)$ describing an entity S. Then $\Sigma, <$ and $\mathcal{L}, <$ are pre-ordered sets.*

We can prove the following:

Proposition 4 *Consider a state context property system $(\Sigma, \mathcal{M}, \mathcal{L}, \mu, \xi)$ describing an entity S. (1) Suppose that $a, b \in \mathcal{L}$ and $p \in \Sigma$. If $a \in \xi(p)$ and $a < b$, then $b \in \xi(p)$. (2) Suppose that $p, q \in \Sigma$ and $a \in \mathcal{L}$. If $q \in \kappa(a)$ and $p < q$ then $p \in \kappa(a)$.*

---

[d] Remark that the state implication and property implication are not defined in a completely analogous way. Indeed, then we should have written $p < q \Leftrightarrow \xi(p) \subset \xi(q)$. That we have chosen to define the state implication the other way around is because historically this is how intuitively is thought about states implying one another.

Proof: (1) We have $p \in \kappa(a)$ and $\kappa(a) \subset \kappa(b)$. This proves that $p \in \kappa(b)$ and hence $b \in \xi(p)$. (2) We have $a \in \xi(q)$ and $\xi(q) \subset \xi(p)$ and hence $a \in \xi(p)$. This shows that $p \in \kappa(a)$. □

It is possible to prove that the morphisms of the category **SCOP** conserves the two implications.

Proposition 5 *Consider two state context property systems* $(\Sigma, \mathcal{M}, \mathcal{L}, \mu, \xi)$ *and* $(\Sigma', \mathcal{M}', \mathcal{L}', \mu', \xi')$ *describing entities* S *and* S'*, and a morphism* (m, l, n) *between these two state context. For* $p', q' \in \Sigma'$ *and* $a, b \in \mathcal{L}$ *we have:*

$$p' < q' \Leftrightarrow m(p') < m(q') \tag{28}$$

$$a < b \Leftrightarrow n(a) < n(b) \tag{29}$$

Proof: Suppose that $p' < q'$. This means that $\xi'(q') \subset \xi'(p')$. Consider $a \in \xi(m(q'))$. Using 18 implies that $n(a) \in \xi(q')$. But then $n(a) \in \xi(p')$ and again using 18 we have $a \in \xi'(m(p'))$. This means that we have shown that $\xi(m(q')) \subset \xi(m(q'))$. As a consequence we have $m(p') < m(q')$. This proves one of the implications of 28. Suppose now that $m(p') < m(q')$, which implies $\xi(m(q')) \subset \xi(m(p'))$. Consider $a \in \xi'(q')$. Form 18 follows that $n(a) \in \xi(m(q'))$ and hence also $n(a) \in \xi(m(p'))$. Again from 18 follows that $a \in \xi'(p')$. So we have shown that $\xi'(q') \subset \xi'(p')$, and as a consequence $p' < q'$. Formula 29 is proven in a completely analogous way. □

5 Experiments And Preparations

Some contexts are used to perform an experiment on the entity under consideration and other context are used to prepare a state of the entity. We want to study these types of context more carefully, because they will play an important role in further operational foundations of the formalism. Let us analyze what requirements are to be fulfilled for a context to be an experiment.

For a context that is an experiment the context will change under influence of the state in such a way that from the new context we can determine the outcome of the experiment. What are then outcomes of an experiment?

5.1 Outcomes of Experiments

For a context to play the role of an experiment it must be possible to identify outcomes of the experiment. We will denote outcomes by $x, y, z, \ldots$, and the set of possible outcomes corresponding to an experiment $e \in \mathcal{M}$, the entity being in state $p \in \Sigma$, by $O(e, p)$. Obviously, the set of all possible outcomes of the experiment e is then given by $\cup_{p \in \Sigma} O(e, p) = O(e)$.

If we consider the experimental practice in different scientific domains there does not seem to be a standard way to identify outcomes. In general the description of an outcome of an experiment e on an entity S in state p is linked to the state of the entity after the effect of the context, hence to the type of change that has been provoked by the experiment, and also to the experiment itself, and to the new context that arises after the experiment has been performed.

This means that, if we consider a context e that we want to use as an experiment, and suppose that the entity is in state p, and that q is a possible state that the entity can change to under context e, and f is a possible context after e has worked on p, then a possible outcome $x(f, q, e, p)$ for e will occur. But it might well be that another possible state r that the entity might evolve to under context e, identifies the same outcome $x(f, r, e, p) = x(f, q, e, p)$ for e. This is the case when state q only differs from state r in aspects that are not relevant for the physical quantity measured by the experiment e. Let us give an example to explain what we mean.

Suppose that we consider a classical physics entity S that is a point particle located motionless on a line that we have coordinated by the set of real numbers $\mathbb{R}$. The state of the particle in classical physics is described by its position u and its momentum mv, where m is its mass and v its velocity, hence by the vector $(u, mv) \in \mathbb{R}^2$. Our experiment e consists of making a picture of the particle. On the picture we can read off the coordinate where the particle is, hence its position. The set of possible outcomes $O(e)$ for this experiment is a part of the set of real numbers $\mathbb{R}$, namely the points described by the coordinate u, the position coordinate of the particle. The context e is all what takes place when we make the picture, but without the picture being taken, which means that for e the film in the camera has not been exposed. This context e changes after the experiment into f where the film has been exposed. The experiment e is just an observation, not provoking any change on the state. If for example outcome $x \in \mathbb{R}$ occurs, we know that the position u of the particle equals x. The experiment not only does not change the state of the particle, it is also a deterministic context. Indeed $O(e, (u, mv)) = \{u\}$ for all states $(u, mv) \in \mathbb{R}^2$. In this example the experiment gathers knowledge about the state of the entity that we did not have before we performed the experiment.

Let us consider the example of a one dimensional quantum particle, described by a state that is a wave function $\psi(x)$, element of $L^2(\mathbb{R})$ such that

$$\int \|\psi(x)\|^2 dx = 1$$

The context related to a position measurement is described by a self-adjoint operator with a set of spectral projection operators that are the characteristic functions χ_Ω of subsets $\Omega \subset \mathbb{R}$. The probability for the quantum particle to be located in the subset $\Omega \subset \mathbb{R}$ by the effect of the context is given by

$$\int_\Omega \|\psi(x)\|^2 dx$$

and, if the particle is located in the subset Ω, the state $\psi(x)$ is changed into the state

$$\frac{1}{\sqrt{\int_\Omega \|\psi(x)\|^2 dx}} \cdot \chi_\Omega \circ \psi(x) \tag{30}$$

In this new state, given by (30), the probability to be located in Ω, if the position context is applied again to the quantum particle, is given by:

$$\int \|\frac{1}{\sqrt{\int_\Omega \|\psi(x)\|^2 dx}} \cdot \chi_\Omega \circ \psi(x)\|^2 dx = 1$$

That is the reason that we can consider the subset Ω as an outcome for the position experiment.

Definition 20 (Experiment) *Consider a state context property system $(\Sigma, \mathcal{M}, \mathcal{L}, \mu, \xi)$ describing an entity S. We say that a context $e \in \mathcal{M}$ is an experiment if for $p \in \Sigma$ there exists a set $O(e,p)$ and a map:*

$$x : \mathcal{M} \times \Sigma \times \mathcal{M} \times \Sigma \to O(e,p) \tag{31}$$

$$(f, q, e, p) \mapsto x(f, q, e, p) \tag{32}$$

where $x(f,q,e,p)$ is the outcome of experiment e for the entity in state p, and where (e,p) has changed to (f,q). We further have that:

$$\mu(f, q, e, p) \neq \{0\} \tag{33}$$

expressing that an outcome for e is only possible if the transition probability from (e,p) to (f,q) is different from $\{0\}$. We denote $q = P^e_x(p)$. The probability for the experiment e to give outcome x if the entity is in state p equals

$$\mu(e, P^e_x(p), e, p) \tag{34}$$

For $x \in O(e,p) \cap O(e,r)$ we have:

$$P^e_x(p) = P^e_x(r) \tag{35}$$

The definition of an experiment that we have given here is still very general. As we saw already, in classical physics an experiment is usually an observation, which is a much more specific type of experiment. An observation just 'observes' the state of the entity that is there without provoking any kind of change. Also in quantum mechanics an experiment is much less general than the definition that we have given here.

5.2 Contexts and Experiments of the First Kind

In quantum mechanics an experiment provokes a change of state such that the state after the experiment is an eigenstate of this experiment, and the outcome is identified by means of this eigenstate. Experiments with this property have been called experiments of the first kind in physics.

Definition 21 (Contexts of the First Kind) *Consider a state context property system $(\Sigma, \mathcal{M}, \mathcal{L}, \mu, \xi)$ describing an entity S. We say that a context $e \in \mathcal{M}$ is a context of the first kind if for $p \in \Sigma$, we have that (e,p) changes to (f,q), with $f \in \mathcal{M}$ and $q \in \Sigma$, where q is an eigenstate of f.*

This means that for a context of the first kind, when f is being applied again and again to the entity, its state will remain q. This means that the entity has been changed into a stable state that no longer changes under influence of context. This is a perfect situation to be able to identify an outcome of an experiment by means of this eigenstate.

Definition 22 (Experiment of the First Kind) *Consider a state context property system $(\Sigma, \mathcal{M}, \mathcal{L}, \mu, \xi)$ describing an entity S. We say that $e \in \mathcal{M}$ is an experiment of the first kind if e is an experiment and a context of the first kind. This means that for $p \in \Sigma$ we have that $P^e_x(p)$ is an eigenstate of e, in the sense that $\mu(e, P^e_x(p), e, P^e_x(p)) = \{1\}$, and the experiment e makes occur the outcome x with probability equal to 1.*

An experiment of the first kind pushes each state of the entity into an eigenstate of this experiment. This is the way that experiments act in quantum mechanics. Let us consider again the example of the experiment that measures the position of a quantum particle. The state $\psi(x)$ is changed to the state

$$\frac{1}{\sqrt{\int_\Omega \|\psi(x)\|^2 dx}} \cdot \chi_\Omega \circ \psi(x)$$

if we test whether the outcome is the interval $\Omega \subset \mathbb{R}$. If we test again whether the outcome is in the subset Ω, the state does not change any longer, because

$$\chi_\Omega(\frac{1}{\sqrt{\int_\Omega \|\psi(x)\|^2 dx}} \cdot \chi_\Omega \circ \psi(x)) = \frac{1}{\sqrt{\int_\Omega \|\psi(x)\|^2 dx}} \cdot \chi_\Omega \circ \psi(x)$$

The set of outcomes for a quantum entity has the structure of the set of all subsets of another set, this other set being the spectrum of the self-adjoint operator that represents the experiment in quantum mechanics. But there is more. Let us consider two consecutive measurements of position, once in the subset Ω_1 and second in the subset Ω_2, such that $\Omega_2 \subset \Omega_1 \subset \mathbb{R}$. First, for Ω_1 the state $\psi(x)$ changes to

$$\frac{1}{\sqrt{\int_{\Omega_1} \|\psi(x)\|^2 dx}} \cdot \chi_{\Omega_1} \circ \psi(x)$$

with probability

$$\int_{\Omega_1} \|\psi(x)\|^2 dx$$

Remark first that:

$$\int_{\Omega_2} \|\frac{1}{\sqrt{\int_{\Omega_1} \|\psi(x)\|^2 dx}} \cdot \chi_{\Omega_1} \circ \psi(x)\|^2 dx = \frac{\int_{\Omega_2} \|\psi(x)\|^2 dx}{\int_{\Omega_1} \|\psi(x)\|^2 dx}$$

This means that for the second change of state the probability equals to:

$$\frac{\int_{\Omega_2} \|\psi(x)\|^2 dx}{\int_{\Omega_1} \|\psi(x)\|^2 dx}$$

and the state

$$\frac{1}{\sqrt{\int_{\Omega_1} \|\psi(x)\|^2 dx}} \cdot \chi_{\Omega_1} \circ \psi(x)$$

is changed to

$$\sqrt{\frac{\int_{\Omega_1} \|\psi(x)\|^2 dx}{\int_{\Omega_2} \|\psi(x)\|^2 dx}} \cdot \chi_{\Omega_2} \circ \frac{1}{\sqrt{\int_{\Omega_1} \|\psi(x)\|^2 dx}} \cdot \chi_{\Omega_1} \circ \psi(x)$$

Taking into account that $\chi_{\Omega_2} \circ \chi_{\Omega_1} \circ \psi(x) = \chi_{\Omega_2} \circ \psi(x)$ because $\Omega_2 \subset \Omega_1$ we have that the final state is:

$$\frac{1}{\int_{\Omega_2} \|\psi(x)\|^2 dx} \cdot \chi_{\Omega_2} \circ \psi(x)$$

This is exactly the state that we would have found if we would immediately have measured the position with a test to see that the quantum particle is localized in the subset Ω_2. And also the probabilities multiply, namely the probability to find the position in subset Ω_2 with a direct test on Ω_2 is the product of the probability to find it in Ω_1 with a test on Ω_1, with the probability to find it in Ω_2 after it had already been tested for Ω_1. The process that we have identified here for the position measurement of a quantum entity is generally true for all quantum mechanical measurements. Quantum mechanical measurement have a kind of cascade structure. Let us introduce this structure for a general experiment and call it a cascade experiment. To be able to do so we need to define what we mean by the product of two subsets of the interval $[0, 1]$ and also what we mean by 1 minus this subset for a subset of $[0, 1]$.

Definition 23 *Suppose that* $A, B, C \in \mathcal{P}([0,1])$ *of the interval* $[0,1]$. *We define:*

$$1 - A = \{1 - x \mid x \in A\} \tag{36}$$

$$B \cdot C = \{x \cdot y \mid x \in B, y \in C\} \tag{37}$$

Obviously $1 - A \in \mathcal{P}([0,1])$ whenever $A \in \mathcal{P}([0,1])$ and $A \cdot B \in \mathcal{P}([0,1])$ whenever $A, B \in \mathcal{P}([0,1])$.

Definition 24 (Cascade Experiment) *Consider a state context property system* $(\Sigma, \mathcal{M}, \mathcal{L}, \mu, \xi)$ *describing an entity* S. *An experiment* $e \in \mathcal{M}$ *is a cascade experiment if there exists a set* E *such that the set of outcomes* $O(e)$ *of* e *is a subset* $\mathcal{P}(E)$, *hence* $O(e) \subset \mathcal{P}(E)$. *For* $p \in \Sigma$, *and* $x, y, z, t \in O(e)$ *such that* $x \subset y$ *and* $z \cup t = E$ *and* $z \cap t = \emptyset$, *we have:*

$$P_y^e(P_x^e(p)) = P_x^e(p) \tag{38}$$

$$\mu(e, P_x^e(p), e, P_x^e(p)) = \{1\} \tag{39}$$

$$\mu(e, P_x(p), e, p) = \mu(e, P_x(p), e, P_y(p)) \cdot \mu(e, P_y(p), e, p) \tag{40}$$

$$\mu(e, P_z(p), e, p) = 1 - \mu(e, P_t(p), e, p) \tag{41}$$

We call E *the spectrum of the experiment* e.

Note that the elements of the spectrum E for an experiment e are not necessarily outcomes of the experiment e. Indeed, for example for the position measurement of a free quantum particle, the spectrum of the position operator is a subset of the set of real numbers $\mathbb{R}$, namely the spectrum of the self-adjoint operator corresponding to this position measurement, but none of the numbers of the spectrum is an outcome. Only subsets of this spectrum with measure different from zero are outcomes, because the spectrum is continuous.

5.3 An Extra Condition For the Morphisms

When some of the contexts are experiments we can derive from the merological covariance situation an extra condition to be fulfilled for the morphisms of **SCOP**.

Definition 25 *Consider two state context property systems* $(\Sigma, \mathcal{M}, \mathcal{L}, \mu)$ *and* $(\Sigma', \mathcal{M}', \mathcal{L}', \mu')$. *If* $e \in \mathcal{M}$ *is an experiment, then also* $l(e) \in \mathcal{M}'$ *is an experiment, and for* $p' \in \Sigma'$ *there exists a bijection* k

$$k : O(e, m(p')) \rightarrow O(l(e), p') \tag{42}$$

which expresses that we use the same outcomes whether we experiment on the big entity S' *or on the subentity* S.

5.4 Preparations

Context are also used to prepare the state of an entity. For a context to function as a preparation it is necessary that it brings the entity in a specific state under its influence.

Definition 26 (Preparation) *Consider a state context property system* $(\Sigma, \mathcal{M}, \mathcal{L}, \mu, \xi)$ *describing an entity* S. *We say that a context* $e \in \mathcal{M}$ *is a preparation if there exists a state* $p \in \Sigma$ *such that* $R(e) = \{p\}$. *We call* p *the state prepared by the context* e.

So a context is a preparation if it provokes a change such that each state of the entity is brought to one and the same state. This is then the prepared state.

6 Meet Properties and Join States

Suppose we consider a set of properties $(a_i)_i \in \mathcal{L}$. It is very well possible that there exist states of the entity S in which all the properties a_i are actual. This is in fact always the case if $\cap_i \kappa(a_i) \neq \emptyset$. Indeed, if we consider $p \in \cap_i \kappa(a_i)$ and S in state p, then all the properties a_i are actual. If there corresponds a new property with the situation where all properties a_i of a set $(a_i)_i$ and no other are actual, we will denote such a new property by $\wedge_i a_i$, and call it a 'meet property' of all a_i. Clearly we have $\wedge_i a_i$ is actual for S in state $p \in \Sigma$ iff a_i is actual for all i for S in state p. This means that we have $\wedge_i a_i \in \xi(p)$ iff $a_i \in \xi(p)$ $\forall i$.

Suppose now that we consider a set of states $(p_j)_j \in \Sigma$ of the entity S. It is very well possible that there exist properties of the entity such that these properties are actual if S is in any one of the states p_j. This is in fact always the case if $\cap_j\xi(p_j) \neq \emptyset$. Indeed suppose that $a \in \cap_j\xi(p_j)$. Then we have that $a \in \xi(p_j)$ for each one of the states p_j, which means that a is actual if S is in any one of the states p_j. If it is such that there corresponds a new state to the situation where S is in any one of the states p_j, we will denote such new state by $\vee_j p_j$ and call a 'join state' of all p_j. We can see that a property $a \in \mathcal{L}$ is actual for S in a state $\vee_j p_j$ iff this property a is actual for S in any of the states p_j[e]. The existence of meet properties and join states gives additional structure to Σ and $\mathcal{L}$.

Definition 27 (Property Completeness) *Consider a state context property system $(\Sigma, \mathcal{M}, \mathcal{L}, \mu, \xi)$ describing an entity S. We say that we have 'property completeness' iff for an arbitrary set $(a_i)_i, a_i \in \mathcal{L}$ of properties there exists a property $\wedge_i a_i \in \mathcal{L}$ such that for an arbitrary state $p \in \Sigma$:*

$$\wedge_i a_i \in \xi(p) \Leftrightarrow a_i \in \xi(p)\ \forall\ i \tag{43}$$

Such a property $\wedge_i a_i$ is called a meet property of the set of properties $(a_i)_i$.

Definition 28 (State Completeness) *Consider a state context property system $(\Sigma, \mathcal{M}, \mathcal{L}, \mu, \xi)$ describing an entity S. We say that we have 'state completeness' iff for an arbitrary set of states $(p_j)_j, p_j \in \Sigma$ there exists a state $\vee_j p_j \in \Sigma$ such that for an arbitrary property $a \in \mathcal{L}$:*

$$\vee_j p_j \in \kappa(a) \Leftrightarrow p_j \in \kappa(a)\ \forall\ j \tag{44}$$

Such a state $\vee_j p_j$ is called a join state of the set of states $(p_j)_j$.

The following definition explains why we have introduced the concept completeness.

Definition 29 (Complete Pre-ordered Set) *Suppose that $Z, <$ is a pre-ordered set. We say that Z is a complete pre-ordered set iff for each subset $(x_i)_i, x_i \in Z$ of elements of Z there exists an infimum and a supremum in Z[f].*

Proposition 6 *Consider a state context property system $(\Sigma, \mathcal{M}, \mathcal{L}, \mu, \xi)$ describing an entity S, and suppose that we have property completeness and state completeness. Then $\Sigma, <$ and $\mathcal{L}, <$ are complete pre-ordered sets.*

---

[e] A join state and meet property are not unique. But two join states and two meet properties corresponding to the same sets are equivalent. We remark that we could also try to introduce join properties and meet states. It is however a subtle but deep property of reality, that this cannot be done on the same level. We will understand this better when we study in the next section more of the operational aspects of the formalism. We will see there that only meet properties and join states can be operationally defined in the general situation.

[f] An infimum of a subset $(x_i)_i$ of a pre-ordered set Z is an element of Z that is smaller than all the x_i and greater than any element that is smaller than all x_i. A supremum of a subset $(x_i)_i$ of a pre-ordered set Z is an element of Z that is greater than all the x_i and smaller than any element that is greater than all the x_i.

Proof: Consider an arbitrary set $(a_i)_i, a_i \in \mathcal{L}$. We will show that $\wedge_i a_i$ is an infimum. First we have to proof that $\wedge_i a_i < a_k \ \forall \ k$. This follows immediately from (43) and the definition of $<$ given in (26). Indeed, from this definition follows that we have to prove that $\kappa(\wedge_i a_i) \subset \kappa(a_k) \ \forall \ k$. Consider $p \in \kappa(\wedge_i a_i)$. From (4) follows that this implies that $\wedge_i a_i \in \xi(p)$. Through (43) this implies that $a_k \in \xi(p) \ \forall \ k$. If we apply (4) again this proves that $p \in \kappa(a_k) \ \forall \ k$. So we have shown that $\kappa(\wedge_i a_i) \subset \kappa(a_k) \ \forall \ k$. This shows already that $\wedge_i a_i$ is a lower bound for the set $(a_i)_i$. Let us now show that it is a greatest lower bound. So consider another lower bound, a property $b \in \mathcal{L}$ such that $b < a_k \ \forall \ k$. Let us show that $b < \wedge_i a_i$. Consider $p \in \kappa(b)$, then we have $p \in a_k \ \forall \ k$ since b is a lower bound. This gives us that $a_k \in \xi(p) \ \forall \ k$, and as a consequence $\wedge_i a_i \in \xi(p)$. But this shows that $p \in \kappa(\wedge_i a_i)$. So we have proven that $b < \wedge_i a_i$ and hence $\wedge_i a_i$ is an infimum of the subset $(a_i)_i$. Let us now prove that $\vee_j p_j$ is a supremum of the subset $(p_j)_j$. The proof is very similar, but we use (44) in stead of (43). Let us again first show that $\vee_j p_j$ is an upper bound of the subset $(p_j)_j$. We have to show that $p_l < \vee_j p_j \ \forall \ l$. This means that we have to prove that $\xi(\vee_j p_j) \subset \xi(p_l) \ \forall \ l$. Consider $a \in \xi(\vee_j p_j)$, then we have $\vee_j p_j \in \kappa(a)$. From (44) it follows that $p_l \in \kappa(a) \ \forall \ l$. As a consequence, and applying (4), we have that $a \in \xi(p_l) \ \forall \ l$. Let is now prove that it is a least upper bound. Hence consider another upper bound, meaning a state q, such that $p_l < q \ \forall \ l$. This means that $\xi(q) \subset \xi(p_l) \ \forall \ l$. Consider now $a \in \xi(q)$, then we have $a \in \xi(p_l) \ \forall \ l$. Using again (4), we have $p_l \in \kappa(a) \ \forall \ l$. From (44) follows then that $\vee_j p_j \in \kappa(a)$ and hence $a \in \xi(\vee_j a_j)$.

We have shown now that $\wedge_i a_i$ is an infimum for the set $(a_i)_i, a_i \in \mathcal{L}$, and that $\vee_j p_j$ is a supremum for the set $(p_j)_j, p_j \in \Sigma$. It is a mathematical consequence that for each subset $(a_i)_i, a_i \in \mathcal{L}$, there exists also a supremum in $\mathcal{L}$, let is denote it by $\vee_i a_i$, and that for each subset $(p_j)_j, p_j \in \Sigma$, there exists also an infimum in Σ, let us denote it by $\wedge_j p_j$. They are respectively given by $\vee_i a_i = \wedge_{x \in \mathcal{L}, a_i \prec x \forall i} \ x$ and $\wedge_j p_j = \vee_{y \in \Sigma, y \prec p_j \forall j} \ y$[g]. □

For both $\mathcal{L}$ and Σ it can be shown that this implies that there is at least one minimal and one maximal element. Indeed, an infimum of all elements of $\mathcal{L}$ is a minimal element of $\mathcal{L}$ and an infimum of the empty set is a maximal element of $\mathcal{L}$. In an analogous way a supremum of all elements of Σ is a maximal element of Σ and a supremum of the empty set is a minimal element of Σ. Of course there can be more minimal and maximal elements. If a property $a \in \mathcal{L}$ is minimal we will express this by $a \approx 0$ and if a property $b \in \mathcal{L}$ is maximal we will express this by $b \approx I$. An analogous notation will

[g] We remark that the supremum for elements of $\mathcal{L}$ and the infimum for elements of Σ, although they exists, as we have proven here, have no simple operational meaning.

be used for the maximal and minimal states.

When there is property completeness and state completeness we can specify the structure of the maps ξ and κ somewhat more after having introduced the concept of 'property state' and 'state property'.

Proposition 7 *Consider a state context property system $(\Sigma, \mathcal{M}, \mathcal{L}, \mu, \xi)$ describing an entity S, and suppose that we have property completeness and state completeness. For $p \in \Sigma$ we define the 'property state' corresponding to p as the property $s(p) = \wedge_{a \in \xi(p)} a$. For $a \in \mathcal{L}$ we define the 'state property' corresponding to a as the state $t(a) = \vee_{p \in \kappa(a)} p$. We have two maps :*

$$\begin{array}{c} t : \mathcal{L} \rightarrow \Sigma \; a \mapsto t(a) \\ s : \Sigma \rightarrow \mathcal{L} \; p \mapsto s(p) \end{array} \tag{45}$$

and for $a, b \in \mathcal{L}$, and $(a_i)_i, a_i \in \mathcal{L}$ and $p, q \in \Sigma$ and $(p_j)_j, p_j \in \Sigma$ we have :

$$\begin{array}{c} a < b \Leftrightarrow t(a) < t(b) \\ p < q \Leftrightarrow s(p) < s(q) \\ t(\wedge_i a_i) \approx \wedge_i t(a_i) \\ s(\vee_j p_j) \approx \vee_j s(p_j) \end{array} \tag{46}$$

Proof: Suppose that $p < q$. Then we have $\xi(q) \subset \xi(p)$. From this follows that $s(p) = \wedge_{a \in \xi(p)} a < \wedge_{a \in \xi(q)} a = s(q)$. Suppose now that $s(p) < s(q)$. Take $a \in \xi(q)$, then we have $s(q) < a$. Hence also $s(p) < a$. But this implies that $a \in \xi(p)$. Hence this shows that $\xi(q) \subset \xi(p)$ and as a consequence we have $p < q$. Because $\wedge_i a_i < a_k \; \forall \; k$ we have $t(\wedge_i a_i) < t(a_k) \; \forall k$. This shows that $t(\wedge_i a_i)$ is a lower bound for the set $(t(a_i))_i$. Let us show that it is a greatest lower bound. Suppose that $p < t(a_k) \; \forall \; k$. We remark that $t(a_k) \in \kappa(a_k)$. Then it follows that $p \in \kappa(a_k) \; \forall \; k$. As a consequence we have $a_k \in \xi(p) \; \forall \; k$. But then $\wedge_i a_i \in \xi(p)$ which shows that $p \in \kappa(\wedge_i a_i)$. This proves that $p < t(\wedge_i a_i)$. So we have shown that $t(\wedge_i a_i)$ is a greatest lower bound and hence it is equivalent to $\wedge_i t(a_i)$. □

Proposition 8 *Consider a state context property system $(\Sigma, \mathcal{M}, \mathcal{L}, \mu, \xi)$ describing an entity S, and suppose that we have property completeness and state completeness. For $p \in \Sigma$ we have $\xi(p) = [s(p), +\infty] = \{a \in \mathcal{L} \mid s(p) < a\}$. For $a \in \mathcal{L}$ we have $\kappa(a) = [-\infty, t(a)] = \{p \in \Sigma \mid p < t(a)\}$.*

Proof: Consider $b \in [s(p), +\infty]$. This means that $s(p) < b$, and hence $b \in \xi(p)$. Consider now $b \in \xi(p)$. Then $s(p) < b$ and hence $b \in [s(p), +\infty]$. □

If p is a state such that $\xi(p) = \emptyset$, this means that there is no property actual for the entity being in state p. We will call such states 'improper' states. Hence a 'proper' state is a state that makes at least one property actual. In an analogous way, if $\kappa(a) = \emptyset$, this means that there is no state that makes the property a actual. Such a property will be called an 'improper' property. A 'proper' property is a property that is actual for at least one state.

Definition 30 (Proper States and Properties) *Consider a state context property system $(\Sigma, \mathcal{M}, \mathcal{L}, \mu, \xi)$ describing an entity S. We call $p \in \Sigma$ a 'proper' state iff $\xi(p) \neq \emptyset$. We call $a \in \mathcal{L}$ a 'proper' property iff $\kappa(a) \neq \emptyset$. A state $p \in \Sigma$ such that $\xi(p) = \emptyset$ is called an 'improper' state, and a property $a \in \mathcal{L}$ such that $\kappa(a) = \emptyset$ is called an 'improper' property.*

It easily follows from proposition 8 that when there is property completeness and state completeness there are no improper states ($I \approx \wedge\emptyset \in \xi(p)$) and no improper properties ($0 \approx \vee\emptyset \in \kappa(a)$). Let us find out how the morphism behave in relation with 'meet' and 'join'.

Proposition 9 *Consider two state context property systems $(\Sigma, \mathcal{M}, \mathcal{L}, \mu, \xi)$ and $(\Sigma', \mathcal{M}', \mathcal{L}', \mu', \xi')$ describing entities S and S' that are property and state complete, and a morphism (m, l, n) between $(\Sigma, \mathcal{M}, \mathcal{L}, \mu, \xi)$ and $(\Sigma', \mathcal{M}', \mathcal{L}', \mu', \xi')$. For $(a_i)_i \in \mathcal{L}$ and $(p'_j)_j \in \Sigma'$ we have:*

$$n(\wedge_i a_i) \approx \wedge_i n(a_i) \tag{47}$$

$$m(\vee_j p'_j) \approx \vee_j m(p'_j) \tag{48}$$

Proof: Because $\wedge_i a_i < a_j \ \forall j$ we have $n(\wedge_i a_i) < n(a_j) \ \forall j$. From this follows that $n(\wedge_i a_i) < \wedge_i n(a_i)$. There remains to prove that $\wedge_i n(a_i) < n(\wedge_i a_i)$. Suppose that $\wedge_i n(a_i) \in \xi'(p')$. Then $n(a_j) \in \xi'(p') \ \forall j$, which implies that $a_j \in \xi(m(p')) \ \forall j$. As a consequence we have $\wedge_i a_i \in \xi(m(p'))$. From this follows that $n(\wedge_i a_i) \in \xi'(p')$. So we have proven that $\wedge_i n(a_i) < n(\wedge_i a_i)$. Formula 48 is proven in an analogous way. □

7 Operationality

We have introduced states, contexts and properties for a physical entity. In this section we analyze in which way operationality introduces connections between these concepts.

7.1 *Testing Properties and Operationality*

Experiments can be used to measure many things, and in this sense they can also be used to test properties. Let us explain how this works. Consider an experiment $e \in \mathcal{M}$ and a property $a \in \mathcal{L}$. If there exists a subset $A \subset O(e)$ of the outcome set of e, such that the property a is actual iff the outcome of e is contained in A, we say that e tests the property a.

Definition 31 (Test of a Property) *Consider a state context property system $(\Sigma, \mathcal{M}, \mathcal{L}, \mu, \xi)$ describing an entity S. If for a property $a \in \mathcal{L}$ there is* *an experiment $e \in \mathcal{M}$, and a subset $A \subset O(e)$ of the outcome set of e, such that:*

$$a \in \xi(p) \Leftrightarrow O(e,p) \subset A \tag{49}$$

We say that e is a 'test' for the property a.

If all the properties of the entity that we consider can be tested by an experiment, we say that we have operationality, or that our entity is an operational entity.

Definition 32 (Operational Entity) *Consider a state context property system $(\Sigma, \mathcal{M}, \mathcal{L}, \mu, \xi)$ describing an entity S. If for each property $a \in \mathcal{L}$ there is an experiment $e \in \mathcal{M}$ that tests this property, and if moreover the experiments to test the properties are such that for two properties $a, b \in \mathcal{L}$ we have at least experiments $e, f \in \mathcal{M}$ that test respectively a and b such that $O(e) \cap O(f) = \emptyset$, we say that the entity S is an operational entity.*

Of course, for an operational entity, it is not necessary to give the set of properties apart, they can be derived from the rest of the mathematical structure. We have done this explicitly in [51] for the case where all experiment contexts are yes/no-experiments. It is possible to generalize the construction of [51]. For this reason we need to introduce what we will call a state context system.

Definition 33 (The Category SCO) *A state context system $(\Sigma, \mathcal{M}, \mu)$ consists of two sets Σ and $\mathcal{M}$, and a function*

$$\mu : \mathcal{M} \times \Sigma \times \mathcal{M} \times \Sigma \to \mathcal{P}([0,1]) \tag{50}$$

The sets Σ and $\mathcal{M}$ play the role of the set of states and the set of contexts of an entity S, and the function μ describes the transition probability. Consider two state context systems $(\Sigma, \mathcal{M}, \mu)$ and $(\Sigma', \mathcal{M}', \mu')$. A morphism is a couple of functions (m, l) such that:

$$m : \Sigma' \to \Sigma \tag{51}$$

$$l : \mathcal{M} \to \mathcal{M}' \tag{52}$$

and the following formula is satisfied:

$$\mu(f, m(q'), e, m(p')) = \mu(l(f), q', l(e), p') \tag{53}$$

Further we have that if $e \in \mathcal{M}$ is an experiment, then also $l(e) \in \mathcal{M}'$ is an experiment and for $p' \in \Sigma'$ we have a bijection k:

$$k : O(e, m(p')) \to O(l(e), p') \tag{54}$$

We denote the category of state context systems and their morphisms by **SCO**.

For such a state context system we can construct the set of properties that are testable by experiments of $\mathcal{M}$, and this will deliver us a state context property system for an operational entity. Let us see how this works.

Definition 34 *Suppose that we have a state context system* $(\Sigma, \mathcal{M}, \mu)$. *We define:*

$$\mathcal{L} = \{A \mid \exists\, e \in \mathcal{M},\ e \text{ experiment and } A \subset O(e)\} \tag{55}$$

$$\xi\,:\,\Sigma \to \mathcal{P}(\mathcal{L}) \tag{56}$$

$$A \in \xi(p) \Leftrightarrow O(e,p) \subset A \tag{57}$$

and call $(\Sigma, \mathcal{M}, \mathcal{L}, \mu, \xi)$ *the state context property system related to* $(\Sigma, \mathcal{M}, \mu)$.

Proposition 10 *Suppose that* $(\Sigma, \mathcal{M}, \mathcal{L}, \mu, \xi)$ *and* $(\Sigma', \mathcal{M}', \mathcal{L}', \mu', \xi')$ *are the state context property systems related to the state context systems* $(\Sigma, \mathcal{M}, \mu)$ *and* $(\Sigma', \mathcal{M}', \mu)'$. *A morphism* (m, l) *between* $(\Sigma, \mathcal{M}, \mu)$ *and* $(\Sigma', \mathcal{M}', \mu)'$ *determines a morphism* (m, l, n) *between* $(\Sigma, \mathcal{M}, \mathcal{L}, \mu, \xi)$ *and* $(\Sigma', \mathcal{M}', \mathcal{L}', \mu', \xi')$.

Proof: Consider $A \in \mathcal{L}$. This means that $\exists\, e \in \mathcal{M}$ where e is an experiment, and $A \subset O(e)$. We know that $l(e) \in \mathcal{M}'$ is also an experiment, and if we consider $k(A)$, where k is the bijection of (54) we have $k(A) \subset O(l(e)) = k(O(e))$. This means that $k(A) \in \mathcal{L}'$. Let us define:

$$n : \mathcal{L} \to \mathcal{L}' \tag{58}$$

$$A \mapsto k(A) \tag{59}$$

Take $p' \in \Sigma'$ and $A \in \mathcal{L}$. We have $A \in \xi(m(p')) \Leftrightarrow O(e, m(p')) \subset A \Leftrightarrow k(O(e, m(p'))) \subset k(A) \Leftrightarrow O(l(e), p') \subset n(A) \Leftrightarrow n(A) \in \xi'(p')$. This means that (m, l, n) is a morphism between $(\Sigma, \mathcal{M}, \mathcal{L}, \mu, \xi)$ and $(\Sigma', \mathcal{M}', \mathcal{L}', \mu', \xi')$.

We introduce in the next section specific types of contexts and states that make it possible to test the meet property and deliver a join state for our entity.

7.2 Product Contexts and Product States

Suppose that we consider a set of contexts $(e_i)_i$. In general it will be possible for only one of these contexts to be realized together with an entity S. We can however consider the following operation: we choose one of the contexts of the set $(e_i)_i$ and realize this context together with the entity S. We can consider this operation together with the set of contexts $(e_i)_i$ as a new context. Let us denote it as $\Pi_i e_i$ and call it the product context of the set $(e_i)_i$. It is interesting to note that the product of different contexts gives rise to indeterminism. In earlier work we have been able to prove that the quantum type of indeterminism is exactly due to the fact that each experiment is the product of some hidden experiments. We have called this explanation of the quantum probability structure the 'hidden measurement' approach[18,19,32]. In [51] we analyze in detail how we have to introduce the product experiment in a mathematical way, and it is shown how a subset probability is necessary for this purpose.

Definition 35 (Product Context) *Consider a state context property system $(\Sigma, \mathcal{M}, \mathcal{L}, \mu, \xi)$ describing an entity S. Suppose we have a set of contexts $(e_i)_i \in \mathcal{M}$. The product context $\Pi_i e_i$ is defined in the following way. For $p, q \in \Sigma$ and $f \in \mathcal{M}$ we have:*

$$\mu(f, q, \Pi_i e_i, p) = \cup_i \mu(f, q, e_i, p) \tag{60}$$

Proposition 11 *Consider a state context property system $(\Sigma, \mathcal{M}, \mathcal{L}, \mu, \xi)$ describing an entity S. For the product context $\Pi_i e_i$ of a set of contexts $(e_i)_i \in \mathcal{M}$ we have for $p \in \Sigma$:*

$$R(\Pi_i e_i, p) = \cup_i R(e_i, p) \tag{61}$$

Proof: Suppose now that $q \in R(\Pi_i e_i, p)$. This means that there exist $f \in \mathcal{M}$ such that $\mu(f, q, \Pi_i e_i, p) \neq \{0\}$. Hence $\cup_i \mu(f, q, e_i, p) \neq \{0\}$, which means that there is at least j such that $\mu(f, q, e_j, p) \neq \{0\}$. This shows that $q \in R(e_j, p)$, and hence $q \in \cup_i R(e_i, p)$. On the contrary, suppose that $q \in \cup_i R(e_i, p)$. This means that there is at least one j such that $q \in R(e_j, p)$. Hence there exist $f \in \mathcal{M}$ such that $\mu(f, q, e_j, p) \neq \{0\}$. As a consequence we have $\cup_i \mu(f, q, e_i, p) \neq \{0\}$, and hence $\mu(f, q, \Pi_i e_i, p) \neq \{0\}$, which shows that $q \in R(\Pi_i e_i, p)$. This proves (61). □

Proposition 12 *Consider a state context property system $(\Sigma, \mathcal{M}, \mathcal{L}, \mu, \xi)$ describing an entity S. Suppose that $(e_i)_i$ is a set of experiments. We have:*

$$O(\Pi_i e_i, p) = \cup_i O(e_i, p) \tag{62}$$

Proof: Suppose that $x \in O(\Pi_i e_i, p)$. This means that there exist $q \in \Sigma$ and $f \in \mathcal{M}$ such that $\mu(f, q, \Pi_i e_i, p) \neq \{0\}$, and x occurs whenever p changes into q. Because of (60) we have $\cup_i \mu(f, q, e_i, p) \neq \{0\}$. This means that there is at least one j such that $\mu(f, q, e_j, p) \neq \{0\}$. Hence x is a possible outcome of e_j that occurs when p is changed into q by e_j. As a consequence we have $x \in O(e_j, p)$, and hence $x \in \cup_i O(e_i, p)$. On the contrary, suppose now that $x \in \cup_i O(e_i, p)$. This means that there is at least one j such that $x \in O(e_j, p)$. This means that there exist $q \in \Sigma$ and $f \in \mathcal{M}$ such that $\mu(f, q, e_j, p) \neq \{0\}$, and x is the outcome that occurs when e_j changes the state p to the state q. As a consequence we have $\cup_i \mu(f, q, e_i, p) \neq \{0\}$, which means that $\mu(f, q, \Pi_i e_i, p) \neq \{0\}$, which means that x also occurs when $\Pi_i e_i$ changes the state p to q. Hence $x \in O(\Pi_i e_i, p)$. This proves (62). □

Suppose that we consider a set of states $(p_i)_i \in \Sigma$. Then it is possible to consider a situation where the entity is in one of these states, but we do not know which one, as a new state, that we will call the product state $\Pi_i p_i$ of the set of states $(p_i)_i$. More specifically we define the product state as the state that is prepared by a product of contexts that are preparations.

Definition 36 (Product State) *Consider a state context property system $(\Sigma, \mathcal{M}, \mathcal{L}, \mu, \xi)$ describing an entity S. Suppose that $(p_i)_i \in \Sigma$ is a set of states. The product state $\Pi_i p_i$ is defined in the following way, for $e, f \in \mathcal{M}$ and $q \in \Sigma$ we have:*

$$\mu(f, q, e, \Pi_i p_i) = \cup_i \mu(f, q, e, p_i) \tag{63}$$

Proposition 13 *Consider a state context property system $(\Sigma, \mathcal{M}, \mathcal{L}, \mu, \xi)$ describing an entity S. For the product state $\Pi_i p_i$ of a set of states $(p_i)_i \in \Sigma$ we have for $e \in \mathcal{M}$:*

$$R(e, \Pi_i p_i) = \cup_i R(e, p_i) \tag{64}$$

Proof: Suppose that $q \in R(e, \Pi_i p_i)$. This means that there exists $f \in \mathcal{M}$ such that $\mu(f, q, e, \Pi_i p_i) \neq \{0\}$. Hence $\cup_i \mu(f, q, e, p_i) \neq \{0\}$. This means that there is at least one j such that $\mu(f, q, e, p_j) \neq \{0\}$. Hence $q \in R(e, p_j)$ which shows that $q \in \cup_i R(e, p_i)$. On the contrary, suppose now that $q \in \cup_i R(e, p_i)$. This means that there is at least one j such that $q \in R(e, p_j)$. Hence there exists $f \in \mathcal{M}$ such that $\mu(f, q, e, p_j) \neq \{0\}$. As a consequence we have $\cup_i \mu(f, q, e, p_j) \neq \{0\}$, and hence $\mu(f, q, e, \Pi_i p_i) \neq \{0\}$. This shows that $q \in R(e, \Pi_i p_i)$. □

7.3 Meet Properties and Join States

We can show that product contexts test meet properties while product states are join states.

Proposition 14 *Consider a state context property system $(\Sigma, \mathcal{M}, \mathcal{L}, \mu, \xi)$ describing an entity S. Consider a set of properties $(a_i)_i \in \mathcal{L}$. Suppose that we have experiments $(e_i)_i$ available, such that experiment e_j tests property a_j, and such that $O(e_j) \cap O(e_k) = \emptyset$ for $j \neq k$, then the product experiment $\Pi_i e_i$ tests the meet property $\wedge_i a_i$.*

Proof: Experiment e_i tests property a_i. This means there exists $A_i \subset O(e_i)$ such that $a_i \in \xi(p) \Leftrightarrow O(e, p) \subset A_i$. Consider now the product experiment $\Pi_i e_i$. We will prove that $\Pi_i e_i$ tests the property $\wedge_i a_i$. Consider $A = \cup_i A_i$. Then we have $\cup_i A_i \subset \cup_i O(e_i) = O(\Pi_i e_i)$. We have $O(\Pi_i e_i, p) = \cup_i O(e_i, p)$ and, since $O(e_i) \cap O(e_j) = \emptyset$ for $i \neq j$ we also have $O(e_i, p) \cap O(e_j, p) = \emptyset$ for $i \neq j$. This means that $\cup_i O(e_i, p) \subset \cup_i A_i \Leftrightarrow O(e_j, p) \subset A_j \ \forall j$. Consider the property a tested as follows by $\Pi_i e_i$: $a \in \xi(p) \Leftrightarrow O(\Pi_i e_i, p) \subset A$. Then $a \in \xi(p) \Leftrightarrow a_j \in \xi(p)\ \forall j$, which proves that $a = \wedge_i a_i$. □

Proposition 15 *Consider a state context property system $(\Sigma, \mathcal{M}, \mathcal{L}, \mu, \xi)$ describing an operational entity S. Suppose that we have a set of states $(p_i)_i \in \Sigma$, then the product state $\Pi_i p_i$ is a join state of the set of states $(p_i)_i$.*

Proof: Suppose that we have $\Pi_i p_i \in \kappa(a)$ for $a \in \mathcal{L}$. Since the entity is operational we have an experiment $e \in \mathcal{M}$ that tests the property a. This means that there exists $A \subset O(e)$ such that $a \in \xi(p) \Leftrightarrow O(e,p) \subset A$. From $\Pi_i p_i \in \kappa(a)$ follows that $a \in \xi(\Pi_i p_i)$. Hence $O(e, \Pi_i p_i) \subset A$. As a consequence we have $\cup_i O(e, p_i) \subset A$, which shows that $O(e, p_j) \subset A\ \forall j$, and hence $a \in \xi(p_j)\ \forall j$. From this follows that $p_j \in \kappa(a)\ \forall j$. On the contrary, suppose now that $p_j \in \kappa(a)\ \forall j$, where e is again an experiment that tests the property $a \in \mathcal{L}$. Then there exists $A \subset O(e)$ such that $O(e, p_j) \subset A\ \forall j$. As a consequence we have that $\cup_i O(e, p_i) \subset A$. Hence $O(e, \Pi_i p_i) \subset A$. From this follows that $\Pi_i p_i \in \kappa(a)$. This means that we have proven that $\Pi_i p_i$ is a join state of the set of states $(p_i)_i$. □

Definition 37 (Operational Completeness) *Consider a state context property system $(\Sigma, \mathcal{M}, \mathcal{L}, \mu, \xi)$ describing an operational entity S. We will say that the state context property system is operationally complete if for any set $(e_i)_i \in \mathcal{M}$ of contexts the product context $\Pi_i e_i \in \mathcal{M}$ and for any set of states $(p_i)_i$ the product state $\Pi_i p_i \in \Sigma$.*

Theorem 1 *Any operationally complete state context property system $(\Sigma, \mathcal{M}, \mathcal{L}, \mu, \xi)$ satisfies property completeness and state completeness.*

Let us introduce the category with elements the operationally complete state context property systems.

8 Conclusion

There remains a lot of work to make the formalism that we put forward into a full grown theory. Potentially however such a theory will be able to describe dynamical change as well as change by a measurement in a unified way. Both are considered to be contextual change. Certainly for application in other fields of reality this generality will be of value. As for applications to physics it will be interesting to reconsider the quantum axiomatics and reformulate the essential axioms within the formalism that is proposed here. This project has been elaborated already within the earlier axiomatic approaches, which means that part of the work is translation into the more general scheme that we present here. However, because the basic concepts are different it will not just be a translation of earlier results. In the years to come we plan to engage in this enterprise.

References

1. D. Aerts, "Foundations of quantum physics: a general realistic and operational approach", *Int. J. Theor. Phys.* **38**, 289 - 358 (1999).
2. D. Aerts, "Quantum mechanics: structures, axioms and paradoxes", in *Quantum Mechanics and the Nature of Reality,* eds. Aerts, D. and Pykacz, J., Kluwer Academic, Dordrecht (1999).
3. C. Piron, *Foundations of Quantum Physics,* Reading, Mass., W. A. Benjamin (1976).
4. D. Aerts, *The One and the Many: Towards a Unification of the Quantum and the Classical Description of One and Many Physical Entities,* Doctoral Thesis, Brussels Free University (1981).
5. D. Aerts, "Description of many physical entities without the paradoxes encountered in quantum mechanics", *Found. Phys.* **12**, 1131-1170 (1982).
6. D. Aerts, D. "Classical theories and nonclassical theories as a special case of a more general theory", *J. Math. Phys.* **24**, 2441-2453 (1983).
7. D. Aerts, "The description of one and many physical systems", in *Foundations of Quantum Mechanics,* eds. C. Gruber, A.V.C.P. Lausanne, 63 (1983).
8. C. Piron, "Recent developments in quantum mechanics", *Helv. Phys. Acta* **62**, 82 (1989).
9. C. Piron, *Mècanique Quantique: Bases et Applications,*, Press Polytechnique de Lausanne (1990).
10. C. Randall and D. Foulis, "A mathematical setting for inductive reasoning", in *Foundations of Probability Theory, Statistical Inference, and Statistical Theories of Science III,* eds. W. L. Harper and C. A. Hooker, Kluwer Academic, Dordrecht, 169 (1975).
11. C. Randall and D. Foulis, "The operational approach to quantum mechanics", in *Physical Theories as Logico-Operational structures,* ed. C. A. Hooker, Kluwer Academic, Dordrecht, Holland, 167 (1978).
12. D. Foulis and C. Randall, "What are quantum logics and what ought they to be?" in *Current Issues in Quantum Logic,* eds. E. Beltrametti and B. van Fraassen, Kluwer Academic, Dordrecht, 35 (1981).
13. C. Randall and D. Foulis, D., "Operational statistics and tensor products", in *Interpretations and Foundations of Quantum Theory,* ed. H. Neumann, B.I. Wissenschaftsverslag, Bibliographisches Institut, Mannheim, 21 (1981).
14. D. Foulis, C. Piron and C. Randall, "Realism, operationalism, and quan-

tum mechanics", *Found. Phys.* **13**, 813 (1983).
15. C. Randall and D. Foulis, "Properties and operational propositions in quantum mechanics", *Found. Phys.* **13**, 835 (1983).
16. D. Aerts and B. D'Hooghe, "The problem of submeasurements in quantum mechanics", submitted to *Foundations of Science*.
17. D. Aerts, D., "A possible explanation for the probabilities of quantum mechanics and a macroscopical situation that violates Bell inequalities", in *Recent Developments in Quantum Logic*, eds. Mittelstaedt, P., et al., Grundlagen der Exacten Naturwissenschaften, vol.6, Wissenschaftverlag, Bibliographisches Institut, Mannheim, 235-251 (1985).
18. D. Aerts, "A possible explanation for the probabilities of quantum mechanics", *J. Math. Phys.* **27**, 202-210 (1986).
19. D. Aerts, "The origin of the non-classical character of the quantum probability model", in *Information, Complexity, and Control in Quantum Physics*, eds. Blanquiere, A., Diner, S. and Lochak, G., Springer-Verlag, Wien-New York, 77-100 (1987).
20. D. Aerts, "The description of separated systems and quantum mechanics and a possible explanation for the probabilities of quantum mechanics", in *Micro-physical Reality and Quantum Formalism*, eds. van der Merwe, A., et al., Kluwer Academic Publishers, 97-115 (1988).
21. D. Aerts and S. Aerts, "The hidden measurement formalism: quantum mechanics as a consequence of fluctuations on the measurement", in *Fundamental Problems in Quantum Physics II*, eds. Ferrero, M. and van der Merwe, A., Kluwer Academic, Dordrecht (1997).
22. D. Aerts, S. Aerts, B. Coecke, B. D'Hooghe, T. Durt and F. Valckenborgh, "A model with varying fluctuations in the measurement context", in *Fundamental Problems in Quantum Physics II*, eds. Ferrero, M. and van der Merwe, A., Kluwer Academic, Dordrecht (1997).
23. D. Aerts, "The hidden measurement formalism: what can be explained and where paradoxes remain", *Int. J. Theor. Phys.* **37**, 291-304 (1998)
24. S. Aerts, "Interactive probability models: inverse problems on the sphere, *Int. J. Theor. Phys.* **37**, 1 (1998).
25. D. Aerts, S. Aerts, T. Durt and O. Lévêque, "Quantum and classical probability and the epsilon-model, *Int. J. Theor. Phys.* **38**, 407-429 (1999).
26. D. Aerts, B. Coecke and S. Smets, "On the origin of probabilities in quantum mechanics: creative and contextual aspects", in *Metadebates on Science*, eds. Cornelis, G., Smets, S., and Van Bendegem, J.P., Kluwer Academic, Dordrecht (1999).
27. S. Aerts, "Hidden measurements from contextual axiomatics", this volume.

28. T. Durt and B. D'Hooghe, "The classical limit of the lattice-theoretical orthocomplementation in the framework of the hidden-measurement approach", this volume.
29. T. Durt, J. Baudon, R. Mathevet, J. Robert and B. Viaris de Lesegno, "Memory effects in atomic interferometry: a negative result", this volume.
30. D. Aerts, "A macroscopical classical laboratory situation with only macroscopical classical entities giving rise to a quantum mechanical probability model", in *Quantum Probability and Related Topics, Volume VI*, ed. Accardi, L., World Scientific Publishing Company, Singapore, 75-85 (1991).
31. D. Aerts, "Quantum structures due to fluctuations of the measurement situations", *Int. J. Theor. Phys.* **32**, 2207-2220 (1993).
32. D. Aerts, "Quantum structures, separated physical entities and probability", *Found. Phys.* **24**, 1227-1259 (1994).
33. D. Aerts and S. Aerts, "Applications of quantum statistics in psychological studies of decision processes", *Found. Sc.* **1**, 85-97 (1994).
34. D. Aerts and S. Aerts, "Application of quantum statistics in psychological studies of decision processes", in *Topics in the Foundation of Statistics*, eds. Van Fraassen B., Kluwer Academic, Dordrecht (1997).
35. D. Aerts, S. Aerts, J. Broekaert and L. Gabora, "The violation of Bell inequalities in the macroworld", *Found. Phys.* **30**, 1387-1414 (2000), lanl archive ref: *quant-ph/0007044*.
36. D. Aerts, J. Broekaert and S. Smets, S., "The liar paradox in a quantum mechanical perspective", *Found. Sc.* **4**, 156 (1999), lanl archive ref: *quant-ph/0007047*.
37. D. Aerts, J. Broekaert and S. Smets, "A quantum structure description of the liar paradox", *Int. J. Theor. Phys.* **38**, 3231-3239 (1999), lanl archive ref: *quant-ph/0106131*.
38. L. Gabora and D. Aerts, "Distilling the essence of an evolutionary process, and implications for a formal description of culture", in *Cultural Evolution*, ed. Kistler, W., Foundation for the Future, Washington (2000).
39. L. Gabora, *Cognitive Mechanisms Underlying the Origin and Evolution of Culture*, Doctoral thesis, CLEA, Brussels Free University (2001).
40. D. Aerts and L. Gabora, "Towards a general theory of evolution", in preparation.
41. D. Aerts, E. Colebunders, A. Van der Voorde and B. Van Steirteghem, "State property systems and closure spaces: a study of categorical equiv-

alence", *Int. J. Theor. Phys.* **38**, 359-385 (1999).
42. D. Aerts, E. Colebunders, A. Van der Voorde and B. Van Steirteghem, "The construct of closure spaces as the amnestic modification of the physical theory of state property systems", to appear in *Applied Categorical Structures.*
43. C. Piron, "Axiomatique quantique", *Helv. Phys. Acta* **37**, 439 (1964).
44. B. Van Steirteghem, *Quantum Axiomatics: Investigation of the Structure of the Category of Physical Entities and Solér's Theorem*, Dissertation for the degree of Bachelor in Science, Brussels Free University (1998).
45. B. Van Steirteghem, "T_0 separation in axiomatic quantum mechanics, *Int. J. Theor. Phys.* **39**, 955 (2000).
46. A. Van der Voorde, "A categorical approach to T_1 separation and the product of state property systems", *Int. J. Theor. Phys.* **39**, 947-953 (2000).
47. A. Van der Voorde, *Separation Axioms in Extension Theory for Closure Spaces and Their Relevance to State Property Systems*, Doctoral Thesis, Brussels Free University (2001).
48. D. Aerts, A Van der Voorde and D. Deses, "Connectedness applied to closure spaces and state property systems", *Journal of Electrical Engineering*, **52**, 18-21 (2001).
49. D. Aerts, A Van der Voorde and D. Deses, "Classicality and connectedness for state property systems and closure spaces", submitted to *International Journal of Theoretical Physics.*
50. D. Aerts and D. Deses, "State property systems and closure spaces: extracting the classical and nonclassical parts", this volume.
51. D. Aerts, "Reality and Probability: Introducing a New Type of Probability Calculus", this volume.
52. G. d'Emma, "On quantization of the electromagnetic field", *Helv. Phys. Acta* **53**, 535 (1980).
53. W. Daniel, "On non-unitary evolution of quantum systems", *Helv. Phys. Acta* **55**, 330 (1982).
54. G. Cattaneo, C. Dalla Pozza, C. Garola and G. Nistico, "On the logical foundations of the Jauch-Piron approach to quantum physics", *Int. J. Theor. Phys.* **27**, 1313 (1988).
55. G. Cattaneo and G. Nistico, "Axiomatic foundations of quantum physics: Critiques and misunderstandings. Piron's question-proposition system", *Int. J. Theor. Phys.* **30**, 1293 (1991).
56. G. Cattaneo and G. Nistico, "Physical content of preparation-question structures and Bruwer-Zadeh lattices", *Int. J. Theor. Phys.* **31**, 1873 (1992).

57. G. Cattaneo and G. Nistico, "A model of Piron's preparation-question structures in Ludwig's selection structures", *Int. J. Theor. Phys.* **32**, 407 (1993).
58. D. Moore, "Categories of Representations of Physical Systems", *Helv. Phys. Acta* **68**, 658 (1995).
59. D. Aerts, "Framework for possible unification of quantum and relativity theories", *Int. J. Theor. Phys.* **35**, , (2399-2416)1996.
60. D. Aerts, "Relativity theory: what is reality?" *Found. Phys.* **26**, 1627-1644 (1996).
61. D. Aerts and B. D'Hooghe, "Operator structure of a non-quantum and a non-classical system", *Int. J. Theor. Phys.* **35**, 2285-2298 (1996).
62. D. Aerts, B. Coecke and B. D'Hooghe, "A mechanistic macroscopical physical entity with a three dimensional Hilbert space quantum description", *Helv. Phys. Acta* **70**, 793-802 (1997).
63. D. Aerts, B. Coecke, T. Durt and F. Valckenborgh, "Quantum, classical and intermediate I: a model on the Poincaré sphere", *Tatra Mt. Math. Publ.*, **10**, 225 (1997).
64. D. Aerts, B. Coecke, T. Durt and F. Valckenborgh, "Quantum, classical and intermediate II: the vanishing vector space structure", *Tatra Mt. Math. Publ.*, **10**, 241 (1997).
65. D. Aerts, "The entity and modern physics: the creation-discovery view of reality", in *Interpreting Bodies: Classical and Quantum Objects in Modern Physics*, ed. E. Castellani, Princeton University Press, Princeton (1998).
66. H. Amira, B. Coecke and I. Stubbe, "How quantales emerge by introducing induction within the operational approach", *Helv. Phys. Acta* **71**, 554 (1998).
67. I. Stubbe, *Quantaloids of operational resolutions and state transitions*, Dissertation for the degree of Bachelor in Science, Brussels Free University (1998).
68. D. Aerts, "The stuff the world is made of: physics and reality", in *Einstein meets Magritte: An Interdisciplinary Reflection*, eds. D. Aerts, J. Broekaert and E. Mathijs, Kluwer Academic, Dordrecht, (1999), lanl archive ref: *quant-ph/0107044*.
69. D. Aerts, J. Broekaert and L. Gabora, "Nonclassical contextuality in cognition: Borrowing from quantum mechanical approaches to indeterminism and observer dependence", *Dialogues in Psychology*, **10** (1999).
70. D. Aerts and B. Coecke, "The creation discovery view: towards a possible explanation of quantum reality", in *Language, Quantum, Music: Selected Contributed Papers of the Tenth International Congress of Logic, Method-*

ology and Philosophy of Science, Florence, August 1995, eds. M. L. Dalla Chiara, R. Giuntini and F. Laudisa, Kluwer Academic, Dordrecht, (1999).

71. C. Piron, "Quanta and relativity: two failed revolutions", in *Einstein Meets Magritte: An Interdisciplinary Reflection*, eds. Aerts, D., Broekaert, J. and Mathijs, E., Kluwer Academic (1999).
72. B. Coecke and I. Stubbe, "Operational resolutions and state transitions in a categorical setting", *Found. Phys. Lett.*, **12**, 29 (1999).
73. B. Coecke and I. Stubbe, "On a duality of quantales emerging from an operational resolution", *Int. J. Theor. Phys.* **38**, 3269 (1999).
74. D. Aerts, "The description of joint quantum entities and the formulation of a paradox", *Int. J. Theor. Phys.* **39**, 485-496 (2000), lanl archive ref: *quant-ph/0105106*.
75. D. Aerts, J. Broekaert and L. Gabora, "Intrinsic contextuality as the crux of consciousness", in *Fundamental Approaches to Consciousness*, ed. K. Yasue, John Benjamins Publishing Company, Amsterdam (2000).
76. D. Aerts and B. Van Steirteghem, "Quantum axiomatics and a theorem of M.P. Soler", *Int. J. Theor. Phys.* **39**, 497-502 (2000), lanl archive ref: *quant-ph/0105107*.
77. B. Coecke and D.J. Moore, "Operational Galois adjunctions", in *Current Research in Operational Quantum Logic*, 195-218, Kluwer Academic Publishers (2000).
78. B. Coecke and I. Stubbe, "State transitions as morphisms for complete lattices", *Int. J. Theor. Phys.* **39**, 601 (2000).
79. D. Aerts, "Quantum Structures and their future importance", *Soft Computing*, **5**, 131 (2001).
80. F. Valckenborgh, *Compound Systems in Quantum Axiomatics: Analysis of Subsystems and Classification of the Various Products for a Categorical Perspective*, doctoral dissertation, Brussels Free University (2001).
81. S. Smets, *The Logic of Physical Properties in Static and Dynamic Perspective*, doctoral dissertation, Brussels Free University (2001).

THE CLASSICAL LIMIT OF THE LATTICE-THEORETICAL ORTHOCOMPLEMENTATION IN THE FRAMEWORK OF THE HIDDEN-MEASUREMENT APPROACH

THOMAS DURT
Foundations of the Exact Sciences (FUND) and Applied Physics and Photonics (TONA), Brussels Free University, Pleinlaan 2, 1050 Brussels, Belgium
E-mail: thomdurt@vub.ac.be

BART D'HOOGHE
Foundations of the Exact Sciences (FUND), Department of Mathematics, Brussels Free University, Pleinlaan 2, 1050 Brussel, Belgium
E-mail: bdhooghe@vub.ac.be

We study the Grib-Zapatrin orthocomplementation for the epsilon model, and show that it does not allow a Boolean orthoclosure in the deterministic limit of the model. We generalize the epsilon model to a model in which the change of state induced by the measurement process is controlled by a continuous parameter such that in the classical limit this change of state vanishes. As a consequence, the Grib-Zapatrin orthoclosure will become Boolean, showing that determinism by itself is not sufficient to ensure a complete classical character for the system. Finally, we use Wooters topology for physical systems to show that this theorem holds in general.

1 Introduction

We developed in Brussels a heuristic model, the ϵ-model, aimed at simulating a continuous transition between a classical (deterministic) and a quantum (probabilistic) regime. The lattice of properties related to this model was intensively studied [1,2,3,4,5,6,7] and it appears that it exhibits a continuous transition between the Boolean and the Hilbertian lattices. In lattice theory, the orthogonality relation is important at least for two reasons:

(A) It is possible to associate to an orthogonality relation what is called an orthocomplementation, as we shall show it. The existence of an orthocomplementation is postulated in many attempts made in order to generalize the classical (Boolean) paradigm. For instance it is an unavoidable element of the representation theorem of Piron [8,9,10]. (B) The orthocomplementation can be interpreted as a generalized logical negation.

We study here the Grib-Zapatrin orthogonality relation in the framework of the ϵ-model, show that it evolves continuously during the classical-quantum transition, and discuss whether the lattice of properties of the system described by the ϵ-model is orthocomplemented by this orthogonality relation. We pay particular attention to the classical limit and show that for the ϵ-model the G-Z complementation is not a classical (i.e., Boolean) complementation in the classical (deterministic) limit $\epsilon = 0$. We solve this problem by generalizing the ϵ-model to a new model, the N-model, for which in the classical limit the measurement process is deterministic and non-disturbing as well.

2 The ϵ-Model

The system can be considered as a generalised spin 1/2 particle. It is well-known that the rays of a 2-dimensional Hilbert space can be represented on the surface of the unit 3-dimensional sphere (the so-called Bloch or Poincaré-sphere). From now on, we shall consider that the state of the system is represented by a point on the surface of this sphere.

2.1 Guiding Principles of the ϵ-Model

The ϵ-model aims at simulating the interaction which occurs during a measurement process. It is based on the following guiding principles:

- The measuring apparatus is essentially classical and interacts with the observed system in a deterministic way. The state of this apparatus undergoes statistical fluctuations,responsible for a dispersion in the results of measurement [11].

- The amplitude of these fluctuations is quantified by a real parameter ϵ comprised between 0 and 1. When ϵ equals 0, the fluctuations vanish, when ϵ equals 1, they are maximal. The classical limit is reached when we can neglect these fluctuations ($\epsilon = 0$); intuitively, this means that the scale of the disturbance caused by the measuring apparatus is smaller than the scale of the observed system. When the scale of the disturbance of the measuring apparatus is larger than the scale of the observed system, ϵ is not negligible anymore, the dispersion of the experimental results increases, and when ϵ equals 1, we recover a distribution of results equal to the quantum one.

2.2 Simple Version of the ϵ-Model

Let us briefly describe how the model works (this is an ultrasimplified version, see [1,3,12] for more sophisticated presentations). We advice the readers who are not familiar with this kind of models to consult [4]. Properties of the model that we shall only briefly sketch in the present work are developed in detail there. The "metaphor of the elastic", that we summarize now, illustrates in an intuitive manner the modelization of the measurement process that we develop in the ϵ -model. The measurement process is assumed to occur in two steps: the particle "falls" on an elastic put between the two Poles p and $-p$ and sticks there for a while; thereafter, the elastic, which is assumed to be fragile in a certain zone[a], breaks at random in this zone, and the part of the elastic on which the particle is located pulls the particle onto the pole to which it is attached. The point where the elastic breaks, is a random variable which is supposedly uncontrollable. This is the hidden variable of our model, and its stochastic nature allows us to reproduce the stochasticity inherent to any quantum measurement process. Let us now give, in an abstract style, a presentation which contains the essential features of the model.

- The ϵ-model describes a generalized spin 1/2 measurement. The state of the generalized spinning particle is represented by a point q of the Poincaré-sphere (a direction). The set of states of the system is thus the surface of the Poincaré sphere. The generalized Stern-Gerlach device is represented by another point p of the Poincaré-sphere and a hidden variable ϕ.

- We assume that the hidden variable ϕ is submitted to uncontrollable fluctuations, so that it is a random variable homogeneously distributed over the interval $[\phi_{min}, \phi_{max}]$ (with $0 \leq \phi_{min} \leq \phi_{max} \leq 1$). The measurement process forces the state to collapse along two possible directions: the direction of the device p, or its opposite $-p$. We shall then say that the spin of the system is "up" or "down" along p. Let us denote by θ the angle between p and q. In virtue of the "metaphor of the elastic" already mentioned, the result of the measurement is determined as follows: If $2\phi - 1$ is strictly less than $cos^2\theta$, i.e., if ϕ is strictly less than $cos^2\frac{\theta}{2}$, the result of the measurement is "spin-up" and the state q collapses on p. If ϕ is larger than or equal to $cos^2\frac{\theta}{2}$, the result of the measurement is "spin-down" and the state q collapses on $-p$.

We obtain after averaging on the variable ϕ, for each value of the pair (ϕ_{min}, ϕ_{max}), a transition probability which depends on the angle between the initial and the final state (i.e., the respective state before and after the measurement

[a] In the interval of the elastic situated between the distances $2\phi_{min}$ and $2\phi_{max}$ from the south pole in our case.

process). The probability $P(p|q)$ of getting spin-up can be shown [12] to depend on the angle θ between p and q as follows:

- It is equal to 1 when $0 \leq \theta < \theta_{up}$, 0 when $\pi \geq \theta \geq \pi - \theta_{down}$, where $cos^2 \frac{\theta_{up}}{2} = \phi_{max}$ and $sin^2 \frac{\theta_{down}}{2} = \phi_{min}$. The subscripts *up* and *down* are justified by the fact that these angles are the extents of the classical zones (probability 0 or 1) around the poles in which the spin measurements are predetermined with a probability equal to hundred percent.
- In between, i.e., in a zone of angular opening θ_{sup} the experiment has two possible results. The subscript *sup* is justified by the fact that θ_{sup} is the extent of the superposition zone (probability neither 0, nor 1) between the classical zones. We have, in the superposition zone, that:

$$P(p|q) = \frac{cos\theta + cos\theta_{down}}{cos\theta_{up} + cos\theta_{down}}. \quad (1)$$

The three angles θ_{sup}, θ_{up}, θ_{down} fulfill the following relation: $\theta_{sup} + \theta_{up} + \theta_{down} = \pi$. We define the classical (deterministic) and quantum limits of the model by imposing that θ_{sup} is respectively equal to 0 and π.

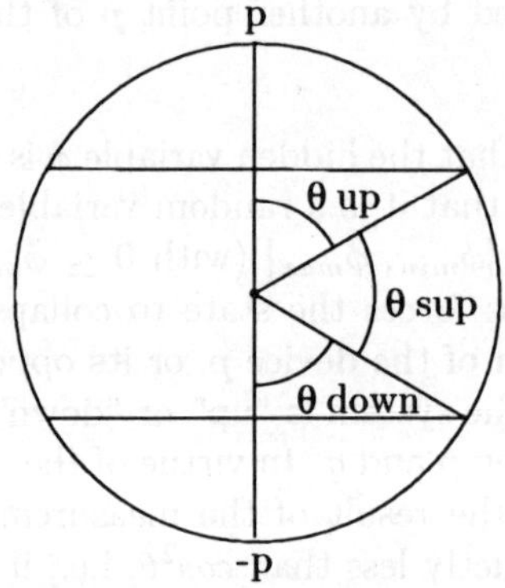

Figure 1. The up and down zones around p.

The classical case corresponds to a purely deterministic situation (the probability of getting spin up takes only values 0 or 1), while in the quantum case we recover the quantum probability associated to the measurement of the spin direction of a spin 1/2 particle [b]. In the symmetric case $\theta_{up} = \theta_{down}$ we have

[b] The idea to associate hidden variables to the apparatus in the quantum context was originally presented in [2].

that $sin(\frac{\theta_{sup}}{2}) = \epsilon$. In the following, we shall study the properties of the G-Z orthogonality relation that we obtain every time we fix the values of θ_{up} and θ_{down}. Other orthogonality relations are studied in [13].

3 Structures on the Set of Properties

3.1 The Cartan Map of the Lattice of Properties of a Physical System

The lattice of properties of the ϵ-model and its Cartan map were intensively studied in other publications [4,6,7,12] so that we shall only sketch here the most important properties related to this concept. The so-called Cartan map of the lattice of properties of a physical system can be illustrated by the following example. Let us assume that one uses a list of properties in order to classify a human population. It could be a list of the kind: male human being, between 7 and 77 years old, with brown eyes, a red car, and so on. The formulation of a new property selects in the set of persons defined by the enumeration of the previous properties a more restricted subset. For a given set of properties, the set of all the subsets of the population that one can delimitate by arbitrary lists of properties taken inside this set reflects the "topology" induced by this set of properties inside the considered population. It is representative of this set. Conversely, the knowledge of all the subsets of the population that one can delimitate by arbitrary lists of properties taken inside this set of properties allows us in principle to recover these properties. We obtained thus a one to one correspondence between a set of properties and a set of subsets of the population.

If we transpose this idea into physics, with, instead of the human population, the set of states of a system, and, instead of an arbitrary set of properties, the set of the properties that one can deduce from experimentation (for instance, from yes-no measurements like a Stern-Gerlach experiment), one obtains the Cartan map of the lattice of properties of the system.

The Cartan map (denoted μ) plays a similar role between the lattice of properties of a system (denoted $\mathcal{L}$) and a set of subsets of the set of states (denoted Σ). At this level, we shall not define in detail the lattice of (physical) properties of a system. All what we need to know is that the lattice of properties is a set of (physical) properties which fulfills particular conditions, and that a (physical) property is defined by the set of states for which this property is true or actual (see [14,8,10] for more details). If one considers a given lattice of properties, its Cartan map is the set of all the subsets of the state space of the system that one can obtain by enumerating arbitrary lists composed by properties chosen in this lattice. The Cartan map reflects thus the "topology" induced by this set of properties on the set of states of the system under consideration. In general, for systems which present some physical interest and relevance (this is related to a property called "fullness of states" [14]), the Cartan map $\mu(\mathcal{L})$ is in bijective relation with the lattice of properties of the system $\mathcal{L}$, so that it is sufficient to study the Cartan map in order to know the lattice of properties. The axiomatic formulation of lattice theory can be duplicated at the level of the Cartan map, which is a set of subsets of the set of states (denoted Σ), and this often simplifies the approach. For instance, the concept of atomicity, translated at the level of the Cartan map expresses that the set of properties which characterizes our knowledge of the system is extended enough, so that for each state chosen in the set of states it is possible to find a list of properties which are simultaneously fulfilled by this state only [4].

It is useful for what follows to note that, for practical reasons (like the existence of an upper and of a lower bound in the lattice for instance), one always considers that the set of properties of the system contains the "always true" and the "always false" properties, which are realized for all the elements of the set of states and for none of them respectively. Another important aspect of the lattice theory (at least in the Geneva formulation [14,8,9,10]) is that when the properties are deduced from yes-no experiments (and this will always be the case in the framework of the ϵ-model), all the relevant information is contained in the so-called eigenstate sets, the sets of states for which the results of a given measurement can be deterministically foreseen (with probability one or zero). For instance, in the ϵ-model, the eigenstate sets of the property "the spin is up (down) when we measure it along the direction p" are spherical sectors of opening θ_{up} (θ_{down}) along p ($-p$).

3.2 The Cartan Map of the Lattice of Properties of the ϵ-Model

It is possible, in the Geneva approach, to build (see [14] for a rigorous proof) the Cartan map of the lattice of properties of the system, when we know the set of states Σ of a given system and the eigenstate sets of all the experiments that we could perform on it. $\mu(\mathcal{L})$ is, according to the discussions of the previous section, a set of subsets of the set of states. In the case of the ϵ-model, we shall consider the situations for which the fluctuations of the hidden parameters are not controllable experimentally, so that θ_{up} and θ_{down} are constant, but in which the experimenter can choose arbitrarily the direction of the generalized spin measurement described in the section 1. The given of an arbitrary value of the pair (θ_{up}, θ_{down}) defines thus all the experiments that one can perform on the system and all the yes-no properties associated to them. The formalism allows us to define precisely the Cartan map of the lattice of properties for each couple $(\theta_{up}, \theta_{down})$. This map is a set of subsets of the set of states so to say of the Poincaré sphere (also denoted Σ) in this case. One obtains (see [12,6,7]) that this map consists of all the subsets of the set of states which are arbitrary intersections of spherical sectors of which the angular opening is equal to $max(\theta_{up}, \theta_{down})$, centered around arbitrary directions on the sphere, plus the whole sphere, plus the empty set.

An important property of the Cartan map of the lattice of properties, which is intimately connected with the lattice-theoretical "completeness" axiom [14,12], is that this Cartan map is closed under intersection. It is easy to check this property in the case of the ϵ-model, essentially because, if we consider the intersections of sets of spherical sectors centered around arbitrarily chosen directions, the intersection of such sets is still such a set. It is trivial to include in the proof the case where one of the sets is the empty set or the whole sphere.

We shall now study another set of subsets of the set of states which is also closed under intersection, and that one can deduce, as it was shown by Birkhoff [15], from the given of an orthogonality relation, the so-called bi-orthogonal set.

3.3 The Bi-Orthogonal Set

Let us consider a set X, $(X \subset \mathcal{P}(\Sigma))$ partially ordered by set-inclusion in Σ, then X is said to be orthocomplemented if there exists a bijection $'$, from X to X, such that $\forall A, B \in X$:

$$\begin{cases} i) \;\; A \subset B \Rightarrow B' \subset A' \\ ii) \;\; (A')' = A \\ iii) \; A \cap A' = \emptyset \end{cases}$$

Note that, instead of saying that $\mu(\mathcal{L})$, the Cartan map of the lattice of properties, is orthocomplemented according to the previous definition, one can equivalently say that the lattice $\mathcal{L}$ of properties itself is orthocomplemented.

Following Birkhoff [15], let us now show how one can build an orthocomplemented set of subsets of the set of states Σ provided we know a relation between the elements of Σ which is antireflexive and symmetrical.

Let us suppose that we know a relation $*$ between the states, such that

$$\forall p, q \in \Sigma : p * q \Rightarrow q * p \text{ and } \nexists p \in \Sigma : p * p. \tag{2}$$

Then, we can associate to any subset K of Σ its orthogonal defined by:

$$K^* = \{p \in \Sigma : p * q, \forall q \in K\} \tag{3}$$

The set-orthogonal relation defined here will allow us to build an orthocomplementation inside Σ, thanks to the so-called bi-orthogonal set $\mathcal{F}_{orth}$:

Definition 1 *The bi-orthogonal set, denoted $\mathcal{F}_{orth}$, is the set of subsets of Σ which are preserved by a double application of the set-orthogonal relation (or bi-orthogonal application): $\mathcal{F}_{orth} = \{F \in \mathcal{P}(\Sigma) : (F^*)^* = F\}$.*

Following Birkhoff [15], one can show [12] that the bi-orthogonal set $\mathcal{F}_{orth}$ is orthocomplemented under the set-orthogonality.

Theorem 1 *$\mathcal{F}_{orth}$ is orthocomplemented under *, the set-orthogonality, which is a permutation of it.*

One can also prove (see e.g. [12]) that the bi-orthogonal set is closed under intersection:

Theorem 2 *The bi-orthogonal set $\mathcal{F}_{orth}$ is closed under intersection.*

As we mentioned it already, it is also possible to show that the Cartan map of the lattice of properties $\mu(\mathcal{L})$ is closed under intersection, but this property is not essential for our purposes so that we shall not develop this aspect of the formalism. Nevertheless, other deep structural analogies exist between the Cartan map of the set of properties of a physical system and bi-orthogonal sets. Among others, it can be shown [16] that the lattices encountered in classical logics and in quantum logics, that we shall call the Boolean and the Hilbertian lattices respectively, are orthocomplemented. Hereby we intend that the Cartan map of a Boolean lattice and the Cartan map of a quantum lattice are orthocomplemented. One can show more: every set inside these Cartan maps is equal to its bi-orthogonal under a well chosen orthogonality relation. We shall now mention, without proof, some general results obtained in lattice theory in the case of classical and quantum systems (see [17,18,19]).

3.4 The Cartan Map of the Lattice of Properties of Boolean and Quantum Systems

Definition 2 *A system is said to be Boolean if the Cartan map of its lattice of properties is the power set $P(\Sigma)$ of the set of states.*

Definition 3 *A system is said to be quantum if its set of states is in a bijective relation with the set of rays of a Hilbert space H and if the Cartan map of its lattice of properties is the set of all Hilbertian subspaces of H, so to say, all the subsets of H which are closed under linear superposition of their elements.*

For a Boolean lattice, the orthocomplementation is the Boolean or set-theoretical complementation: the orthogonal of a set is its Boolean complement, so to say: two states are orthogonal iff they are different.

Sometimes, this complementation is associated to the classical logical negation. Note that some other properties of F_{orth} can be deduced as a corollary of the already proven theorems [15,20,12], which, formally, present some analogy with the properties of the logical negation in classical logics. We give them without proof (see [14,12] for the extended proofs):

For all $F, F_i \in F_{orth}$:

$$\begin{cases} a)\ (\bigcap_i F_i)^* = cl_{orth}(\bigcup_i F_i^*) \\ b)\ (\bigcup_i F_i)^* = \bigcap_i F_i^* \\ c)\ cl_{orth}(F \cup F^*) = \Sigma \end{cases}$$

Remark that, although often in lattice theory classical means Boolean, this will not necessarily be the case in the framework of the ϵ-model where classical (in our notation) means deterministic, and where it can be shown that deterministic measurements are not always associated to a Boolean lattice of properties and vice versa.

It is possible to show in the quantum case that the set orthogonal to a state p, which is now identified with a one dimensional subspace of the Hilbert space H (a ray), is the subspace of H for which the Hilbert product with p is equal to zero.

4 Physically Defined Orthogonality Relations

4.1 The Grib-Zapatrin Orthogonality Relation

Grib and Zapatrin proposed the following orthogonality relation [21,12,22].

Definition 4 *Two states p, q are G-Z-orthogonal iff no experiment exists which induces a transition from one of the states to the other. In other words, the probability of transition from one state to the other is always zero.*

Note that the G-Z-relation is given in a form symmetrical in p, q, and that it is antireflexive, because in the lattice theory, the "always true" experiment which induces a transition of any state onto itself with probability 1 is always assumed to exist a priori, as we mentioned it in the second section.

We recover the standard Hilbertian relation in the quantum case, as we will show now. The probability of transition is given in quantum mechanics by the modulus squared of the Hilbert product between the initial state and the state after the projection. Two states are thus G-Z-orthogonal iff they are H-orthogonal, so that the G-Z-orthogonality is equivalent with the H-orthogonality.

The definition of the G-Z-orthogonality implicitly presupposes some kind of collapse ("jump" between different states). The hypothesis of the existence of a collapse is of quantum inspiration and is not always realized, so that the concept of the G-Z orthogonality relation is not always applicable. Nevertheless, in the ϵ-model, we recover the quantum property that after a spin measurement along a direction, the state of the system aligns itself (collapses) along this direction or its antipode.

Remark also that the G-Z orthogonality relation presented here requires the knowledge of the "non-superposition" states (which realize an outcome with probability 1 or 0) only since the G-Z orthogonality is based on the impossible transitions (probability 0). This illustrates that quantum logics in the Geneva approach can be considered as an attempt to build a mathematical structure which generalizes the classical (Boolean) structure but is still built on the basis of the "actual, absolutely true and certain" propositions which survive in probabilistic theories[c].

Let us now study the G-Z orthogonality relation in the framework of the ϵ-model.

4.2 The G-Z Orthogonality Relation and the ϵ-Model

Before discussing the orthogonality, it is useful to recall some properties of the ϵ-model:

- The measurement process forces the state to collapse along two possible directions: the direction of the device p, or its opposite $-p$. We shall then say that the spin of the system is "up" or "down".
- The probability of getting spin-up $P(p|q)$ depends on the angle θ between p and q only and yields 1 when $0 \leq \theta \leq \theta_{up}$, 0 when $\pi \geq \theta \geq \pi - \theta_{down}$.
- In between two results are possible, in a zone of angular opening θ_{sup} (with in the symmetric case $\sin\left(\frac{\theta_{\text{sup}}}{2}\right) = \epsilon$).

If we apply the definition of the G-Z-orthogonality, we obtain the following:

Theorem 3 *Two states p and q are G-Z-orthogonal iff they make an angle superior or equal to* π - $min(\theta_{down}, \theta_{up})$ *when* $\epsilon \neq 0$. *In the classical case* ($\theta_{sup} = 0$), *the state q is G-Z-orthogonal to p when q belongs to the intersection of a closed sector of opening* θ_{down} *with an open sector of opening* θ_{up}, *both around* $-p$.

Proof: If we consider a state p on the sphere, two experiments could induce a transition on it: the first one with the direction of the experimental device

[c]This is also true for what concerns the Birkhoff-von Neumann approach [16,15].

parallel to p, the second one with this direction anti-parallel to p. The state q is then G-Z-orthogonal to p if the probability of having the result "up" is zero in the first case, one in the second. This means that q belongs to the intersection of the eigenstate set of the outcome "up" around $-p$ with the eigenstate set of the outcome "down" around p.

When $\epsilon \neq 0$, q must then belong to a closed sector around $-p$ of opening $min(\theta_{down}, \theta_{up})$. If we permute the role of p and q in the definition, we shall recover the same condition, essentially because we can write this last condition as $\theta(p, q) \geq \pi - min(\theta_{down}, \theta_{up})$, which is symmetrical in p, q. The state-G-Z-orthogonals are thus closed spherical sectors of opening $min(\theta_{down}, \theta_{up})$ around the antipode of the state.

When $\epsilon = 0$, the proof is entirely similar, excepted that we must consider open spherical caps with angular opening θ_{up} and closed spherical caps of angular opening θ_{down}. □

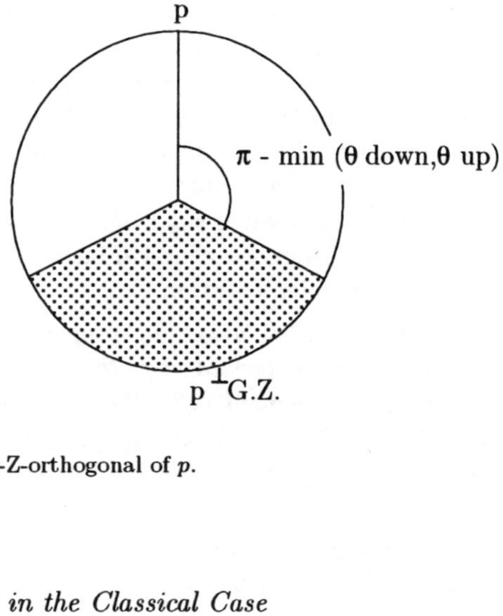

Figure 2. The G-Z-orthogonal of p.

4.3 The G-Z Orthogonality Relation in the Classical Case

It is obvious, from the previous theorems that in the quantum case ($\epsilon = 1$), the definition of G-Z orthogonality implies that two states are orthogonal iff they are antipodal. This corresponds, after the Pauli mapping from the sphere to the two-dimensional Hilbert space, to a ray H-orthogonal to this state, as it must, because this is a special case (dimension 2) of a general property fulfilled by the G-Z orthogonality relations.

The classical case ($\epsilon = 0$) can be treated analogously (see [12] for a rigorous proof), excepted that we must now be careful because of the presence of open and closed spherical sectors simultaneously. Let us discuss the G-Z orthogonality relation in the classical case.

We must distinguish two cases:

- When $\theta_{up} > \theta_{down}$, the state q is G-Z-orthogonal to p when q belongs to a closed sector of opening θ_{down} around $-p$.
- When $\theta_{up} \leq \theta_{down}$, the state q is G-Z-orthogonal to p when q belongs to a open sector of opening θ_{up} around $-p$.

Beside this, the eigenclosure structure is generated by closed sectors of opening θ_{down} and open sectors of opening θ_{up}.

In the first case, when $\theta_{up} > \theta_{down}$, it is generated by open sectors of opening θ_{up}. The eigenclosure structure and the G-Z-ortho-closure-structure are then equal if and only if the state-orthogonals (closed sectors of opening θ_{down}) are equal to open sectors of opening θ_{up}, which is clearly impossible.

In the second case, when $\theta_{up} \leq \theta_{down}$, it is generated by closed sectors of opening θ_{down}. The eigenclosure structure and the G-Z-ortho-closure-structure are then equal if and only if the state-orthogonals (open sectors of opening θ_{up}) are equal to closed sectors of opening θ_{down}, which is also impossible, even when $\theta_{up} = \theta_{down} = \pi/2$.

In conclusion, no classical situation is orthocomplemented for the Grib-Zapatrin complementation.

Theorem 4 *The lattice of properties deduced from an ϵ-probability distribution characterized by the given of the angles $\theta_{up}, \theta_{down}$ is G-Z-orthocomplemented iff the ϵ-probability distribution is symmetrical ($min(\theta_{down}, \theta_{up}) = max(\theta_{down}, \theta_{up})$) when $\epsilon \neq 0$. In the classical case ($\theta_{sup} = 0$), the lattice of properties is never orthocomplemented.*

The Aerts and the ϵ orthocomplemented structures which have been studied in [12,13] are more general (θ_{up} continuously covers the interval $[0, \pi]$, from the quantum to and with the classical situation) than the G-Z-orthocomplemented ones (θ_{up} covers the interval $[0, \pi/2[$, and stays far from the classical case). This is not so astonishing if we note, as we already remarked, that the G-Z definition of the orthogonality is of quantum inspiration. This is emphasized in the following theorem, which shows that the G-Z orthogonality relation does not make sense in a classical context:

Theorem 5 *If at least one of the experiments that one can carry out on the system sends a state on a different state with non zero probability, the G-Z-orthoclosure structure associated to the system is not Boolean.*

Proof: It is a direct consequence of the definition of the G-Z orthogonality that the two states mentioned in the hypothesis are not G-Z orthogonal but are different. The G-Z orthogonality is then not the Boolean orthogonality (which is the relation of difference). □

The previous theorem shows that in the framework of the ϵ model the Boolean limit is not well defined. Actually, the deep reason for this failure is not due to the definition of Grib and Zapatrin because intuitively it seems to be very natural that in the classical limit no experiment will induce a transition from one state onto a different state, in which case the Grib-Zapatrin orthogonality relation will be the difference relation, as it must be in a classical context. Obviously, the problem comes from the ϵ model itself in which, even in the classical limit, non-negligible state-transitions are still likely to occur, depending on the choice that we perform regarding the direction of the measurement axis. Therefore, it would be interesting to reconsider the G-Z orthogonality relation in a situation for which the classical limit is accompanied by the vanishing of the magnitude of the transitions induced by the measurement. Such a situation is described by the model discussed in the next section, which is a simplified version of the model studied extensively in [23].

5 The N-Model: A Model with Vanishing State Transitions in the Classical Limit

5.1 The N-Model

The system consists of a point p on the unit sphere in three dimensions. The experiments e_u^N are defined as follows. Let us consider the measurement axis $[-u, u]$ and divide the sphere in N spherical sections of opening $\theta_N = \frac{\pi}{N}$ by defining $N+1$ circles C_k parallel to the equator: $C_k := \{p_k \mid \theta_{up_k} = k\theta_N\}\,, k = 0, \ldots, N$. For $k = 0$ the circle C_0 reduces to the point u and for $k = N$ the circle C_N reduces to the point $-u$. All these circles are parallel and equidistant to the equator. u is the north pole and $-u$ is the south pole. The half great circle containing $u, -u$ and p is denoted by $C_{u,-u,p}$. Let us consider the situation in which the entity is prepared in a state represented by the point p, lying in a region B_n defined by the circles C_n and C_{n+1}, i.e., $B_n = \{q \mid n\theta_N \leq \theta_{uq} \leq (n+1)\,\theta_N\}$. The intersection of $C_{u,-u,p}$ with B_n is a small longitudinal arc along the sphere and is denoted by C_p. The intersection of the longitudinal arc C_p with the parallel circle C_n is denoted by p_n and the intersection of C_p with the parallel circle C_{n+1} is denoted by p_{n+1}. The measurement proceeds by selecting in the arc C_p randomly a point q. If $\theta_{uq} \geq \theta_{up}$ the measurement induces a state transition towards the state represented by the point p_n and if $\theta_{uq} < \theta_{up}$ the particle ends up in p_{n+1}. We define the set of outcomes of the experiment e_u^N by identifying a state transition towards the state p_n with an outcome $S_n = \cos(n\theta_N)$ for the measurement. Depending on the distribution function $f(q)$ of the random point q, one obtains the probability of the outcomes S_n and S_{n+1}. For instance, if the distribution function $f(q)$ of q is uniform, then the probabilities for outcome S_n, respectively outcome S_{n+1}, is given by:

$$P(S_n) = \frac{\theta_{n+1} - \theta}{\theta_{n+1} - \theta_n} = \frac{N}{\pi}(\theta_{n+1} - \theta)$$

and

$$P(S_{n+1}) = \frac{\theta - \theta_n}{\theta_{n+1} - \theta_n} = \frac{N}{\pi}(\theta - \theta_n)$$

One could also choose a non-uniform distribution function $f(q)$, resulting in another transition probability. For instance, if one would choose the distribution function $f(q)$

$$f(q) = \frac{N}{2}\sin[N(\theta - \theta_n)]$$

one obtains the probability

$$P(S_n) = \frac{\int_{\theta}^{\theta_{n+1}} dq f(q)}{\int_{\theta_n}^{\theta_{n+1}} dq f(q)} = \cos^2\left[\frac{N}{2}(\theta - \theta_n)\right]$$

In the case $N = 1$, the N-model reduces to the sphere model [11], which is a model for the spin properties of a spin-$\frac{1}{2}$ quantum particle. Therefore we prefer to use this distribution function $f(q)$ for the N-model.

5.2 The G-Z Orthogonality Relation and the N-Model

According to a previous definition, the states p and q are G-Z-orthogonal (denoted as $p \perp_{GZ} q$) iff there does not exist a single experiment which induces a state transition from one of the states to the other.

Let us consider two states p and q. The only experiments which could induce a change of state towards p are the experiments e_u^N for which $\theta_{pu} = k\theta_N, k \in \{0, \ldots, N\}$. The only states for which such a measurement e_u^N could induce a change of state towards the state p are the states q along the longitudinal arc for which $\varphi_q = \varphi_p, (k-1)\theta_N < \theta_q < (k+1)\theta_N$. Therefore, we can restrict ourselves to the directions u in the plane defined by o, p and q such that $\theta_{pu} = k\theta_N, k \in \{0, \ldots, N\}$. Since the model is symmetric under rotations over θ_N, the discussion can be restricted to the study of the experiment e_u^N with $k = 0$, i.e., $e_u^N = e_p^N$. The following theorem holds:

Theorem 6 *The states p and q are Grib-Zapatrin orthogonal iff $\theta_{pq} \geq \theta_N$*

Proof: Following the previous discussion it suffices to consider the experiment e_p^N only. Results obtained for this specific experiment hold by reasons of symmetry for all other experiments also. Then if $\theta_{pq} \geq \theta_N$ it follows that $q \in eig_{e_p^N}(\cos(k\theta_N) \mid k \in \{1, \ldots, N\})$ and no transition towards p can occur. On the other hand, if $\theta_{pq} < \theta_N = \frac{\pi}{N}$ then the transition probability $P(p \mid q) = \cos^2\left[\frac{N}{2}\theta_{pq}\right]$ is non-zero, which proves the theorem. □

Following the argument that the orthoclosure is generated by making intersections of state orthogonals, it follows that the orthoclosure is generated by making intersections of spherical caps with opening angle $\pi - \theta_N$. Since in the classical limit $N \longrightarrow \infty$ the angle $\theta_N \longrightarrow 0$, the G-Z state orthogonal will be given by the set-theoretic complement of the state. These generate a Boolean structure by intersection and the lattice of properties is Boolean and therefore the N-model in its limit $N \longrightarrow \infty$ can be called classical.

5.3 Classical Limit for the Orthoclosure of the N-Model

Since in the classical limit $N \longrightarrow \infty$ the angle $\theta_N \longrightarrow 0$, the G-Z state orthogonal will be given by the set-theoretic complement of the state. These generate a Boolean structure by intersection. Therefore the generated Grib-Zapatrin orthoclosure will be Boolean in the classical limit $N \longrightarrow \infty$ of the N-model, in contrast with the epsilon model. The reason for this is that the transition probability will only be non-zero in a region which decreases as N increases, and in the limit vanishes. Clearly, in such a case the Grib-Zapatrin orthogonality relation reduces to the relation of difference. Let us generalize these results in the next section.

6 The G-Z Orthogonality Relation for a General Physical System

Before formulating the general theorem for a general physical entity, we mention that it is possible to define a topology on the set of states in a physical way in terms of the available experiments [24].

6.1 Wooters' Topology Defined by the Set of Available Experiments

Let us show how the set of available experiments allows us to define an operational topology on the set of states, and illustrate this procedure on the

N-model. In Wooters [24] it is shown how the set of experiments allows to define a topology on the set of states, in the sense that distance is defined by how many experimentally distinguishable states there are between the two states. States are called distinguishable if the differences in statistical fluctuations are large enough to conclude that these states are different. Wooters has given (we have changed his notations to make the formula conform with the rest of the notations in this paper) the following distance function on the set of states, in terms of the transition probability P:

$$d(q_1, q_2) = \int dq \frac{\left|\frac{dP}{dq}\right|}{2\sqrt{P(1-P)}}$$

In this formula the shortest possible path between the two states is required, i.e., the path with the smallest value for the integral. For the N-model, this means that one has to follow a path along a great circle on the sphere. Therefore, we can choose coordinates such that $\varphi_{q_1} = \varphi_{q_2} = \varphi_q$, and we can define a state with just the θ-coordinate only. Since the transition probability is given by

$$P(\theta) = \cos^2\left(\frac{N}{2}(\theta - \theta_k)\right)$$

Wooters' formula yields:

$$d(\theta_1, \theta_2) = \int d\theta \frac{\left|\frac{dP}{d\theta}\right|}{2\sqrt{P(1-P)}} = \frac{N}{2}\left|\theta_2 - \theta_1\right|$$

which is a distance function proportional to the standard angular distance on the sphere. This is a natural result due to the spherical symmetry of the model. As a consequence, the Wooters' topology for the N-model is isomorphic with the standard topology on the sphere.

6.2 *General Theorem*

Provided we have such a topology at our disposal, we can define physically the concept of neighborhood, so that following theorem can be formulated:

Theorem 7 (Main Theorem) *If the transition probability is symmetric and if departing from a state p, all experimentally induced transitions will send p on a final state q that remains confined inside a neighborhood of p, and if the size of this neighborhood goes to zero in the classical limit, then the Grib-Zapatrin orthoclosure becomes Boolean in this limit.*

Proof: Since by symmetry of the state transition probability the set of states for which a state transition towards a state p is possible vanishes to the singleton p in the limit, the Grib-Zapatrin orthogonality relation reduces to the relation of difference. Hence it follows immediately that the orthoclosure is Boolean in this limit. □

Although this theorem is very natural and essentially nothing more than a reformulation of the Grib-Zapatrin orthogonality relation, this theorem is useful since it shows the possibility of a continuous dependence of the Grib-Zapatrin orthogonality relation in terms of the physical defined topology.

The theorem confirms the idea that non-perturbability of the experiment induces a Boolean structure on the orthoclosures, and this shows that the classical regime is characterized not only by the condition of determinism of the experiments, as it was the case for the epsilon model, but also by the condition of non-perturbability of the experiment. In fact, if the experiments are ideal, then determinism follows from non-perturbability:

Theorem 8 *If the experiments are ideal, non-perturbability implies determinism.*

Proof: Let us recall that an experiment is called ideal (see e.g., [9]), if a measurement induces a state transition towards an eigenstate of the observed outcome. Therefore, if an ideal experiment does not change the state of the system, the state prior to the measurement has to be an eigenstate of the observed outcome, from which determinism follows. □

Since in most cases it is assumed that experiments are ideal (otherwise one cannot even make experiments since a 'preparation' experiment would be impossible), this shows that non-perturbability is a stronger condition than determinism, and it is easy to understand why the non-perturbability of the N-model in its classical limit leads to Boolean closure structures, while the deterministic limit of the ϵ-model fails to do so for the Grib-Zapatrin orthogonality relation.

Acknowledgments

This work was realized in the framework of a project of the Flemish Funds for Scientific Research (FWO). Some part of it were already presented in reference [12]. It was partially written during an academic visit in Gdansk, realized in the framework of the bilateral Flemish-Polish project 127/E-335/S/2000. The authors would like to thank J. Pykacz for his comments which helped to improve the quality of this manuscript. T. Durt and B. D' Hooghe are

Postdoctoral Fellows of the Fund for Scientific Research - Flanders, Belgium (FWO-Vlaanderen).

References

1. D. Aerts, T. Durt and B. Van Bogaert, "A physical example of quantum fuzzy sets and the classical limit", *Tatra Mountains Math. Publ.*, **1**, 5 (1992).
2. D. Aerts, T. Durt and B. Van Bogaert, "Indeterminism, nonlocality and the classical limit", in *Foundations of Modern Physics*, ed. T. Hyvonen, World Scientific, Singapore (1993).
3. D. Aerts and T. Durt, "Quantum, classical and intermediate, a measurement model", in *Foundations of Modern Physics*, eds. Laurikainen, K. V., Montonen, C. and Sunnaborg, K., Editions Frontieres, Gives Sur Yvettes, France (1994).
4. D. Aerts and T. Durt, "Quantum, classical and intermediate, an illustrative example, *Found. Phys.*, **24**, 1407 (1994).
5. D. Aerts, S. Aerts, B. Coecke, B. D'Hooghe, T. Durt and F. Valckenborgh, "A model with varying fluctuations in the measurement context", in *New Developments on Fundamental Problems in Quantum Physics*, eds. M. Ferrero and A. van der Merwe, Kluwer Academic, Dordrecht (1997).
6. D. Aerts, B. Coecke, T. Durt and F. Valckenborgh, "Quantum, classical and intermediate I: a model on the Poincaré sphere", *Tatra Mt. Math. Publ.*, **10**, 225 (1997).
7. D. Aerts, B. Coecke, T. Durt and F. Valckenborgh, "Quantum, classical and intermediate II: the vanishing vector space structure", *Tatra Mt. Math. Publ.*, **10**, 241 (1997).
8. C. Piron, "Axiomatique quantique", *Helv. Phys. Acta*, **37**, 439 (1964).
9. C. Piron, *Foundations of Quantum Physics*, W. A. Benjamin, Inc. (1976).
10. C. Piron, *Mécanique Quantique, Bases et Applications*, Presses Polytechnique et Universitaire Romandes, Laussane (1990).
11. D. Aerts, "A possible explanation for the probabilities of quantum mechanics", *J. Math. Phys.*, **27**, 203 (1986).
12. T. Durt, *From quantum to classical, a toy model*, Doctoral Dissertation, Free University of Brussels (1996).
13. T. Durt, "Orthogonality relations: from classical to quantum", in *Quantum Structures and the Nature of Reality*, eds. D. Aerts and J. Pykacz, Kluwer Academic, Dordrecht, 207 (1999).
14. D. Aerts, *The One and the Many*, Doctoral Dissertation, Free University

of Brussels (1981).
15. G. Birkhoff, *Lattice Theory, third edition*, Amer. Math. Soc., Colloq. Publ. **Vol. XXV**, Providence.
16. G. Birkhoff and J. von Neumann, "The logic of quantum mechanics", *Annals of Mathematics*, **37**, 823 (1936).
17. E. Beltrametti and G. Cassinelli, *The Logic of Quantum Mechanics*, Addison-Wesley Publishing Company (1981).
18. J. von Neumann, *Grundlehren, Math. Wiss. XXXVIII* (1932).
19. J. von Neumann, *Mathematische Grundlagen der Quanten-Mechanik*, Springer-Verlag, Berlin (1932).
20. H. H. Crapo and G. C. Rota, "Geometric lattices", in *Trends in Lattice Theory*, ed. J. C. Abbott, Van Nostrand-Reinhold, New York (1970).
21. D. Aerts, T. Durt, A.A. Grib, B. Van Bogaert and R.R. Zapatrin, "Quantum structures in macroscopic reality", *Int. J. Theor. Phys.*, **32**, 489 (1993).
22. A. A. Grib and R. R. Zapatrin, *Int. Journal of Theor. Phys.*, **29**, 113 (1990).
23. B. D'Hooghe, *From quantum to classical: A study of the effect of varying fluctuations in the measurement context and state transitions due to experiments*, Doctoral Dissertation, Brussels Free University (2000).
24. W. K. Wooters, "Statistical distance and Hilbert space", *Phys. Rev.* **D 23**, 357 (1981).

STATE PROPERTY SYSTEMS AND CLOSURE SPACES: EXTRACTING THE CLASSICAL EN NONCLASSICAL PARTS

DIEDERIK AERTS

Center Leo Apostel (CLEA) and Foundations of the Exact Sciences (FUND), Brussels Free University, Krijgskundestraat 33, 1160 Brussels, Belgium
E-mail: diraerts@vub.ac.be

DIDIER DESES

Department of Mathematics (TOPO), Brussels Free University, Pleinlaan 2, 1050 Brussels, Belgium
E-mail: diddesen@vub.ac.be

In [1] an equivalence of the categories **SP** and **Cls** was proven. The category **SP** consists of the state property systems [2] and their morphisms, which are the mathematical structures that describe a physical entity by means of its states and properties [3,4,5,6,7,8]. The category **Cls** consists of the closure spaces and the continuous maps. In earlier work it has been shown, using the equivalence between **Cls** and **SP**, that some of the axioms of quantum axiomatics are equivalent with separation axioms on the corresponding closure space. More particularly it was proven that the axiom of atomicity is equivalent to the T_1 separation axiom [9]. In the present article we analyze the intimate relation that exists between classical and nonclassical in the state property systems and disconnected and connected in the corresponding closure space, elaborating results that appeared in [10,11]. We introduce classical properties using the concept of super selection rule, i.e. two properties are separated by a superselection rule iff there do not exist 'superposition states' related to these two properties. Then we show that the classical properties of a state property system correspond exactly to the clopen subsets of the corresponding closure space. Thus connected closure spaces correspond precisely to state property systems for which the elements 0 and I are the only classical properties, the so called pure nonclassical state property systems. The main result is a decomposition theorem, which allows us to split a state property system into a number of 'pure nonclassical state property systems' and a 'totally classical state property system'. This decomposition theorem for a state property system is the translation of a decomposition theorem for the corresponding closure space into its connected components.

1 State Property Systems and Closure Spaces

The general approaches to quantum mechanics make use of mathematical structures that allow the description of pure quantum entities and pure classical entities, as well as mixtures of both. In this article we study the Geneva-Brussels approach, where the basic physical concepts are the one of state and property of a physical entity [3,4,5,6,7,8]. Traditionally the collection of properties is considered to be a complete lattice, partially ordered by the implication of properties, with an orthocomplementation, representing the quantum generalization of the 'negation' of a property. A state is represented by the collections of properties that are actual whenever the entity is in this state. We mention however that in these earlier approaches [3,4,5,6,7,8] the mathematical structure that underlies the physical theory had not completely been identified. To identify the mathematical structure in a complete way, the structure of a state property system was introduced in [2].

Suppose that we consider a physical entity S, and we denote its set of states by Σ and its set of properties by $\mathcal{L}$. The state property system corresponding to this physical entity S is a triple $(\Sigma, \mathcal{L}, \xi)$, where Σ is the set of states of S, $\mathcal{L}$ the set of properties of S, and ξ a map from Σ to $\mathcal{P}(\mathcal{L})$, that makes correspond to each state $p \in \Sigma$ the set of properties $\xi(p) \in \mathcal{P}(\mathcal{L})$ that are actual if the entity S is in state p. Some additional requirements, that express exactly how the physicists perceives a physical entity in relation with its states and properties, are satisfied in a state property system. Let us introduce the formal definition of a state property system and then explain what these additional requirements mean.

Definition 1 (State Property System) *A triple $(\Sigma, \mathcal{L}, \xi)$ is called a state property system if Σ is a set, $\mathcal{L}$ is a complete lattice and $\xi : \Sigma \to \mathcal{P}(\mathcal{L})$ is a function such that for $p \in \Sigma$, 0 the minimal element of $\mathcal{L}$ and $(a_i)_i \in \mathcal{L}$, we have:*

$$0 \notin \xi(p) \tag{1}$$

$$a_i \in \xi(p)\ \forall i \Rightarrow \wedge_i a_i \in \xi(p) \tag{2}$$

and for $a, b \in \mathcal{L}$ we have:

$$a < b \Leftrightarrow \forall r \in \Sigma : a \in \xi(r) \text{ then } b \in \xi(r) \tag{3}$$

We demand that $\mathcal{L}$, the set of properties, is a complete lattice. This means that the set of properties is partially ordered, with the physical meaning of the partial order relation $<$ being the following: $a, b \in \mathcal{L}$, such that $a < b$ means that whenever property a is actual for the entity S, also property b is actual for the entity S. If $\mathcal{L}$ is a complete lattice, it means that for an arbitrary family of properties $(a_i)_i \in \mathcal{L}$ also the infimum $\wedge_i a_i$ of this family is a property. The property $\wedge_i a_i$ is the property that is actual iff all of the properties a_i are actual. Hence the infimum represents the logical 'and'. The minimal element 0 of the lattice of properties is the property that is never actual (*e.g.* the physical entity does not exist). Requirement (1) expresses that a property that is in the image by ξ of an arbitrary state $p \in \Sigma$ can never be the 0 property. Requirement (2) expresses that if for a state $p \in \Sigma$ all the properties a_i are actual, this implies that for this state p also the 'and' property $\wedge_i a_i$ is actual. Requirement (3) expresses the meaning of the partial order relation that we gave already: $a < b$ iff whenever p is a state of S such that a is actual if S is in this state, then also b is actual if S is in this state.

Along the same lines, just traducing what the physicist means when he imagines the situation of two physical entities, of which one is a sub entity of the other, the morphisms of state property systems can be deduced. More concretely, suppose that S is a sub entity of S'. Then each state p' of S' determines a state p of S, namely the state p where the sub entity S is in when S' is in state p'. This defines a map $m : \Sigma' \to \Sigma$. On the other hand, each property a of S determines a property a' of S, namely the property of the sub entity, but now conceived as a property of the big entity. This defines a map $n : \Sigma \to \Sigma'$. Suppose that we consider now a state p' of S', and a property a of S, such that $a \in \xi(m(p'))$. This means that the property a is actual if the sub entity S is in state $m(p')$. This state of affairs can be expressed equally by stating that the property $n(a)$ is actual when the big entity is in state p'. Hence, as a basic physical requirement of merological covariance we should have:

$$a \in \xi(m(p')) \Leftrightarrow n(a) \in \xi'(p') \tag{4}$$

This all gives rise to the following definition of morphism for state property systems.

Definition 2 (Morphisms of State Property Systems) *Suppose that* $(\Sigma, \mathcal{L}, \xi)$ *and* $(\Sigma', \mathcal{L}', \xi')$ *are state property systems then*

$$(m, n) : (\Sigma', \mathcal{L}', \xi') \to (\Sigma, \mathcal{L}, \xi)$$

is called an **SP**-*morphism if* $m : \Sigma' \to \Sigma$ *and* $n : \mathcal{L} \to \mathcal{L}'$ *are functions such that for* $a \in \mathcal{L}$ *and* $p' \in \Sigma'$ *:*

$$a \in \xi(m(p')) \Leftrightarrow n(a) \in \xi'(p') \tag{5}$$

Using the previous definitions we can use these concept to generate a category of state property systems, in the mathematical sense.

Definition 3 (The Category SP) *The category of state property systems and their morphisms is denoted by* **SP**.

Definition 4 (The Cartan Map) *If* $(\Sigma, \mathcal{L}, \xi)$ *is a state property system then its Cartan map is the mapping* $\kappa : \mathcal{L} \to \mathcal{P}(\Sigma)$ *defined by :*

$$\kappa : \mathcal{L} \to \mathcal{P}(\Sigma) : a \mapsto \kappa(a) = \{p \in \Sigma \mid a \in \xi(p)\} \tag{6}$$

It was amazing to be able to prove (see [1]) that this category of states property systems and its morphisms is equivalent to a category which arises as a generalization of the category of topological spaces and continuous maps, namely to the category of closure spaces and the continuous maps. We will now introduce this category of closure spaces.

A topological space consists of a set X, and a collection of 'open' subsets, such that X is open, any union of open subsets is again open and any finite intersection of open subsets is again open. A subset of X is called closed if it's complement is open. Therefore we have that in a topological space the empty set is closed, any intersection of closed sets is closed and any finite union of closed sets is again closed. Hence a topological space is also defined by it's closed sets.

In mathematics the concept topological space is very useful and arises in many different areas. However there are occasions when we 'almost' have a topological space. Let's take the following example. Consider the plane $\mathbb{R}^2$ and the collection of all convex subsets of $\mathbb{R}^2$ (A is convex if the segment between any two points of A lies completely within A). Clearly $\emptyset$ is convex and every intersection of convex sets is again convex. However a finite union of convex sets does not need to be convex. Hence the convex subsets of the plane can 'almost' be considered as closed sets, but they do not form a topological space. To be able to consider such structures one has introduced the notion of closure spaces.

Definition 5 (Closure Space) *A closure space $(X, \mathcal{F})$ consists of a set X and a family of subsets $\mathcal{F} \subseteq \mathcal{P}(X)$ satisfying the following conditions:*

$$\emptyset \in \mathcal{F}$$

$$(F_i)_i \in \mathcal{F} \Rightarrow \cap_i F_i \in \mathcal{F}$$

The closure operator corresponding to the closure space $(X, \mathcal{F})$ is defined as

$$cl : \mathcal{P}(X) \to \mathcal{P}(X) : A \mapsto \bigcap \{F \in \mathcal{F} \mid A \subseteq F\} \tag{7}$$

If $(X, \mathcal{F})$ and $(Y, \mathcal{G})$ are closure spaces then a function $f : (X, \mathcal{F}) \to (Y, \mathcal{G})$ is called a continuous map if $\forall B \in \mathcal{G} : f^{-1}(B) \in \mathcal{F}$. The category of closure spaces and continuous maps is denoted by **Cls**.

The following theorem shows how we can associate with each state property system a closure space and with each morphism a continuous map, hence we get the categorical equivalence described in [1].

Theorem 1 *The correspondence F :* **SP** $\longrightarrow$ **Cls** *consisting of*
(1) the mapping

$$|\mathbf{SP}| \to |\mathbf{Cls}|$$

$$(\Sigma, \mathcal{L}, \xi) \mapsto F(\Sigma, \mathcal{L}, \xi) = (\Sigma, \kappa(\mathcal{L}))$$

(2) for every pair of objects $(\Sigma, \mathcal{L}, \xi), (\Sigma', \mathcal{L}', \xi')$ *of* **SP** *the mapping*

$$\mathbf{SP}((\Sigma', \mathcal{L}', \xi'), (\Sigma, \mathcal{L}, \xi)) \to \mathbf{Cls}(F(\Sigma', \mathcal{L}', \xi'), F(\Sigma, \mathcal{L}, \xi))$$
$$(m, n) \mapsto m$$

is a covariant functor.

We can also connect a state property system to a closure space and a morphism to a continuous map.

Theorem 2 *The correspondence* $G : \mathbf{Cls} \longrightarrow \mathbf{SP}$ *consisting of*
(1) the mapping

$$|\mathbf{Cls}| \to |\mathbf{SP}|$$
$$(\Sigma, \mathcal{F}) \mapsto G(\Sigma, \mathcal{F}) = (\Sigma, \mathcal{F}, \bar{\xi})$$

where $\bar{\xi} : \Sigma \to \mathcal{P}(\mathcal{F}) : p \mapsto \{F \in \mathcal{F} \mid p \in F\}$
(2) for every pair of objects $(\Sigma, \mathcal{F}), (\Sigma', \mathcal{F}')$ *of* **Cls** *the mapping*

$$\mathbf{Cls}((\Sigma', \mathcal{F}'), (\Sigma, \mathcal{F})) \to \mathbf{SP}(G(\Sigma', \mathcal{F}'), G(\Sigma, \mathcal{F}))$$
$$m \mapsto (m, m^{-1})$$

is a covariant functor.

Theorem 3 (Equivalence of SP and Cls) *The functors*

$$F : \mathbf{SP} \to \mathbf{Cls}$$

$$G : \mathbf{Cls} \to \mathbf{SP}$$

establish an equivalence of categories.

The above equivalence is a very powerful tool for studying state property systems. It states that the lattice $\mathcal{L}$ of properties can be seen as the lattice of closed sets of a closure space on the states Σ, conversely every closure space on X can be considered as a set of states (X) and a lattice of properties (the lattice of closed sets).

Recall that closure spaces are in fact a generalization of topological spaces, hence a number of topological properties have been generalized to closure spaces. Moreover with the previous equivalence, a concept which can be defined using closed sets on a closure space can be translated in an equivalent concept for state property systems. At first sight this translation does not need to be meaningful in the context of physical systems. However it turned out that many such translations actually coincided with well known physical concepts.

We shall give one example which was studied in [9]. A topological space is called T_1 if the following separation axiom is satisfied. For every two points x, y there are open sets which contain x resp. y but do not contain y resp. x. This is equivalent to stating that all singletons are closed sets. Hence the following definition.

Definition 6 (T_1 Closure Space) *A closure space $(X, \mathcal{F})$ is a T_1 closure space iff $\forall x \in X : \{x\} \in \mathcal{F}$.*

In the theory of state property systems, or more general of property lattices the concept of atomistic lattice is quite fundamental. In [9] it was proven that using the equivalence between state property systems and closure spaces both concepts are in fact related.

Definition 7 (Atomistic State Property System) *Let $(\Sigma, \mathcal{L}, \xi)$ be a state property system. Then the map s_ξ maps a state p to the strongest property it makes actual, i.e.*

$$s_\xi : \Sigma \rightarrow \mathcal{L} : p \mapsto \wedge \xi(p) \tag{8}$$

T.F.A.E.

(1) $\xi : \Sigma \rightarrow \mathcal{P}(\mathcal{L})$ is injective and $\forall p \in \Sigma : s_\xi(p)$ is an atom of $\mathcal{L}$.

(2) $\forall p, q \in \Sigma : \xi(p) \subset \xi(q) \Rightarrow p = q$

(3) $F(\Sigma, \mathcal{L}, \xi) = (\Sigma, \kappa(\mathcal{L}))$ is a T_1 closure space.

If a state property system satisfies one, and hence all of the above conditions it is called an atomistic state property system, in this case $\mathcal{L}$ is a complete atomistic lattice.

If we write $\mathbf{Cls_1}$ for the full subcategory of $\mathbf{Cls}$ given by T_1 closure spaces, and $\mathbf{SP_a}$ for the full subcategory of $\mathbf{SP}$ given by the atomistic state property systems, then the general equivalence can be reduced.

Theorem 4 (Equivalence of $\mathbf{SP_a}$ and $\mathbf{Cls_1}$) *The functors*

$$F : \mathbf{SP_a} \rightarrow \mathbf{Cls_1}$$

$$G : \mathbf{Cls_1} \rightarrow \mathbf{SP_a}$$

establish an equivalence of categories.

For a more extensive study of separation axioms and their relation with state property systems we refer to [12]. In the present text our final aim is to use the described equivalence to translate the concept of connectedness in closure spaces into terms of state property systems. It will give us a means to distinguish 'classical' and 'quantum mechanical' properties of a physical entity. First we will need a more precise concept of classical property.

2 Super Selection Rules

In this section we start to distinguish the classical aspects of the structure from the quantum aspects. We all know that the concept of superposition state is very important in quantum mechanics. The superposition states are the states that do not exist in classical physics and hence their appearance is one of the important quantum aspects. To be able to define properly a superposition state we need the linearity of the set of states. On the level of generality that we work now, we do not necessarily have this linearity, which could indicate that the concept of superposition state cannot be given a meaning on this level of generality. This is however not really true: the concept can be traced back within this general setting, by introducing the idea of 'superselection rule'. Two properties are separated by a superselection rule iff there do not exist 'superposition states' related to these two properties. This concept will be the first step towards a characterization of classical properties of a physical system.

Definition 8 (Super Selection Rule) *Consider a state property system* $(\Sigma, \mathcal{L}, \xi)$. *For* $a, b \in \mathcal{L}$ *we say that* a *and* b *are separated by a super selection rule, and denote* $a\ ssr\ b$, *iff for* $p \in \Sigma$ *we have:*

$$a \vee b \in \xi(p) \Rightarrow a \in \xi(p) \text{ or } b \in \xi(p) \tag{9}$$

We again use the equivalence between state property systems and closure spaces to translate the concept of 'separation by a superposition rule' into a concept for the closed sets of a closure space. Amazingly we find that properties that are 'separated by a superselection rule' (i.e. they are 'classical' properties in a certain sense) correspond to closed sets that also behave in a classical way, where classical now refers to classical topology.

Theorem 5 *Consider a state property system* $(\Sigma, \mathcal{L}, \xi)$ *and its corresponding closure space* $\mathcal{F} = \kappa(\mathcal{L})$. *For* $a, b \in \mathcal{L}$ *we have:*

$$a\ ssr\ b \Leftrightarrow \kappa(a \vee b) = \kappa(a) \cup \kappa(b) \Leftrightarrow \kappa(a) \cup \kappa(b) \in \mathcal{F} \tag{10}$$

Proof: Suppose that $a, b \in \mathcal{L}$ such that $a\ ssr\ b$. If $p \in \kappa(a \vee b)$, then $a \vee b \in \xi(p)$. Then it follows that $a \in \xi(p)$ or $b \in \xi(p)$. So we have $p \in \kappa(a)$ or $p \in \kappa(b)$, which shows that $p \in \kappa(a) \cup \kappa(b)$. This proves that $\kappa(a \vee b) \subseteq \kappa(a) \cup \kappa(b)$. We obviously have the other inclusion and hence $\kappa(a \vee b) = \kappa(a) \cup \kappa(b)$. It follows immediately that $\kappa(a) \cup \kappa(b) \in \mathcal{F}$. Conversely, if $\kappa(a) \cup \kappa(b) \in \mathcal{F}$, then there exists a property $c \in \mathcal{L}$ such that $\kappa(c) = \kappa(a) \cup \kappa(b)$. From $\kappa(a) \subseteq \kappa(c)$ it follows that $a < c$, and in a similar way we have $b < c$. So it follows that $a \vee b < c$. As a consequence we have $\kappa(a \vee b) \subseteq \kappa(c) = \kappa(a) \cup \kappa(b)$. Since $\kappa(a) \cup \kappa(b) \subseteq \kappa(a \vee b)$, we have $\kappa(a \vee b) = \kappa(a) \cup \kappa(b)$. Consider now an arbitrary $p \in \Sigma$ such that $a \vee b \in \xi(p)$. Then $p \in \kappa(a \vee b) = \kappa(a) \cup \kappa(b)$. As a consequence $p \in \kappa(a)$ or $p \in \kappa(b)$. This proves that $a \in \xi(p)$ or $b \in \xi(p)$ which shows that a ssr b. □

This theorem shows that the properties that are separated by a super selection rule are exactly the ones that behave also classically within the closure system. In the sense that their set theoretical unions are closed. This also means that if our closure system reduces to a topology, and hence all finite unions of closed subsets are closed, all finite sets of properties are separated by super selection rules.

Corollary 1 *Let $(\Sigma, \mathcal{L}, \xi)$ be a state property system. T.F.A.E.:*

(1) Every two properties of $\mathcal{L}$ are separated by a super selection rule.

(2) The corresponding closure space $(\Sigma, \kappa(\mathcal{L}))$ is a topological space.

A state property satisfying one, and hence both of the above conditions will be called a 'super selection classical' state property system or 's-classical' state property system. The full subcategory of **SP** *given by the s-classical state property systems will be written as* $^{sc}\mathbf{SP}$.

Hence the equivalence between state property systems and closure space can be reduced to an equivalence between s-classical state property systems, in which no two properties have 'superposition states' related to them, and topological spaces.

Theorem 6 (Equivalence of $^{sc}\mathbf{SP}$ and Top) *The functors*

$$F : {}^{sc}\mathbf{SP} \to \mathbf{Top}$$

$$G : \mathbf{Top} \to {}^{sc}\mathbf{SP}$$

establish an equivalence of categories.

3 D-classical Properties

We are ready now to introduce the concept of a 'deterministic classical property' or 'd-classical property'. To make clear what we mean by this we have to explain shortly how properties are tested. For each property $a \in \mathcal{L}$ there exists a test α, which is an experiment that can be performed on the physical entity under study, and that can give two outcomes, 'yes' and 'no'. The property a tested by the experiment α is actual iff the state p of S is such that we can predict with certainty (probability equal to 1) that the outcome 'yes' will occur for the test α. If the state p of S is such that we can predict with certainty that the outcome 'no' will occur, we test in some way a complementary property of the property a, let us denote the complementary property by a^c. Now we have three possibilities: (1) the state of S is such that α gives 'yes' with certainty; (2) the state of S is such that α gives 'no' with certainty; and (3) the state of S is such that neither the outcome 'yes' nor the outcome 'no' is certain for the experiment α. The third case represents the situations of 'quantum indeterminism'. That is the reason that a property a tested by an experiment α where the third case is absent will be called a 'deterministic classical' property or 'd-classical' property.

Definition 9 (D-classical Property) *Consider a state property system $(\Sigma, \mathcal{L}, \xi)$. We say that a property $a \in \mathcal{L}$ is a 'deterministic classical property' or 'd-classical' property, if there exists a property $a^c \in \mathcal{L}$ such that $a \vee a^c = I$, $a \wedge a^c = 0$ and a ssr a^c.*

Remark that for every state property system $(\Sigma, \mathcal{L}, \xi)$ the properties 0 and I are d-classical properties. Note also that if $a \in \mathcal{L}$ is a d-classical property, we have for $p \in \Sigma$ that $a \in \xi(p) \Leftrightarrow a^c \notin \xi(p)$ and $a \notin \xi(p) \Leftrightarrow a^c \in \xi(p)$. This follows immediately from the definition of a d-classical property.

Theorem 7 *Consider a state property system $(\Sigma, \mathcal{L}, \xi)$. If $a \in \mathcal{L}$ is a d-classical property, then a^c is unique and is a d-classical property. We will call it the complement of a. Further we have:*

$$(a^c)^c = a$$
$$a < b \Rightarrow b^c < a^c$$
$$\kappa(a^c) = \kappa(a)^C$$

Proof: Suppose that we have another property $b \in \mathcal{L}$ such that $a \vee b = I$, $a \wedge b = 0$ and a ssr b. Consider an arbitrary state $p \in \Sigma$ such that $a^c \in \xi(p)$. This means that $a \notin \xi(p)$. We have however $a \vee b \in \xi(p)$, which implies, since a ssr b, that $a \in \xi(p)$ or $b \in \xi(p)$. As a consequence we have $b \in \xi(p)$. This means that we have proven that $a^c < b$. In a completely analogous way we can show that also $b < a^c$, which shows that a^c is unique. Obviously a^c is a d-classical property. Then the idempotency follows from the fact that a is the complement of a^c and from the uniqueness of the complement. Consider $a < b$ and an arbitrary state $p \in \Sigma$ such that $b^c \in \xi(p)$. This means that $b \notin \xi(p)$, which implies that $a \notin \xi(p)$. As a consequence we have $a^c \in \xi(p)$. So we have shown that $b^c < a^c$. Further we have $p \in \kappa(a^c)$ iff $a^c \in \xi(p)$. From the above mentioned remark this is equivalent with $a \notin \xi(p)$. and $p \notin \kappa(a)$ which is the same as saying that $p \in \kappa(a)^C$. So we have $\kappa(a^c) = \kappa(a)^C$. ☐

Definition 10 (Connected Closure Space) *A closure space $(X, \mathcal{F})$ is called connected if the only clopen (i.e. closed and open) sets are $\emptyset$ and X.*

We shall see now that these subsets that make closure systems disconnected are exactly the subsets corresponding to d-classical properties.

Theorem 8 *Consider a state property system* $(\Sigma, \mathcal{L}, \xi)$ *and its corresponding closure space* $(\Sigma, \kappa(\mathcal{L}))$. *For* $a \in \mathcal{L}$ *we have:*

$$a \text{ is d} - \text{classical } \Leftrightarrow \kappa(a) \text{ is clopen} \tag{11}$$

Proof: ¿From the previous propositions it follows that if a is d-classical, then $\kappa(a)$ is clopen. So now consider a clopen subset $\kappa(a)$ of Σ. This means that $\kappa(a)^C$ is closed, and hence that there exists a property $b \in \mathcal{L}$ such that $\kappa(b) = \kappa(a)^C$. We clearly have $a \wedge b = 0$ since there exists no state $p \in \Sigma$ such that $p \in \kappa(a)$ and $p \in \kappa(b)$. Since $\Sigma = \kappa(a) \cup \kappa(b)$ we have $a \vee b = I$. Further we have that for an arbitrary state $p \in \Sigma$ we have $a \in \xi(p)$ or $b \in \xi(p)$ which shows that a ssr b. This proves that $b = a^c$ and that a is d-classical. □

This means that the d-classical properties correspond exactly to the clopen subsets of the closure system.

Corollary 2 *Let* $(\Sigma, \mathcal{L}, \xi)$ *be a state property system. T.F.A.E.*
(1) The properties 0 *and* I *are the only d-classical ones.*
(2) $F(\Sigma, \mathcal{L}, \xi) = (\Sigma, \kappa(\mathcal{L}))$ *is a connected closure space.*

We now introduce 'completely quantum mechanical' or pure nonclassical state property systems, in the sense that there are no (non-trivial) d-classical properties.

Definition 11 (Pure Nonclassical State Property System) *A state property system* $(\Sigma, \mathcal{L}, \xi)$ *is called a pure nonclassical state property system if the properties* 0 *and* I *are the only d-classical properties.*

Theorem 9 *Let* $(\Sigma, \mathcal{F})$ *be a closure space. T.F.A.E.*
(1) $(\Sigma, \mathcal{F})$ *is a connected closure space.*
(2) $G(\Sigma, \mathcal{F}) = (\Sigma, \mathcal{F}, \bar{\xi})$ *is a pure nonclassical state property system.*

Proof: Let $(\Sigma, \mathcal{F})$ be a connected state property system. Then $\emptyset$ and Σ are the only clopen sets in $(\Sigma, \mathcal{F})$. Since the Cartan map associated to ξ is given by $\kappa : \mathcal{F} \to \mathcal{P}(\Sigma) : F \mapsto F$, we have $\kappa(\emptyset) = \emptyset$ and $\kappa(\Sigma) = \Sigma$. Applying proposition 8, we find that $\emptyset$ and Σ are the only d-classical properties of $\mathcal{F}$. Conversely, let $G(\Sigma, \mathcal{F}) = (\Sigma, \mathcal{F}, \bar{\xi})$ be a pure nonclassical state property system. Then by corollary 2, $(\Sigma, \mathcal{F}) = FG(\Sigma, \mathcal{F})$ is a connected closure space. □

If we define $\mathbf{SP_Q}$ as the full subcategory of **SP** where the objects are the pure nonclassical state property systems and we define $\mathbf{Cls_C}$ as the full subcategory of **Cls** where the objects are the connected closure spaces, then the previous propositions and theorem 3 imply an equivalence of the categories $\mathbf{SP_Q}$ and $\mathbf{Cls_C}$.

Theorem 10 (Equivalence of $\mathbf{SP_Q}$ and $\mathbf{Cls_C}$) *The functors*

$$F : \mathbf{SP_Q} \to \mathbf{Cls_C}$$

$$G : \mathbf{Cls_C} \to \mathbf{SP_Q}$$

establish an equivalence of categories.

Again we have found using the equivalence 3 that a physical concept (i.e. nonclassicality) translates to a known topological property (i.e. connectedness). In the next section we will use topological methods to construct a decomposition of a state property system into pure nonclassical components.

4 Decomposition Theorem

As for topological spaces, every closure space can be decomposed uniquely into connected components. In the following we say that, for a closure space $(X, \mathcal{F})$, a subset $A \subseteq X$ is connected if the induced subspace is connected. It can be shown that the union of any family of connected subsets having at least one point in common is also connected. So the component of an element $x \in X$ defined by

$$K_{\mathbf{Cls}}(x) = \bigcup\{A \subseteq X \mid x \in A, A \text{ connected }\} \tag{12}$$

is connected and therefore called the connection component of x. Moreover, it is a maximal connected set in X in the sense that there is no connected subset of X which properly contains $K_{\mathbf{Cls}}(x)$. From this it follows that for closure spaces $(X, \mathcal{F})$ the set of all distinct connection components in X form a partition of X. So we can consider the following equivalence relation on X : for $x, y \in X$ we say that x is equivalent with y iff the connection components $K_{\mathbf{Cls}}(x)$ and $K_{\mathbf{Cls}}(y)$ are equal. Further we remark that the connection components are closed sets.

In the following we will try to decompose state property systems similarly into different components.

Theorem 11 *Let $(\Sigma, \mathcal{L}, \xi)$ be a state property system and let $(\Sigma, \kappa(\mathcal{L}))$ be the corresponding closure space. Consider the following equivalence relation on Σ :*

$$p \sim q \Leftrightarrow K_{\mathbf{Cls}}(p) = K_{\mathbf{Cls}}(q) \tag{13}$$

with equivalence classes $\Omega = \{\omega(p) | p \in \Sigma\}$. If $\omega \in \Omega$ we define the following :

$$\Sigma_\omega = \omega = \{p \in \Sigma \mid \omega(p) = \omega\}$$

$$s(\omega) = s(\omega(p)) = a, \text{ such that } \kappa(a) = \omega(p)$$

$$\mathcal{L}_\omega = [0, s(\omega)] = \{a \in \mathcal{L} \mid 0 \leq a \leq s(\omega)\} \subset \mathcal{L}$$
$$\xi_\omega \,:\, \Sigma_\omega \rightarrow \mathcal{P}(\mathcal{L}_\omega) : p \mapsto \xi(p) \cap \mathcal{L}_\omega$$

then $(\Sigma_\omega, \mathcal{L}_\omega, \xi_\omega)$ *is a state property system.*

Proof: Since $\mathcal{L}_\omega$ is a sublattice (segment) of $\mathcal{L}$, it is a complete lattice with maximal element $I_\omega = s(\omega)$ and minimal element $0_\omega = 0$. Let $p \in \Sigma_\omega$. Then $0 \notin \xi(p)$. So $0 \notin \xi(p) \cap \mathcal{L}_\omega = \xi_\omega(p)$. If $a_i \in \xi_\omega(p)$, $\forall i$, then $a_i \in \mathcal{L}_\omega$ and $a_i \in \xi(p)$, $\forall i$. Hence $\wedge a_i \in \mathcal{L}_\omega \cap \xi(p) = \xi_\omega(p)$. Finally, let $a, b \in \mathcal{L}_\omega$ with $a <_\omega b$ and let $r \in \Sigma_\omega$. If $a \in \xi_\omega(r)$, then $a \in \mathcal{L}_\omega$ and $a \in \xi(r)$, thus $b \in \mathcal{L}_\omega$ and $b \in \xi(r)$. So $b \in \xi_\omega(r)$. Conversely, if $a, b \in \mathcal{L}_\omega$ and $\forall r \in \Sigma_\omega : a \in \xi_\omega(r) \Rightarrow b \in \xi_\omega(r)$ then we consider a q such that $a \in \xi(q)$ (q must be in Σ_ω by definition of $\mathcal{L}_\omega$). Then $a \in \xi_\omega(q)$ implies that $b \in \xi_\omega(q)$. So $b \in \xi(q)$ and $a < b$. Thus $a <_\omega b$. □

Moreover we can show that the above introduced state property systems $(\Sigma_\omega, \mathcal{L}_\omega, \xi_\omega)$ have no proper d-classical properties, and hence are pure non-classical state property systems.

Theorem 12 *Let* $(\Sigma, \mathcal{L}, \xi)$ *be a state property system. If* $\omega \in \Omega$, *then* $(\Sigma_\omega, \mathcal{L}_\omega, \xi_\omega)$ *is a pure nonclassical state property system.*

Proof: If a is classical element of $\mathcal{L}_\omega$, then $\kappa(a)$ must be a clopen set of the associated closure space $(\Sigma_\omega, \kappa(\mathcal{L}_\omega))$ which is a connected subspace of $(\Sigma, \kappa(\mathcal{L}))$. Hence there are no proper classical elements of $\mathcal{L}_\omega$. □

Theorem 13 *Let* $(\Sigma, \mathcal{L}, \xi)$ *be a state property system. If we introduce the following :*

$$\Omega = \{\omega(p) \mid p \in \Sigma\}$$
$$\mathcal{C} = \{\vee s(\omega_i) \mid \omega_i \in \Omega\}$$
$$\eta \,:\, \Omega \rightarrow \mathcal{P}(\mathcal{C}) : \omega = \omega(p) \mapsto \xi(p) \cap \mathcal{C}$$

then $(\Omega, \mathcal{C}, \xi)$ *is an atomistic state property system.*

Proof: First we remark that η is well defined because if $\omega(p) = \omega(q)$, then $\xi(p) \cap \mathcal{C} = \xi(q) \cap \mathcal{C}$. Indeed, if $\vee s(\omega_i) \in \xi(p)$ then $p \in \kappa(\vee s(\omega_i)) = cl(\cup \omega_i)$ in the corresponding closure space $(\Sigma, \kappa(\mathcal{L}))$. Since $cl(\cup\omega_i)$ is not connected we have that $K_{\mathbf{Cls}}(p) = \omega(p) = \omega(q) \subset cl(\cup\omega_i)$ so $q \in cl(\cup\omega_i) = \kappa(\vee s(\omega_i))$ and $\vee s(\omega_i) \in \xi(q)$. Now, since $\mathcal{C}$ is a sublattice of $\mathcal{L}$ it is a complete lattice with $1_\mathcal{C} = 1$ and $0_\mathcal{C} = 0$. By definition $\mathcal{C}$ is generated by its atoms $\{s(\omega) \mid \omega \in \Omega\}$. Clearly $0 \notin \eta(\omega(p))$ because $0 \notin \xi(p)$. If $a_i \in \eta(\omega(p)) = \xi(p) \cap \mathcal{C}$, $\forall i$, then $\wedge a_i \in \xi(p) \cap \mathcal{C} = \eta(\omega(p))$. Finally, let $a, b \in \mathcal{C}$ with $a <_\mathcal{C} b$. Let $\omega(p) \in \Omega$ with $a \in \eta(\omega(p))$. Thus $a \in \xi(p)$. $a <_\mathcal{C} b$ implies $a < b$. So we have $b \in \xi(p) \cap \mathcal{C} = \eta(\omega(p))$. Conversely, let $a, b \in \mathcal{C}$ and assume that $\forall p \in \Sigma$: $a \in \eta(\omega(p)) \Rightarrow b \in \eta(\omega(p))$. Then we have $\forall p \in \Sigma : a \in \xi(p) \Rightarrow b \in \xi(p)$.

Thus $a < b$ and $a <_{\mathcal{C}} b$. In order to prove that $(\Omega, \mathcal{C}, \eta)$ is an atomistic state property, we show that η is injective. So consider $p, q \in \Sigma$ such that $\omega(p) \neq \omega(q)$. Since $p \in \omega(p) = \kappa(s(\omega(p)))$, we have $s(\omega(p)) \in \xi(p) \cap \mathcal{C}$ and since $q \notin \omega(p) = \kappa(s(\omega(p)))$ we have $s(\omega(p)) \notin \xi(q) \cap \mathcal{C}$. This implies that $\xi(p) \cap \mathcal{C} \neq \xi(q) \cap \mathcal{C}$, i.e. $\eta(\omega(p)) \neq \eta(\omega(q))$. Thus η is injective and $(\Omega, \mathcal{C}, \eta)$ is an atomistic state property system. □

Theorem 14 *$(\Omega, \mathcal{C}, \eta)$ is a totally classical state property system, in the sense that the only pure nonclassical segments (i.e. segments with no proper classical elements) are trivial, i.e. $\{0, s(\omega)\}$.*

Proof: Suppose $[0, a]$ is a pure nonclassical segment of $\mathcal{C}$, then in the corresponding closure space $(\Sigma, \kappa(\mathcal{L}))$ the subset $\kappa(a)$ is connected hence $\kappa(a) \subset \omega$ for some $\omega \in \Omega$, hence $a < s(\omega)$. Since $s(\omega)$ is an atom, $a = s(\omega)$. Thus $[0, a] = \{0, s(\omega)\}$. □

Corollary 3 *The closure space associated with $(\Omega, \mathcal{C}, \eta)$ is a totally disconnected closure space.*

Summarizing the previous results we get:

Theorem 15 (decomposition theorem) *Any state property system $(\Sigma, \mathcal{L}, \xi)$ can be decomposed into:*

- *a number of pure nonclassical state property systems $(\Sigma_\omega, \mathcal{L}_\omega, \xi_\omega), \omega \in \Omega$*
- *and a totally classical state property system $(\Omega, \mathcal{C}, \eta)$*

Thus the decomposition of a closure space into its maximal connected components yields a way to decompose a state property system $(\Sigma, \mathcal{L}, \xi)$ into pure nonclassical state property systems $(\Sigma_\omega, \mathcal{L}_\omega, \xi_\omega), \omega \in \Omega$. In the context of closure spaces the maximal connected components are subspaces of the given space. However we do not yet have that the pure nonclassical state property systems $(\Sigma_\omega, \mathcal{L}_\omega, \xi_\omega)$ are subsystems of $(\Sigma, \mathcal{L}, \xi)$. To show this we introduce a new concept of subsystem.

5 Closed Subspaces and *ap*-Subsystems

Definition 12 (AP-subsystem) *Let $(\Sigma, \mathcal{L}, \xi)$ be a state property system and let $a \in \mathcal{L}$. Consider the following:*

- $\Sigma' = \kappa(a)$
- $\mathcal{L}' = [0, a]$
- $\xi' = \xi_{|\Sigma'}$

We now have a new state property system $(\Sigma', \mathcal{L}', \xi')$ which we shall call an 'actual property' (ap-) subsystem of $(\Sigma, \mathcal{L}, \xi)$ generated by a .

The name 'actual property' subsystem comes from the physical interpretation of this construction: give a property a of the physical system, we consider only those states Σ' for which a is always actual.

Theorem 16 *Let $(\Sigma', \mathcal{L}', \xi')$ be an ap-subsystem of $(\Sigma, \mathcal{L}, \xi)$, generated by a. Consider the corresponding closure spaces $(\Sigma', \kappa(\mathcal{L}'))$ and $(\Sigma, \kappa(\mathcal{L}))$, we have that $(\Sigma', \kappa(\mathcal{L}'))$ is a closed subspace of $(\Sigma, \kappa(\mathcal{L}))$.*

Proof: Follows immediately from the definition. □

Theorem 17 *Consider a closed subspace $(\Sigma', \mathcal{F}')$ of the closure space $(\Sigma, \mathcal{F})$, we have that $(\Sigma', \mathcal{F}', \bar{\xi}')$ is an ap-subsystem of $(\Sigma, \mathcal{F}, \bar{\xi})$ generated by Σ'.*

Proof: Follows immediately from the definition. □

¿From the above two theorems we see that *ap*-subsystems correspond exactly to closed subspaces of the associated closure space.

Any closed subspace Σ' of a closure space $(\Sigma, \mathcal{F})$ induces in a natural way a canonical inclusion map:

$$i : (\Sigma', \mathcal{F}') \to (\Sigma, \mathcal{F})$$

which in turn, by the functional equivalence between the category of closure spaces and state property systems gives a morphism:

$$(i, i^{-1}) : (\Sigma', \mathcal{F}', \bar{\xi}') \to (\Sigma, \mathcal{F}, \xi)$$

Theorem 18 *Let $(\Sigma', \mathcal{L}', \xi')$ be an ap-subsystem of $(\Sigma, \mathcal{L}, \xi)$, generated by a. We now define the following maps:*

$$m : \Sigma' \to \Sigma : p \mapsto p$$

$$n : \mathcal{L} \to \mathcal{L}' : c \mapsto a \wedge c$$

then $(m, n) : (\Sigma', \mathcal{L}', \xi') \to (\Sigma, \mathcal{L}, \xi)$ is a morphism in the category of state property systems which reduces to the canonical inclusion between the underlying closure spaces.

Proof: We have to show that for $c \in \mathcal{L}$ and $p' \in \Sigma' : c \in \xi(m(p')) \Leftrightarrow n(c) \in \xi'(p')$. Let's start with $c \in \xi(m(p')) \Leftrightarrow c \in \xi(p') \Leftrightarrow c \in \xi'(p')$. Because $\kappa(a) = \Sigma'$ we know that $a \in \xi'(p') = \xi(p')$, therefore $n(c) = c \wedge a \in \xi'(p')$. Conversely, if $n(c) = c \wedge a \in \xi'(p')$ then $p' \in \kappa'(c \wedge a) = \kappa'(c) \cap \kappa'(a) = \kappa'(c) \cap \Sigma' = \kappa'(c)$ therefore $c \in \xi'(p)$. □

We shall apply these results to the pure nonclassical state property systems $(\Sigma_\omega, \mathcal{L}_\omega, \xi_\omega), \omega \in \Omega$ that we have introduced in the previous section. Recall that we started with a state property system $(\Sigma, \mathcal{L}, \xi)$ with associated closure space $(\Sigma, \kappa(\mathcal{L}))$. By means of the connection relation on $(\Sigma, \kappa(\mathcal{L}))$ we obtained a partition $\Omega = \{\omega(p) = K_{\mathbf{Cls}}(p) | p \in \Sigma\}$ of Σ. Moreover each $w \in \Omega$ with $\omega = \omega(p) = K_{\mathbf{Cls}}(p)$ was a closed subset of $(\Sigma, \kappa(\mathcal{L}))$. Hence there was an $a = s(\omega)$ such that $\kappa(a) = \omega$. We will now use this property $a = s(\omega)$ to create an *ap*-subsystem.

$$\Sigma' = \kappa(a) = \omega$$

$$\mathcal{L}' = [0, a] = [0, s(\omega)]$$

$$\xi' = \xi_{|\Sigma'}^{|\mathcal{L}'} : p' \mapsto \xi(p) \cap \mathcal{L}'$$

We easily see that for an $\omega \in \Omega$ this *ap*-subsystem is in fact $(\Sigma_\omega, \mathcal{L}_\omega, \xi_\omega)$. Let

$$m : \Sigma_\omega \rightarrow \Sigma : p \mapsto p$$

$$n : \mathcal{L} \rightarrow \mathcal{L}_\omega : c \mapsto s(\omega) \wedge c$$

then $(m, n) : (\Sigma', \mathcal{L}', \xi') \rightarrow (\Sigma, \mathcal{L}, \xi)$ is a morphism in the category of state property systems which reduces to the canonical inclusion between the underlying closure spaces. In this way $(\Sigma_\omega, \mathcal{L}_\omega, \xi_\omega), \omega \in \Omega$ is always an *ap*-subsystem of $(\Sigma, \mathcal{L}, \xi)$.

6 The D-classical Part of a State Property System

In this section we want to show how it is possible to extract the d-classical part of a state property system. First of all we have to define the d-classical property lattice related to the entity S that is described by the state property system $(\Sigma, \mathcal{L}, \xi)$.

Definition 13 (D-classical Property Lattice) *Consider a state property system* $(\Sigma, \mathcal{L}, \xi)$. *We call* $\mathcal{C}' = \{\wedge_i a_i | a_i \text{ is a d} - \text{classical property}\}$ *the d-classical property lattice corresponding to the state property system* $(\Sigma, \mathcal{L}, \xi)$.

Theorem 19 *$\mathcal{C}'$ is a complete lattice with the partial order relation and infimum inherited from $\mathcal{L}$ and the supremum defined as follows: for $a_i \in \mathcal{C}'$, $\vee_i a_i = \wedge_{b \in \mathcal{C}', a_i \leq b \, \forall i} \, b$.*

Remark that the supremum in the lattice $\mathcal{C}'$ is not the one inherited from $\mathcal{L}$.

Theorem 20 *Consider a state property system $(\Sigma, \mathcal{L}, \xi)$. Let $\xi'(q) = \xi(q) \cap \mathcal{C}'$ for $q \in \Sigma$, then $(\Sigma, \mathcal{C}', \xi')$ is a state property system which we shall refer to as the d-classical part of $(\Sigma, \mathcal{L}, \xi)$.*

Proof: Clearly $0 \notin \xi'(p)$ for $p \in \Sigma$. Consider $a_i \in \xi'(p)\ \forall i$. Then $a_i \in \xi(p) \cap \mathcal{C}'\ \forall i$, from which follows that $\wedge_i a_i \in \xi(p) \cap \mathcal{C}'$ and hence $\wedge_i a_i \in \xi'(p)$. Consider $a, b \in \mathcal{C}'$. Let us suppose that $a \leq b$ and consider $r \in \Sigma$ such that $a \in \xi'(r)$. This means that $a \in \xi(r) \cap \mathcal{C}'$. From this follows that $b \in \xi(r) \cap \mathcal{C}'$ and hence $b \in \xi'(r)$. On the other hand let us suppose that $\forall r \in \Sigma : a \in \xi'(r)$ then $b \in \xi'(r)$. Since $a, b \in \mathcal{C}'$, this also means that $\forall r \in \Sigma : a \in \xi(r)$ then $b \in \xi(r)$. From this follows that $a \leq b$. □

Since $(\Sigma, \mathcal{C}', \xi')$ is a state property system, it has a corresponding closure space $(\Sigma, \kappa(\mathcal{C}'))$. In order to check some property of this space we introduce the following concepts.

Definition 14 (Weakly Zero-dimensional Closure Space) *Let $(X, \mathcal{F})$ be a closure space and $\mathcal{B} \subset \mathcal{F}$. $\mathcal{B}$ is called a base of $(X, \mathcal{F})$ iff $\forall F \in \mathcal{F} : \exists B_i \in \mathcal{B} : F = \cap B_i$. $(X, \mathcal{F})$ is called weakly zero-dimensional iff there is a base consisting of clopen sets.*

Theorem 21 *The closure space $(\Sigma, \kappa(\mathcal{C}'))$ corresponding to the state property system $(\Sigma, \mathcal{C}', \xi')$ is weakly zero-dimensional.*

Proof: To see this recall that a is classical iff $\kappa(a)$ is clopen in $(\Sigma, \kappa(\mathcal{L}))$, hence $\kappa(\mathcal{C}')$ is a family of closed sets on Σ which consists of all intersections of the clopen sets of $(\Sigma, \kappa(\mathcal{L}))$. □

In general $(\Sigma, \mathcal{C}', \xi')$ does not need to be atomistic, hence it is different from the totally classical state property system $(\Omega, \mathcal{C}, \eta)$ associated with $(\Sigma, \mathcal{L}, \xi)$. To illustrate this we give an example.

Let's consider the following state property system.

$$\Sigma = \{p, q, r, s, t\}$$
$$\mathcal{L} = \{0, a, b, c, d, I\}$$
$$\xi : \Sigma \to \mathcal{P}(\mathcal{L})$$

with $\xi(p) = \xi(q) = \{b, d, I\}, \xi(r) = \{a, d, I\}$ and $\xi(s) = \xi(t) = \{c, I\}$. The structure for the lattice $\mathcal{L}$ is given by figure 1.

The corresponding closure space (see figure 2) is

$$\Sigma = \{p, q, r, s, t\}$$
$$\kappa(\mathcal{L}) = \{\emptyset, \{r\}, \{p, q\}, \{s, t\}, \{p, q, r\}, \Sigma\}$$

Determining the connectedness components in this closure space, we find the following:

$$K_{\mathbf{Cls}}(p) = K_{\mathbf{Cls}}(q) = \{p, q\}$$

$$K_{\mathbf{Cls}}(r) = \{r\}$$

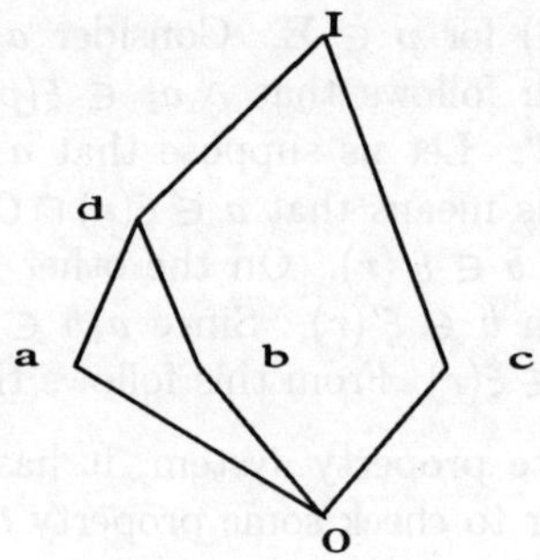

Figure 1. The lattice $\mathcal{L}$

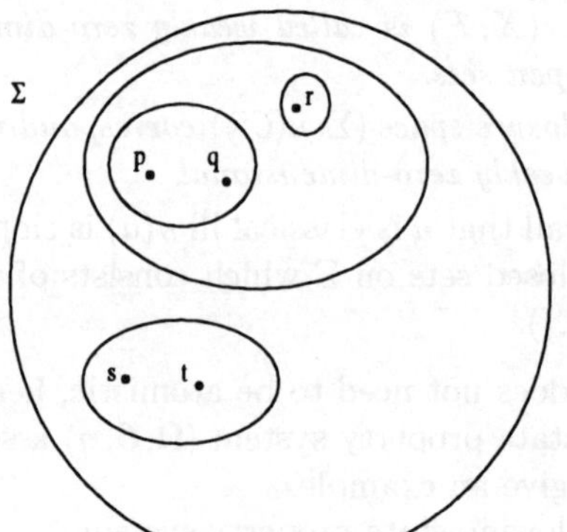

Figure 2. The closure space $\Sigma, \kappa(\mathcal{L})$

$$K_{\mathbf{Cls}}(s) = K_{\mathbf{Cls}}(t) = \{s, t\}$$

We have three pure nonclassical state property systems: $(\Sigma_{\omega_1}, \mathcal{L}_{\omega_1}, \xi_{\omega_1})$, $(\Sigma_{\omega_2}, \mathcal{L}_{\omega_2}, \xi_{\omega_2})$ and $(\Sigma_{\omega_3}, \mathcal{L}_{\omega_3}, \xi_{\omega_3})$.

$$\Sigma_{\omega_1} = \{p, q\}, \ \mathcal{L}_{\omega_1} = [0, b]$$
$$\Sigma_{\omega_2} = \{r\}, \quad \mathcal{L}_{\omega_2} = [0, a]$$
$$\Sigma_{\omega_3} = \{s, t\}, \ \mathcal{L}_{\omega_3} = [0, c]$$

$$\xi_{\omega_1}(p) = \xi_{\omega_1}(q) = \{b\}$$
$$\xi_{\omega_2}(r) = \{a\}$$
$$\xi_{\omega_3}(s) = \xi_{\omega_3}(t) = \{c\}$$

The atomistic totally classical state property system $(\Omega, \mathcal{C}, \eta)$ is given by:

$$\Omega = \{\{p,q\},\{r\},\{s,t\}\}$$
$$\mathcal{C} = \mathcal{L}$$
$$\eta : \Omega \to \mathcal{P}(\mathcal{C})$$

where $\eta(\{p,q\}) = \{b,d,1\}$, $\eta(\{r\}) = \{a,d,1\}$ and $\eta(\{s,t\}) = \{c,1\}$. The classical part is given by $(\Sigma, \mathcal{C}', \xi')$ where

$$\xi'(p) = \xi(p) \cap \mathcal{C}' \text{ for } p \in \Sigma$$

$$\mathcal{C}' = \{0, c, d, I\}$$

Acknowledgments

Part of the research for this article took place in the framework of the bilateral Flemish-Polish project 127/E-335/S/2000. D. Deses is Research Assistent at the FWO Belgium.

References

1. D. Aerts, E. Colebunders, A. Van der Voorde and B. Van Steirteghem, "State property systems and closure spaces: a study of categorical equivalence", *Int. J. Theor. Phys.* **38**, 359-385 (1999).
2. D. Aerts, " Foundations of quantum physics: a general realistic and operational approach", *Int. J. Theor. Phys.* **38**, 289-358 (1999).
3. D. Aerts, "Description of many physical entities without the paradoxes encountered in quantum mechanics", *Found. Phys.* **12**, 1131-1170 (1982).
4. D. Aerts, "Classical theories and non classical theories as special case of a more general theory", *J. Math. Phys.* **24**, 2441-2453 (1983).
5. D. Aerts, "Quantum structures, separated physical entities and probability", *Found. Phys.* **24**, 1227-1259 (1994).
6. C. Piron, *Foundations of Quantum Physics*, W. A. Benjamin, Reading, Mass. (1976).
7. C. Piron, "Recent developments in quantum mechanics", *Helv. Phys. Acta* **62**, 82 (1989).
8. C. Piron, *Mécanique Quantique: Bases et Applications*, Presses Polytechniques et Universitaires Romandes, Lausanne (1990).
9. A. Van der Voorde, "A categorical approach to T_1 separation and the product of state property systems", *Int. J. Theor. Phys.* **39**, 943 (2000).

10. D. Aerts, D. Deses and A. Van der Voorde, "Connectedness applied to closure spaces and state property systems", *Journal of Electrical Engineering*, **52**, 18-21 (2001).
11. D. Aerts, D. Deses and A. Van der Voorde, "Classicality and connectedness for state property systems and closure spaces", submitted to *International Journal of Theoretical Physics*.
12. A. Van der Voorde, *Separation axioms in extension theory for closure spaces and their relevance to state property systems*, PhD. Thesis, Vrije Universiteit Brussel (2001).

HIDDEN MEASUREMENTS FROM CONTEXTUAL AXIOMATICS

SVEN AERTS

Center Leo Apostel (CLEA) and
Department of Mathematics (FUND),
Brussels Free University, Pleinlaan 2,
1050 Brussels, Belgium
E-mail: saerts@vub.ac.be

The hidden measurement approach was originally proposed to show how the tenet of determinism can be upheld in quantum mechanics if one allows the outcome of a measurement to be decided not only by the state of the entity under observation, but also by unkown parameters in the experimental setup. In this paper we show that the structure of hidden measurements is uniquely recovered from three simple axioms and that the state of the entity can be partially defined by the intersection of the sets formed by the closure of each set of measurements that gives rise to a certain outcome.

1 Introduction

In an article by D. Aerts, a proposal to answer the question of the arisal of probabilities in quantum mechanics is given [1]. The author argues that probability enters quantum mechanics because of a lack of knowledge about which measurement was conducted. In a fairly simple and *ad hoc* way the article outlines a procedure that allows one to reproduce any set of probabilities related to a single measurement with n possible alternative (and mutually exclusive) outcomes. Special attention is given in the article to the spin 1/2 case, but we will only concern ourselves with the more general model with n possible outcomes. Let us give a brief description of the procedure as outlined in this article [1].

Suppose, as is done in orthodox quantum mechanics, that we describe a physical system by means of a vector in an n-dimensional Hilbert space $\mathcal{H}$. The n eigenvectors $\{v_1, v_2, ..., v_n\}$ with respective eigenvalues $\{a_1, a_2, ..., a_n\}$, corresponding to an observable A with n possible outcomes, serve as a basis for the state of the entity:

$$w = \sum_{i=1}^{n} \langle w, v_i \rangle v_i$$

The probabilistic rules of orthodox quantum mechanics dictate that the probability of finding the result a_i upon execution of the measurement that corresponds to the observable $\mathcal{A}$ when the entity is in the state w, equals

$$p_w^{\mathcal{A}}(a_i) \equiv p(\mathcal{A} = a_i \mid w) = \mid \langle w, v_i \rangle \mid^2$$

This means that the n-tuple

$$\Pi = (p_w^{\mathcal{A}}(a_1), p_w^{\mathcal{A}}(a_2), ..., p_w^{\mathcal{A}}(a_n))$$

contains all statistical information we can derive from the entity with respect to the observable $\mathcal{A}$ and as such we may use Π as a representation of the statistical state. Because the $p_w^{\mathcal{A}}(a_i)$ are constrained by the requirement $\sum_{i=1}^n p_w^{\mathcal{A}}(a_i) = 1$, we see that the statistical state is an element of the $(n-1)$-simplex $\mathcal{S}_{n-1}$ in $\mathbb{R}^n$ spanned by the canonical base vectors e_i : $\Pi = \sum_i p_w^{\mathcal{A}}(a_i) e_i$. The basic idea of the 'hidden measurement approach' is to associate with each measurement e a set of sub-measurements $e(\lambda)$ such that the measurement $e(\lambda)$ consists of choosing at random one of the λ and performing the measurement $e(\lambda)$ on the entity. The measurements are to be taken classically deterministic, that is, their operation on a fixed state always yields the same result. This is done as follows: take λ to be an n-tuple from the $(n-1)$-simplex:

$$\lambda = (\lambda_1, \lambda_2, ..., \lambda_n), \sum \lambda_i = 1, \lambda_i \geq 0$$

Now call C_i the convex closure of the set $\{e_1, .., e_{i-1}, \Pi, e_{i+1}, ..., e_n\}$. Clearly $\mathcal{S}_{n-1} = \cup_i C_i$. The outcome of the measurement $e(\lambda)$ is determined by λ in the following way: if $\lambda \in C_i$ then the outcome reads a_i. We will not discuss the procedure when the variable λ happens to be chosen on the boundary of one of the C_i as this has a zero probability of occurring and as such does not contribute to the final probabilities. The probability of choosing λ in the simplex C_i is calculated by assuming a uniform and normalized density for λ over $\mathcal{S}_{n-1}$. Hence we obtain

$$p(\lambda \in C_i | w) = m^n(C_i)/m^n(\mathcal{S}_{n-1})$$

where m^n denotes the trace on $\mathcal{S}_{n-1}$ of the Lebesgue measure on $\mathbb{R}^n$, which in this case is simply the weighted n-dimensional volume of the respective simplex. This volume is proportional to both the measure of any of its $(n-2)$-dimensional faces and to the length of the orthogonal projection of its 'height' onto this face. Hence we can easily see that the volumes of the simplexes are proportional to the projections of the statistical state Π onto the base vectors. It is a matter of straightforward determinant calculus to show that

$$m^n(C_i)/m^n(\mathcal{S}_{n-1}) = p_w^{\mathcal{A}}(a_i)$$

and hence we have

$$p(\mathcal{A} = a_i \mid w) = p(\lambda \in C_i|w)$$

The result is deceivingly simple and begs the question what precisely was shown. Its greatest merrit is perhaps that it is difficult to imagine a shorter exposition of the fact that there exist hidden variable models of quantum mechanics if one restricts the latter to measurements related to a single observable. This strength is immediately also the weakness of the exposition: the state of the entity is identified with the statistical state, or the set of probabilities related to a single observable, whereas in quantum mechanics we are able to apply Dirac transformations to calculate the probabilities related to *all* observables. Another question, raised at the end of [1], is how to characterize the measurements that occur in a hidden measurement scheme. In this article we will show how to cope with these two issues at the same time. As we will make extensive use of elementary convex geometry, we will first introduce the most important basic concepts. Readers familiar with elementary convex geometry, in particular the notion of proper and strict separation, can skip the next section.

2 Prerequisites from Convex Geometry

For any pair x_1, x_2 in a vector space X , we call the *open segment* $]x_1, x_2[$ the set of points $x \in X$ satisfying $x = \lambda x_1 + (1 - \lambda)x_2$ with $0 < \lambda < 1$, $\lambda \in \mathbb{R}$, and we call the *closed segment* $[x_1, x_2]$ the set of points $x \in X$ satisfying $x = \lambda x_1 + (1 - \lambda)x_2$ with $0 \leq \lambda \leq 1$, $\lambda \in \mathbb{R}$. A convex set C is a set such that for every pair $x_1, x_2 \in C$, we have that the *segment* $[x_1, x_2] \subset C$. From this definition it follows immediately that $\mathbb{R}^d$ is convex and that the intersection of two convex sets is convex. Therefore, given an arbitrary set $S \subset \mathbb{R}^d$, it makes sense to talk about the smallest convex set that contains K, constructed by taking the intersection of all convex sets that contain the set. In this way we obtain the *closure* or the *convex hull* of the set, denoted by square brackets: $[S] \equiv \bigcap\{C_i \subset \mathbb{R}^d : C_i \text{ is convex}, S \subset C_i\}$. A basic theorem in convex geometry shows that if S is a point set, which will always be the case in this article, this definition reduces to taking special linear combinations. We state the theorem without proof:

Theorem 1 *The* convex hull $[K]$ *of a point set* $K = \{k_1, k_2, ..., k_n\} \subset \mathbb{R}^d$ *equals:*

$$[K] = \{k \in \mathbb{R}^d : k = \sum_{i \in \mathcal{I}} \lambda_i k_i | \lambda_i \geq 0, \sum_{i \in \mathcal{I}} \lambda_i = 1\}$$

Obviously, the closure operation is idempotent: $[[K]] = [K]$. We will also make use of the *affine hull* $[K]_{aff}$ of a point set $K = \{k_1, ..., k_n\} \subseteq \mathbb{R}^d$, which is defined exactly as the convex hull but with the positivity requirement of the λ_i dropped. The *dimension* of the affine hull of a set $dim([K]_{aff})$, equals one less than the minimal number of points that have the same affine hull. A set of n points is called *affinely independent* iff its affine hull has dimension $n-1$, that is iff every proper subset has a smaller affine hull. An (n-1)-dimensional affine set in $\mathbb{R}^n$ is called a *hyperplane.* With $\langle\ ,\ \rangle$ the standard inner product in $\mathbb{R}^n$, we have that the set

$$H = \{x : \langle x, b\rangle = \beta | x, b \in \mathbb{R}^n, \beta \in \mathbb{R}\}$$

represents a hyperplane. It works the other way around too: every hyperplane can be represented in this way, with b and β defined uniquely up to a common non-zero multiple. Hyperplanes are very important because they play a role dual to points in $\mathbb{R}^n$and because of their appearance in separation theorems. Separation is based on the fact that a hyperplane in $\mathbb{R}^n$ divides the space $\mathbb{R}^n$ in two half-spaces in the sense that the complement of the hyperplane in $\mathbb{R}^n$ is the union of two disjoint open convex sets, which we call the open half-spaces associated with the hyperplane H. Stated in terms of the linear equality that defines the hyperplane, this corresponds to the two sets $R_1 = \{x : \langle x, b\rangle \leq \beta \mid x, b \in \mathbb{R}^n, \beta \in \mathbb{R}\}$ and $R_2 = \{x : \langle x, b\rangle \geq \beta \mid x, b \in \mathbb{R}^n, \beta \in \mathbb{R}\}$. If C_1 and C_2 are two sets in $\mathbb{R}^n$, then a hyperplane H is said to *separate* C_1 and C_2 iff C_1 is contained in one of the closed half-spaces associated with H and C_2 lies in the opposite closed half-space. The hyperplane is said to separate C_1 and C_2 *properly* iff H separates C_1 and C_2 and they are not both contained in H. The hyperplane is said to separate C_1 and C_2 *strictly* iff H separates C_1 and C_2 and both have an empty intersection with H. A generalized notion of tangency can be expressed by means of supporting hyperplanes and supporting half-spaces. If C is a convex set in $\mathbb{R}^n$, a *supporting half-space* to C is a closed half-space which contains C and has a point of C in its boundary. A *supporting hyperplane* is a hyperplane which is the boundary of a supporting half-space to C. The *relative interior* $ri(K)$ of a point set K is the set $ri(K) = \{k = \sum_{i\in\mathcal{I}} \lambda_i k_i \mid \lambda_i > 0, \sum_{i\in\mathcal{I}} \lambda_i = 1\}$, and the *boundary* of a convex set $[K]$, denoted ∂K, is the set $[K] \setminus ri(K)$. We are now in a position to state two elementary separation theorems that are intuitively obvious, and yet provide an essential tool in much of convex geometry. We will make extensive use of them in what follows.

Theorem 2 *Let C_1 and C_2 be non-empty convex sets. In order that their exist a hyperplane separating C_1 and C_2 properly, it is necessary and sufficient* *that $ri(C_1)$ and $ri(C_2)$ have no point in common.*

Theorem 3 *Let C_1 and C_2 be disjoint non-empty convex sets and let C_1 be compact and C_2 be closed. Then their exist a hyperplane separating C_1 and C_2 strictly.*

We will omit the proofs; they can be found in any textbook [5,6] on convex geometry. Note that the convex hull of a point set K is compact, since each element of $[K]$ is the image of a single valued, continuous mapping (i.e.. $\sum_{i \in I} \lambda_i k_i$) of a compact set ($\lambda_i \geq 0, \sum_{i \in I} \lambda_i = 1$) into $\mathbb{R}^n$. Also important is a *face* of a convex set C. A face is a convex set $C' \subseteq C$ such that every closed segment in C with a relative interior point in C' has both endpoints in C' The term face is due to the fact that, in case the convex set is a polyhedron, its faces would have exactly the properties listed above. To get some idea of the meaning of the concept of a face in higher dimensions, one can think of a 'maximal flat piece of arbitrary dimension of the boundary' of a convex set. The empty set and C itself are examples of faces of C . The *dimension of a face* $\mathcal{F}$ is defined as the dimension of its affine hull: $dim(\mathcal{F}\ ) \equiv dim([\mathcal{F}]_{aff})$.We call the zero-dimensional faces of C the *extreme points* of C , the one-dimensional faces we call the *edges*. We now turn to the most elementary bodies in convex geometry: simplexes.

Definition 1 *A* d-simplex, *denoted by $\mathcal{S}_d$, is the convex hull of any d+1 affinely independent points in $\mathbb{R}^n, n \geq d$*

The zero-dimensional faces of a simplex are called *vertices* and these are simply the affinely independent points that generate the simplex. The highest dimensional face of a simplex is the simplex itself. A zero-dimensional simplex is a point, a one-dimensional simplex is a line segment, a two dimensional simplex is a triangle and a three-dimensional simplex is a tetrahedron. Simplexes share another nice property: an arbitrary face of a simplex is again a simplex (of course of lower dimension). This inspires us to define a canonical way of partitioning simplexes, that we will make use of at the end of the article:

Definition 2 *Let $\mathcal{S}_n = [k_1, k_2, ..., k_{n+1}]$ be an n-simplex. A family $\mathcal{T}$ of open simplexes $\mathcal{T}^i$, with $F(\mathcal{T}^i)$ the set of faces of $\mathcal{T}^i$, is called a* triangulation *iff*

$$\cup \mathcal{T}^i = \mathcal{S}_n$$

$$i \neq j \Rightarrow \mathcal{T}^i \cap \mathcal{T}^j = \emptyset$$

$$F(\mathcal{T}^i) \subset \mathcal{T}$$

Triangulations can be defined for far more general structures than just simplexes. However, we will only make use of simplex triangulations.

3 Basic Assumptions

We start by assuming the existence of three non-empty sets: a set of *states* Σ, a set of *measurements* $\mathcal{M}$ and a set of *outcomes* X. We will identify the state of an entity as a mathematical object that allows one to derive the statistical properties of an ensemble with a fixed preparation procedure with respect to any set of measurements. Note that according to this definition we can only attribute the concept 'state' to a single entity as the state attributed to a member of an ensemble consisting of identically prepared entities. Note also that a priori the representation of the state may depend on the number and nature of the various measurements one performs. An *experiment* then, consists of choosing a measurement $e \in \mathcal{M}$ and performing it on a system $q \in \Sigma$. Hence an experiment e is an element of the product set $E = \mathcal{M} \times \Sigma$. Every experiment will be assumed to produce a certain physical phenomenon to which a specific *outcome* $x \in X$ is ascribed. How we ascribe an outcome to the physical phenomenon produced by the experiment e, depends on the particular measurement situation, but is in most cases at least to a certain extent arbitrary and decided by the experimenter. Indeed, one may argue that 'reading the outcome of an experiment' ultimately depends on another measurement, and so on *ad infinitum*. However, for reasons of consistency and verifiability, the experimenter has to lay down a set of rules for deciding what outcome occurred as a result of performing the measurement e on the state q. If another experimenter uses the same rules of preparation and measurement and gets exactly the same result, he should assign the very same outcome to the experiment. This is precisely the form in which we will leave this problem: the outcome is whatever the experimenter has decided the outcome to be, on the condition that this obeys the above stated requirement of consistency. Given the fact that any physical measurement is limited in precision, there will always be only a limited number of outcomes possible and, as a consequence, the set of outcomes X is finite. Philosophically, we will now make a bold step and postulate the universe of our experiments to be deterministic. Whenever we perform the same experiment on the same state, we will always obtain the same outcome. One may argue that precisely this classically trivial fact fails in quantum mechanics. But we rescue this particular form of 'quantum-indeterminism' by recognizing that it may very well be beyond the capabilities of the experimenter to control all relevant parameters of the measurement set-up. So whenever we have a statistical pattern in the outcomes, we will interpret the arising probability as a consequence of a very specific lack of knowledge, that we will identify later. But how do a given state and measurement produce the physical phenomenon to which we attribute a specific outcome?

4 Axioms for Contextual Interactions

The physical picture emerging is the following: the state of the entity and the state of the measurement apparatus interact with each other. It is a logical necessity, both in quantum mechanics and in classical physics, that for an experiment to produce an outcome that is related to the entity, the state of the entity under consideration needs to change the state of the measurement apparatus. Obviously, there are important cases for which this relationship is reciprocal in the sense that the state of the entity is also changed by the act of measurement. However, the specific interaction that takes place may depend on many unknown factors. All we get to know are the outcomes! It is this feature that is essential in our approach: that a lack of knowledge about these interaction parameters is responsible for the fluctuations in the outcomes. Of course, the main features of our approach can be traced back immediately to the hidden measurement approach[1,2,3,4]. How will we describe this interaction and the way in which this produces a specific outcome? We will model the differences in the interaction by assuming that with each measurement e there corresponds a set Λ of not directly measurable influences λ, called a *contextual influence* or *context* for short. A contextual influence λ represents a certain *tendency* towards different outcomes. In a sense, it is the sum total of everything that somehow influences the outcome of the measurement. However, contextual influence is not all. It is the context corresponding to a measurement e, together with a state q that determine the measurement outcome. In the founding papers on hidden measurements [1,2,3,4], the contextual influence is called a 'hidden measurement' and assumed to consist out of sub-measurements associated with a measurement. In accordance with this tradition, we could write $e = e(\lambda)$, but for ease of notation, we will drop reference to the measurement e and simply talk about the context λ. This does not give rise to any sort of ambiguity, because throughout the rest of this article, the measurement $e \in \mathcal{M}$ and the state $q \in \Sigma$ are fixed but arbitrary. The use of the word context will be justified at the end of this paper. Mathematically, the requirement of determinism implies that there exists a *single-valued* mapping τ from the set of states and the set of contexts to the set of outcomes. We will also require that our description is complete in the sense that every outcome is necessarily produced by a context and a state. Hence the mapping τ is *surjective*. These two requirements constitute our first axiom. With $\tau : \Lambda \times \Sigma \to X$, we get

Axiom 1 (determinism)

$$\forall(\lambda, q), (\lambda', q') \in \Lambda \times \Sigma : \tau(\lambda, q) \neq \tau(\lambda', q') \Rightarrow (\lambda, q) \neq (\lambda', q')$$

$$\forall x \in X, \exists(\lambda, q) \in \Lambda \times \Sigma : \tau(\lambda, q) = x$$

As we have said, different contextual influences reflect different tendencies towards a certain outcome. Hence a context relates in some way to all outcomes and there should be in each context some indication of just how much this context relates to each separate outcome. Let us consider the set of all linear combinations of the outcomes as a candidate for the set of contexts. Mathematically, this amounts to constructing the free vector space generated by the outcomes. Indeed, let $\mathcal{F}$ be a field and X a non-empty set. We call the *free vector space* F_X over $\mathcal{F}$ with X as basis, the set defined by

$$F_X = \{\sum_{i \in I} \lambda_i x_i | r_i \in \mathcal{F} \text{ and } x_i \in X\}$$

that satisfies the usual rules of addition and multiplication: $rw_i + sw_i = (r+s)w_i$ and $r(sw_i) = (rs)w_i$. Let us postulate that the set Λ should be in the free vector space (over the field of the reals) generated by the set of outcomes: $\Lambda \subset F_X$. To further pin down the structure of Λ, we require that, given a specific context $\sum_{i \in I} \lambda_i x_i$, it is impossible to determine another context $\sum_{i \in I} \lambda'_i x_i$ such that the new context increases (or decreases) its tendency towards *all* the outcomes. The reason why this is the case, is simply because a measurement is going to produce an outcome, regardless of the specific context. Hence it seems superfluous to consider a context that has a greater (or lesser) tendency with respect to *all* outcomes. In other words: only relative tendencies count. But how does a tendency relate to the λ_i? Let us assume that the tendency of a context λ towards a certain outcome x is a fixed and strictly monotone function of the Euclidean distance in F_X between λ and x, and is such that it is impossible for a context λ to have a greater tendency towards *all* outcomes than another context λ'. The strict monotone function of the Euclidean distance means that the set Λ is such that it is impossible to move further away from (or closer to) all the outcomes (base vectors) at the same time. The following theorem shows that this requirement characterizes the points of the convex hull of the outcomes. With $d(\cdot, \cdot)$ the Euclidean distance which is well defined in F_X, we can state the following

Theorem 4 *Let* $\lambda \in \Lambda, \lambda' \in \Lambda$, *and* $X = \{x_1, ..., x_n\}$. *If there do not exist* $\lambda, \lambda' \in \Lambda$ *such that* $d(\lambda, x_i) > d(\lambda', x_i)$ *for all* $i \in 1, ..., n$, *then* $\Lambda \subset [X]$

Proof: First, assume that $\lambda \notin [X]$. Then there exists a hyperplane $\mathcal{A}$ separating λ from $[X]$. Let λ' be the orthogonal projection of λ onto $\mathcal{A}$. For this λ' we have $d(\lambda', x) < d(\lambda, x)$ for every $x \in X$ by the triangle inequality $|\lambda - \lambda'| + |\lambda' - x| \geq |\lambda - x|$. If on the other hand $\Lambda \in [X]$, then the theorem says no $\lambda, \lambda' \in \Lambda$ exist such that $d(\lambda, x_i) > d(\lambda', x_i)$ for all i. We proceed ad absurdum, assuming there do exist $\lambda, \lambda' \in [X]$ such that $d(\lambda, x_i) > d(\lambda', x_i)$ for all i. Let $\mathcal{H}$ be the hyperplane that orthogonally bisects the segment $[\lambda, \lambda']$. If any of the x_i is on the same side as λ with regard to $\mathcal{H}$, then $d(\lambda, x_i) < d(\lambda', x_i)$. Hence all x_i are on the same side as λ' with regard to $\mathcal{H}$. But then λ is separated from Λ, contradicting the assumption. □

Note that the dimension of the free vector space generated by X equals the number of outcomes. Hence we have found that a candidate for the set of contexts is the standard $(n-1)$-simplex in F_X, spanned by the outcome set: $\Lambda = [X] = \mathcal{S}_{n-1}$. However, for reasons which will become clear at the introduction of axiom 3, we need to exclude the outcomes from Λ. Our second axiom then finally reads:

Axiom 2

$$\Lambda = [X] \backslash X$$

With $\mathcal{S}_{n-1} = [x_1, x_2, ..., x_n]$ and with $\mathcal{V}(\mathcal{S}_{n-1})$ the set of vertices of $\mathcal{S}_{n-1}$, Λ can also be written as $\mathcal{S}_{n-1} \backslash \mathcal{V}(\mathcal{S}_{n-1})$. To introduce our third axiom, we need the following definition:

Definition 3 (context pre-order) *We will say a context λ' has less tendency than λ towards outcome x_i and write $\lambda' <_{x_i} \lambda$, if $\lambda' \in [\lambda, x_i[$. A context λ' has a greater tendency than λ towards outcome x_i if $\lambda' \in [\lambda, x_j[$, $j \neq i$*

Clearly we have $\lambda <_{x_i} \lambda$ (reflexivity) and $\lambda <_{x_i} \lambda'$, $\lambda' <_{x_i} \lambda'' \Rightarrow \lambda <_{x_i} \lambda''$ (transitivity) and hence $<_{x_i}$ is indeed a pre-order relation. We will now impose an axiom that completely determines the structure of the context sets.

Axiom 3

$$\lambda, \lambda' \in \Lambda, q \in \Sigma, \tau : \Lambda \times \Sigma \rightarrow X :$$

$$\tau(\lambda, q) \neq x_i \Rightarrow \tau(\lambda', q) \neq x_i, \forall \lambda' : \lambda' <_{x_i} \lambda$$

This axiom states that if a context λ' has less tendency towards outcome x_i than the context λ, and if an experiment on a state q with context λ does not produce x_i, then neither will an experiment on a state q with context λ'. We now see why we excluded the set of outcomes in axiom 2, because otherwise axiom 1, the uniqueness of the outcome mapping τ, fails at the vertices of $\mathcal{S}_{n-1}$. For the same reason we used the half-open interval in the definition of the pre-order relation.

Corollary 1 *If an experiment gives rise to an outcome x_j, then the experiments with a greater tendency with respect to x_j, will give rise to x_j too, $k = 1, ..., j-1, j+1, ..., n$*

Proof: Trivial. □

Corollary 2 $\tau(\lambda, q) = x_j \Rightarrow \tau(\lambda', q) = x_j, \forall \lambda' \in]x_1, ..., x_{j-1}, \lambda, x_{j+1}, ..., x_n[$

In words this says that if an experiment (λ, q) gives rise to an outcome x_j, then experiments (λ', q) with contexts λ' that belong to the open $(n-1)$-simplex $\mathcal{S}_n =]x_1, x_2, ..., x_{j-1}, \lambda, x_{j+1}, ..., x_n[$ will give rise to x_j too.

Proof: Let $V_k = \{v_1, v_2, ..., v_k\}$ be a selection of $k \leq n-1$ distinct elements of $\{x_1, x_2, ..., x_{j-1}, x_{j+1}, ..., x_n\}$. Note that we make no notational distinction between sets and singletons in forming the closure of a set. Hence we write $[V_k, \lambda] \equiv [[V_k], \lambda] = [v_1, v_2, ..., v_k, \lambda]$. We will show that if $\tau(\lambda', q) = x_j, \forall \lambda' \in [V_k, \lambda]$ then $\tau(\lambda', q) = x_j, \forall \lambda' \in [V_{k+1}, \lambda]$, establishing the result by induction. By the first corollary we have $\tau(\lambda', q) = x_j, \forall \lambda' \in [V_1, \lambda] = [v_1, q]$. Next assume $\tau(\lambda', q) = x_j, \forall \lambda' \in [V_k, \lambda]$ and take $w \in [V_{k+1}, q]$. As all v's are affinely independent by definition, we can assume without loss of generality that q is affinely independent too (otherwise the induction step becomes superfluous), $[V_{k+1}, q]$ is in fact a simplex and we can write w as a unique linear combination $w = \alpha z + (1-\alpha)v_{k+1}$ with $0 < \alpha < 1$ where $z \in [V_k, q]$. Stated differently, $w \in]z, v_{k+1}[$. However, because $z \in [V_k, q]$ and v_{k+1} is nothing but one of the $\{x_1, x_2, ..., x_{j-1}, x_{j+1}, ..., x_n\}$, we obtain the desired result by direct application of the axiom. □

But not only the set of all measurements is convex. We can say more: the set of measurements on a fixed state that gives rise to one particular outcome, is convex too. In order to proof this, we will first demonstrate two little lemmas.

Lemma 1 *Suppose two n-simplexes $\mathcal{S}_n$ and $\mathcal{S}'_n$ in $\mathbb{R}^n$ share an (n-1)-face $\mathcal{A}$. A necessary and sufficient condition for $ri(\mathcal{S}_n) \cap ri(\mathcal{S}'_n) \neq \emptyset$ is that the two vertices v and v' of $\mathcal{S}_n$ and $\mathcal{S}'_n$ that do not lie in $\mathcal{A}$, lie in the same half-space generated by $\mathcal{H} = [\mathcal{A}]_{aff}$.*

Proof: Neither v nor v' can be contained in $\mathcal{H}$ for this would violate the assumption that $\mathcal{S}_n$ and $\mathcal{S}'_n$ are n-simplexes (hence, if separated, they are properly separated). If v and v' are in different half-spaces with respect to $\mathcal{H}$, then $\mathcal{H}$ simply becomes a plane of proper separation. For sufficiency, assume v and v' are in the same half-space with respect to $\mathcal{H}$. If there exists a hyperplane that properly separates $\mathcal{S}_n$ and $\mathcal{S}'_n$, it should contain the n vertices of $\mathcal{S}_n$ and $\mathcal{S}'_n$ that are in $\mathcal{H}$ and at least one point in the segment $[v, v']$, a total of $(n+1)$ affinely independent points. But the maximum number of affinely independent points that generate a hyperplane in $\mathbb{R}^n$ equals n. Hence we have $ri(\mathcal{S}_n) \cap ri(\mathcal{S}'_n) \neq \emptyset$. □

Lemma 2 *Given the n-simplex $\mathcal{S}_n = [x_1, x_2, ..., x_n, k]$ in $\mathbb{R}^n, (n > 1)$, that contains three open n-simplexes $\mathcal{S}_n^p =]x_1, ..., x_n, p[, \mathcal{S}_n^q =]x_1, ..., x_n, q[$ and $\mathcal{S}_n^w =]x_1, ..., x_{n-1}, k, w[$ where $w \in [p, q]$. Then $\mathcal{S}_n^w \cap (\mathcal{S}_n^p \cup \mathcal{S}_n^q) \neq \emptyset$*

Proof: Note that all x_i are distinct as they are $n+1$ in number and form the vertices of an n -simplex. If p and q are identical the lemma is trivial, hence assume they are not. First of all, we have that the $(n-2)$-simplex $\mathcal{S}_{n-2} = [x_1, ..., x_{n-1}]$, is a face of $[\mathcal{S}_n^p]$, $[\mathcal{S}_n^q]$ and $[\mathcal{S}_n^w]$ as is obvious from their definition. Construct the hyperplane $\Pi_{n-1} = [x_1, ..., x_{n-1}, w]_{aff}$. The complement of Π_{n-1} in $\mathbb{R}^n$ consists out of two disconnected half-spaces R_1 and R_2. If $p \in \Pi_{n-1}$ then, as $w \in \Pi_{n-1}$, so is q , and the lemma is proven. Hence assume, without loss of generality, that $p \in R_1$ and $q \in R_2$. Because $\mathcal{S}_n^w$ is an n-simplex, $k \notin \Pi_{n-1}$. Take then, for example, $k \in R_2$. Moreover, as Π_{n-1} contains all but 2 of the vertices of $\mathcal{S}_n$ and an interior point w of $\mathcal{S}_n$, it separates x_n and k . In summary, we have $x_n, p \in R_1$ and $k, q \in R_2$ and $w \in \Pi_{n-1}$. It then follows from the first lemma that $\mathcal{S}_n^w \cap \mathcal{S}_n^p = \emptyset$. As both $\mathcal{S}_n^p$ and $\mathcal{S}_n^w$ are convex and do not intersect, they are separated by the hyperplane Π_{n-1}. Now p lies in R_1, so $\mathcal{S}_n^p \subset R_1$. But $\mathcal{S}_n^q$ shares $\mathcal{S}_{n-2}$ with $\mathcal{S}_n^p$ and has a point q in R_2. Hence $\mathcal{S}_n^q$ intersects Π_{n-1}. Let us call the point of intersection r . Now we reapply the first lemma again but this time in the $(n-1)$ simplexes $]x_1, ..., x_{n-1}, r[$ and $]x_1, ..., x_{n-1}, w[$ that both lie in Π_{n-1}. The condition for applying the lemma is fulfilled because, obviously, w and r lie in the same halfplane of Π_{n-1} because both are in the interior of $\mathcal{S}_n$ and the other halfplane of Π_{n-1} is completely exterior to $\mathcal{S}_n$. Hence we have $\mathcal{S}_n^w \cap \mathcal{S}_n^q \neq \emptyset$. If we would have chosen $p \in R_2$ rather than $p \in R_1$ we would have obtained $\mathcal{S}_n^w \cap \mathcal{S}_n^p \neq \emptyset$. Either way we obtain $\mathcal{S}_n^w \cap (\mathcal{S}_n^p \cup \mathcal{S}_n^q) \neq \emptyset$ □

We are now in a position to proof the anticipated theorem:

Theorem 5 (Convexity) *The set of contexts giving rise to the same outcome form a convex set.*

Proof: Let λ_1 and λ_2 be two contexts such that $\tau(\lambda_1, q) = \tau(\lambda_2, q) = x_j$. By the second corollary, we have two simplexes of contexts $\mathcal{S}_{n-1}^{\lambda_1}$and $\mathcal{S}_{n-1}^{\lambda_2}$ that give rise to the same outcome: $\tau(\lambda, q) = x_j, \forall \lambda \in \mathcal{S}_{n-1}^{\lambda_1} \cup \mathcal{S}_{n-1}^{\lambda_2} =]x_1, x_2, ..., x_{j-1}, \lambda_1, x_{j+1}, ..., x_n[\cup]x_1, x_2, ...x_{j-1}, \lambda_2, x_{j+1}, ..., x_n[$. Now let w be an arbitrary element of $[\lambda_1, \lambda_2]$. We proceed ad absurdum. Suppose $\tau(w, q) = x_k \neq x_j$. Then all contexts belonging to $\mathcal{S}_{n-1}^w =]x_1, x_2, ..., x_{k-1}, w, x_{k+1}, ..., x_n[$ give the same outcome. However, the second lemma shows us that $\mathcal{S}_{n-1}^w$ necessarily intersects either $\mathcal{S}_{n-1}^{\lambda_1}$ or $\mathcal{S}_{n-1}^{\lambda_2}$ and this contradicts axiom 1, the single-valuedness of τ. Hence $\tau(w, q) = x_j, \forall$ $w \in [\lambda_1, \lambda_2]$. □

So we have n convex sets, one belonging to each outcome. However, the union of their closure is the simplex Λ. A beautiful theorem of Klee [7] shows that under these conditions, they intersect.

Theorem 6 (Klee) *Let $C_1, C_2, ..., C_n$ be closed convex sets whose union is a convex set. If the intersection of any $(n-1)$ is non-empty, then the intersection of all the C_j is non-empty.*

Proof: Since there exists a closed solid sphere B such that the sets $C_i \cap B$, $(i = 1, ..., n)$ satisfy the hypotheses of the theorem, we may assume that all C_i are compact. This allows the use of the second separation theorem. (Induction) Let C_1 and C_2 be non-empty compact convex sets whose union is convex. If they are disjoint, there exists a hyperplane $\mathcal{H}$ that separates them strictly. As there exist points of $C_1 \cup C_2$ on both sides of $\mathcal{H}$ and $C_1 \cup C_2$ is convex by assumption, then there exist points of $C_1 \cup C_2$ on $\mathcal{H}$, but this is in contradiction with $\mathcal{H}$ *strictly* separating C_1 and C_2. Suppose now that the result holds for $n = r$ convex sets. Put $C = \cap_{j=1}^r C_j$. By hypothesis $C \neq \emptyset$, $C_{r+1} \neq \emptyset$. Suppose these two sets are disjoint, $C_{r+1} \cap C = \emptyset$, then there exists a hyperplane $\mathcal{H}$ that separates them strictly. Writing $C'_j = C_j \cap \mathcal{H}$ we have:

$$\bigcup_{j=1}^{r} C'_j = \bigcup_{j=1}^{r} (C_j \cap \mathcal{H}) \cup (C_{r+1} \cap \mathcal{H}) = \mathcal{H} \cap (\bigcup_{j=1}^{r+1} C_j)$$

Being the intersection of two convex sets; $\bigcup_{j=1}^r C'_j$ is convex. But also the intersection of any $(r-1)$ of the sets C'_j is non-empty, as the intersection of any $(r-1)$ of the sets C_j is non-empty and has a non-empty intersection with both C and C_{r+1} and therefore with $\mathcal{H}$. But then we have $\cap_{j=1}^r C'_j = C \cap \mathcal{H} \neq \emptyset$, contradicting the fact that $\mathcal{H}$ separates C and C_{r+1} strictly. Hence $C \cap C_{r+1} \neq \emptyset$. □

To see the relevance of Klee's theorem to our paper, identify C_i with $[x_1, ..., x_{i-1}, x_{i+1}, ..., x_n]$. The union of all C_i is the $(n-1)$ -simplex, which is certainly convex. Recognize that every two C_i and C_j share at least $[x_1, ..., x_{i-1}, x_{i+1}, .., x_{j-1}, x_{j+1}, ..., x_n]$. Hence the intersection of any $(n-1)$ of the C_i can be written as $\cap_{i \neq k} C_i$ and certainly includes the vertex x_k. The conditions for Klee's theorem are fulfilled and we get that the closure of the convex sets of contexts have a non-empty intersection. It turns out that this intersection is in fact a singleton.

Theorem 7 (Uniqueness) *Let C be a collection of n non-empty closed convex sets $C = \{C_i \mid i = 1, ..., n\}$ with $\cup_i C_i = \mathcal{S}_{n-1} = [x_1, ..., x_n]$ and $ri(C_i) \cap ri(C_j) = \emptyset$. Let H_{ij} be the hyperplane that separates C_i and C_j, and let $C_i \cap C_j \subset H_{ij}$. Then the intersection $\cap_i C_i$ is a singleton.*

Proof: Call σ the intersection of the C_i: $\cap_i C_i = \sigma$. We know by the previous theorem that σ is non-empty, and because it is the intersection of convex sets, it is itself convex. By the first separation theorem and because $ri(C_i) \cap ri(C_j) = \emptyset$, all C_i are properly separated by a set of hyperplanes $H = \{H_{ij} | i, j = 1, ..., n; i \neq j; H_{ij} = H_{ji}$ separates C_i from $C_j\}$, a set of $n(n-1)/2$ hyperplanes in total. Note that $\cap_i C_i = \cap_{i,j} H_{ij}$. We know by corollary 2 that $[x_1, ..., x_{i-1}, x_{i+1}, ..., x_n] \subset C_i$ and likewise $[x_1, ..., x_{j-1}, x_{j+1}, ..., x_n] \subset C_j$. As such we have that $[x_1, ..., x_{i-1}, x_{i+1}, ..., x_{j-1}, x_{j+1}, ..., x_n] \subset H_{ij}$. Hence we have identified $(n-2)$ linear independent points for each hyperplane. With $\dim(H_{ij}) = (n-2)$, identifying one more point in $\mathcal{S}_{n-1}$ that belongs to each H_{ij} effectively fixes the hyperplanes. Let us call this point ρ and choose it to be in σ. Note that ρ, being the intersection of $n(n-1)/2$ hyperplanes of dimension $n-2$, lies in the interior of $\mathcal{S}_{n-1}$ because otherwise (depending on the dimension of the face in which the intersection lies) at least $(n-1)(n-2)/2$ of these hyperplanes will coincide and the remaining $(n-1)$ hyperplanes will separate $\mathcal{S}_{n-1}$ into at most $n-1$ sets C_i, leaving at least one C_i empty, contrary to the assumption that all C_i are non-empty.. Now ρ, being the intersection of hyperplanes, is also an affine space of lower dimension. Hence if ρ is not a singleton, then it is at least a line in $\mathcal{S}_{n-1}$. But then ρ intersects the boundary of $\mathcal{S}_{n-1}$ in at least two points, call one of them y. In the best case y is in the relative interior of a proper face of $\mathcal{S}_{n-1}$, say $y \in]x_1, ..., x_{m-1}, x_{m+1}, ..., x_n[$. Hence y is independent from $[x_1, ..., x_{m-1}, x_{m+1}, ..., x_{r-1}, x_{r+1}, ..., x_n]$ and this y can be chosen to fix the $(n-1)(n-2)/2$ hyperplanes that contain these vertices. But then these hyperplanes coincide, again leaving one of the C_i empty, contrary to the assumption. If y is not in the relative interior of a proper face, then it is in a face of lower dimension, but then even more hyperplanes coincide. Hence the intersection is a singleton in the interior of $\mathcal{S}_{n-1}$. □

The non-emptiness of the C_i is in our case guaranteed by axiom 1, the surjectivity of $\tau(\lambda, q)$. The division of Λ into separate sets takes the form of a special type of triangulation, which is in a sense a simple generalization of a barycentric division, and is in fact affinely isomorphic to it. Imagine a point σ inside an $(n-1)$-simplex $\mathcal{S}_{n-1}$. The simplex $\mathcal{S}_{n-1}$ can be triangulated into simplexes having as vertices σ and $(n-1)$ of the n vertices of $\mathcal{S}_{n-1}$. Each $y \in \mathcal{S}_{n-1}$ obviously lies on some line segment joining ρ with a relative boundary point z of $\mathcal{S}_{n-1}$. As z lies in this boundary, it can be written as a convex combination of $n-1$ vertices of $\mathcal{S}_{n-1}$, say $x_1, x_2, ..., x_{n-1}$. Obviously y and $x_1, x_2, ..., x_{n-1}$ are affinely independent and the simplex they generate contains σ. Hence the following definition makes sense:

Definition 4 *Let $\mathcal{S}_{n-1}$ be an $(n-1)$-simplex and $\sigma \in ri(\mathcal{S}_n)$. We call a σ-*

induced triangulation $\mathcal{T}(\sigma)$ *of* $\mathcal{S}_{n-1}$ *a triangulation* $\mathcal{T}$ *such that every* $(n-1)$-*simplex in* $\mathcal{T}^i$ *contains* σ *as a vertex. We will denote by* $\mathcal{T}_r(\sigma)$ *the elements of* $\mathcal{T}$ *that are r-dimensional,* $r \leq n-1$

We can now easily recognize the partition of Λ, as established by the last theorem, as the (n-1)-dimensional elements $\mathcal{T}_{n-1}(\sigma)$ of the σ-induced triangulation, and these are the relative interiors of the $(n-1)$-simplexes obtained by closure of the set $\{x_1, ..., x_{i-1}, \sigma, x_{i+1}, ..., x_n\}$.

5 Conclusion

It took some effort to realize and show that the three axioms uniquely define a single point in the interior. It is a lot easier to see that, given the point $\sigma \in ri(\mathcal{S}_{n-1})$, we have effectively fixed what context will give rise to what outcome given a fixed state. Imagine now a physical situation in which the context or contextual influence is hidden from us. We do not know and have no way of determining the λ for each experiment. The best we can do is resort to a statistical description of the entity and assume there exists a density function for the contexts. Suppose that the density of the contexts is uniform: every context λ is equally likely. Then the probabilities are fixed, and we could repeat the steps in the introduction, to obtain:

$$p(\mathcal{A} = a_i \mid w) = p(\lambda \in C_i | w) = p_w^{\mathcal{A}}(a_i)$$

So far, we did not discuss what outcome occurs in the rare instance when the context is a member of a lower dimensional element of the σ-induced triangulation. In a probabilistic sense this question does not really matter, as these sets have measure zero with respect to the (n-1)-dimensional elements $\mathcal{T}_{n-1}(\sigma)$ of the triangulation. However, from a mathematical point of view, it could turn out that our set of axioms is contradictory: perhaps the mapping τ cannot satisfy the system of axioms 1, 2 and 3. It turns out this is not the case as we will show by an arbitrary classification of the the lower dimensional elements of the σ-induced triangulation. First fix the state, that is, the point σ. Next, take an arbitrary simplex $]x_1, ..., x_{i-1}, \sigma, x_{i+1}, ..., x_n[$. We know that $\tau(\lambda, q) = x_i$ and assume that this is also the case for the closed simplex (except for the vertices which are excluded from Λ anyway). This set is convex and satisfies all three axioms. Now take any other simplex $]x_1, ..., x_{j-1}, \sigma, x_{j+1}, ..., x_n[$. As this simplex has one (n-2)-dimensional face adjacent to the first simplex, this face is already attributed to the outcome x_i. All contexts λ within other faces of this simplex are attributed to the outcome x_j. This simplex also satisfies all three axioms. The third simplex we choose, $]x_1, ..., x_{k-1}, \sigma, x_{k+1}, ..., x_n[$, will share two (n-2)-dimensional faces, one with each of the simplexes already treated. The other faces are attributed to x_k. We continue in this way untill we reach the last simplex, say $]x_1, ..., x_{z-1}, \sigma, x_{z+1}, ..., x_n[$. This simplex shares all interior (n-2)-dimensional faces with other simplexes, and the mapping τ needs only to be defined in the (n-2)-dimensional face $]x_1, ..., x_{z-1}, x_{z+1}, ..., x_n[$. Attribute the outcome x_z to this face. In this way we have defined the outcome for all contexts within the (n-2)-dimensional faces of the triangulation. In exactly the same manner we can treat the still lower dimensional faces in a way which satisfies our axioms.

We believe the name "contextual influence", or "context" for short, as distinct from the original proposal [1] "hidden measurement" for the λ is justified because the elements of Λ give rise to a contextual axiomatics. To see this, take $\lambda \in C_i =]x_1, ..., x_{i-1}, \sigma, x_{i+1}, ..., x_n[$ so we have $\tau(\lambda, q) = x_i$. Next construct the unique hyperplane that contains λ and is parallel with the hyperplane $[x_1, ..., x_{i-1}, x_{i+1}, ..., x_n]_{aff}$. This is the hyperplane with the equation $x_i = c$. The hyperplane cuts the simplex C_i, (because it separates σ from the x_j, $j = 1, ..., i-1, i+1, ..., n$) but it also cuts all the other simplexes, because it separates x_i from every other simplex than C_i. Hence it is possible to obtain every single outcome by variation of λ within the same hyperplane, that is without changing the coefficient of x_i in λ, which is precisely the meaning of contextuality.

What about the state q? Given the equi-probability of the λ, σ fixes all probabilities with respect to the observable $\mathcal{A}$. But we required that the state q determines all probabilities, and as a consequence σ is a function of q : $\sigma = \sigma(q)$ and every state q must induce a σ, hence this mapping is surjective. So we have obtained what was outlined in the introduction: the state of the entity is no longer identified with the statistical state (instead we have established a surjective function between the two) and we have shown how to characterize the measurements that occur in a hidden measurement scheme by means of three simple axioms.

References

1. D. Aerts, "A possible explanation for the probabilities of quantum mechanics", *J. Math. Phys.*, **27**, 202-210 (1986).
2. D. Aerts,"Quantum structures due to fluctuations of the measurement situation", *Int. J. Theor. Phys.*, **32**, 2207-2220, (1993).
3. D. Aerts, S. Aerts, B. Coecke, B. D'Hooghe, T. Durt and F. Valckenborgh, "A model with varying fluctuations in the measurement context", in *Fundamental Problems in Quantum Physics II*, 7-9, eds. M. Ferrero and A. van der Merwe, Plenum, New York, (1996).

4. D. Aerts, "The hidden measurement formalism: what can be explained and where paradoxes remain", *Int. J. Theor. Phys.*, **37**, 291, (1998).
5. R.Tyrrell Rockafellar, *Convex Analysis*, Princeton Academic Press, New Jersey, (1970).
6. Claude Berge, *Topological Spaces*, Dover Publications, New York, (1997).
7. Frederick A. Valentine, *Convex Sets*, McGraw-Hill, New York, (1964).

MEMORY EFFECTS IN ATOMIC INTERFEROMETRY: A NEGATIVE RESULT

THOMAS DURT

Foundations of the Exact Sciences (FUND) and Applied Physics and Photonics (TONA), Brussels Free University, Pleinlaan 2, 1050 Brussels, Belgium
E-mail: thomdurt@vub.ac.be

JACQUES BAUDON

Laboratoire de Physique des Lasers, Institut Galilee, Univ. de Paris 13, av. J-B Clement F93430, Villetaneuse, France,
E-mail: baudon@galilee.univ-paris13.fr

RENAUD MATHEVET

IRSAMC LCAR UMR5589, Univ P. Sabatier BAT. 3R1B4, 118 route de Narbonne, 31062 Toulouse, France
E-mail: renaud.mathevet@irsamc.ups-tlse.fr

JACQUES ROBERT

Laboratoire de Physique des Lasers, Institut Galilee, Univ. de Paris 13, av. J-B Clement F93430, Villetaneuse, France,
E-mail: robert@galilee.univ-paris13.fr

BRUNO VIARIS DE LESEGNO

IRSAMC LCAR UMR5589, Univ P. Sabatier BAT. 3R1B4, 118 route de Narbonne, 31062 Toulouse, France,
E-mail: Bruno.Viaris@irsamc.ups-tlse.fr

We investigate the possibility of the existence of a memory time inside the measurement device and obtain from experimental data an upper bound on this time, by looking at the distribution of time delays between successive detections performed inside an atomic interferometer.

1 Introduction

The intellectually comfortable image of the world that was elaborated by classical scientists untill the beginning of our century was seriously questioned by the development of the quantum theory. Instead of a world perfectly regulated by deterministic laws, as it is for instance the case in Newton's description of the solar system, the quantum world imposes us the image of a "fuzzy" world, in which nothing is sure but only possible, and in which, as we learn it from Heisenberg's indetermination relations, the extent of certainty that we can reach by making predictions about possible results of measurements carried out on systems, even when they are prepared in a well defined quantum state, is itself intrinsically limited. The Schrödinger cat paradox

[1] shows that the standard interpretation makes it possible to conceive a situation in which, according to the standard interpretation, the most that we "may" say or know about a cat is that this cat is simultaneously dead (with probability 1 - p) and living (with probability p). Such paradoxes illustrate to which extent a probabilistic description of the world is puzzling, when we formulate it in terms of macroscopic systems. Other paradoxes occur if we try to establish a border-line between quantum, probabilistic, and classical, deterministic, systems.

This kind of paradoxes constitutes the so-called measurement problem, which motivated numerous alternative interpretations, among which we would like to distinguish the hidden variable approach. The basic hypothesis made in hidden variable theories is the existence of "extra" degrees of freedom, the so-called hidden variables, which, if they were accurately specified, would allow us to predetermine exactly the result of any kind of measurement performed on a quantum system. In analogy with classical statistical mechanics, the probabilistic features of the quantum theory would then be explainable as a consequence of the statistical distribution of the hidden variables. Viewed so, the hidden variable approach is not totally satisfying, because it does not explain the origin of the probabilistic distribution of the hidden variables. In order to provide a serious alternative to the standard interpretation[a], it is necessary to explain the emergence of the probability distribution of the hidden variables by a convenient randomization process[b]. If we pursue the analogy with classical statistical mechanics, where several attempts were made (in ergodic theory for instance) in order to describe such processes, we come to the conclusion that a major difference must exist between hidden variable theories and the standard quantum theory: in the first approach, the average distribution is not instantaneously reached, an effect that we could measure through

[a]In the present paper, "standard" systematically refers to the so-called orthodox or Copenhagen interpretation.

[b]This is a critics often formulated against Bohm's interpretation [2], according to which the ad hoc distribution of initial positions must be postulated a priori. In the model of Bohm-Vigier [3], for the first time, the concept of a non-instantaneous randomization process of hidden variables was introduced, but unfortunately these ideas were never experimentally tested.

the careful observation of the temporal statistics of quantum phenomena. In standard quantum theory, the measurement process is usually considered to be instantaneous, an hypothesis that is indirectly confirmed by several tests of the experimental violation of Bell's inequalities [4,5,6]. Experiments could thus discriminate between the hidden variable approach and the standard interpretation.

Until now, few attempts were made in this direction (for instance, the Papaliolios experiment [7] performed in order to test Bohm-Bub's model [8], or the Summhammer experiment [9,10] performed in order to test Buonomano's model [11]). We describe them shortly in the first section.

We develop in the second section a simple hidden variable model in which the detector exhibits memory effects. Such effects were not tested in previously mentioned experiments (with light and neutrons). In the third section we present the (negative) results of an attempt realised with neutral atomic beams in the Laboratoire de Physique des Lasers de l' Université de Paris-Nord in order to test the possibility of a memory effect inside the detector.

2 Previous Attempts

2.1 The Papaliolios Experiment

To our knowledge, two experiments were carried out in the past in order to test the possibility of non-standard memory effects[c]. The first of them, realised by Papaliolios in 1967 [7], was aimed at testing the relevance of Bohm-Bub's theory [8]. In this theory, which describes the interaction between a quantum system and a measuring device, it was assumed that hidden variables exist, which complete the description of the state of a quantum system provided by the wave function. The authors suggested the hypohesis according to which these hidden variables would randomize in a time comparable to $\frac{h}{kT}$, where h, k and T are the Planck and the Boltzmann constants and the temperature of the environment respectively.

If we apply the Bohm-Bub theory to the case of a spin 1/2 measurement, in order to describe the passage of a photon through a polarising device, and if we assume that the hidden variable which hypothetically determines the result of a measurement does not randomize instantaneously, but remains "frozen" during a typical time τ_M, where the index M indicates the existence of persistent memory effects, "carried" by the photon, non-standard correlations must appear (departure from the Malus law [12,7]) between successive

[c]We mean hereby memory effects that cannot be explained in the standard theory. Such effects also exist and will be commented upon in the following.

measurements carried out within a very short time interval (shorter than τ_M). In order to check these effects, Papaliolios let pass low intensity light pulses prepared in a given polarization state through two successive, very close to each other, polarisers. If the distance between these two polarisers is smaller than the randomization time of the hidden variable times the speed of light, new correlations are predicted by the Bohm-Bub theory.

Papaliolios did not observe any such effect even for extremely short distances, and considered this negative result as the proof of the non-existence of Bohm-Bub-like variables[d].

2.2 The Summhammer Experiment.

The goal of Summhammer's experiment [9,10] was to check the validity of the local hidden variable theory of Buonomano [11]. The idea of Buonomano was the following: the violation of Bell's inequalities [4] shows that particles are in some manner informed about the setting of distant apparatusses. It is possible to conciliate this phenomenon with Einsteinian causality provided we assume that, (a) particles are localised, and (b) this process of information is realised by the particles themselves. The particles would spread in their surroundings (thanks to a kind of quantum "flavour" of "pheromon") some information about the properties of the external world that they encounter during their way. By doing so, they could (a posteriori) inform the next incoming particle about the obstacles along the way. This next particle could inform the following particle about the nature of the obstacles to be met in the future and so on, untill the information reaches the source. The aim of this hypothesis was to explain how sources of entangled particles could "foresee" what would be the configurations of distant measuring devices. In order to check these ideas, Summhammer used a low-intensity neutronic interferometer in which non-local properties appear: a single neutron is spread among two arms which are separated by a distance much larger than the coherence lenght of the neutron. Afterwards, the bundles originating from both arms are superposed, and interfere according to the standard quantum predictions. Thanks to a shutter, Summhammer opened and closed at random one arm within the interferometer in order to permit a sudden and random switching from interfering to non-interfering conditions and vice cersa. Then, if the theory of Buonomano was right and that a particle could "collect" information stored by the previous particles about the internal configuration of the

[d]Of course, the existence of hidden variables with an extremely short, undetectable, memory time is not excluded by the experience [13], but the existence of such variables constitutes an ad hoc hypothesis, totally unfalsifiable!

interferometer, it would get misled because in the meanwhile the experimental set up would have been changed. If Buonomano's hypothesis would have been correct, the genuinely predicted probabilities would have been disturbed by memory-effects. Summhammer realised the experiment and did not see any discrepancy with standard quantum predictions, even directly after the opening or the closing of the shutter, which constitutes a clear negative result and proves unambiguously the non-existence of the Buonomano type hidden-variables, for so far as these would be observable.

Note that the memory effects considered by Buonomano are localised effects, with a short spatial range and eventually a long temporal range, which connect a large amount of particles, while the effects predicted in Bohm-Bub's theory concern individual particles only and are assumed to have a short temporal range (τ_M).

3 A New Hidden Variable Model

3.1 Memory Effects Inside the Detector

Tutsch [14] criticised the relevance of the negative results obtained by Papaliolios by noting that, after all, it could be that the hidden variable is not attached to the quantum system only, but that its behavior (and thus its randomization) also depend on the measuring apparatus, in which case the experiment of Papaliolios is not concluding. In the framework of axiomatic probability, the distinction between hidden variables inside the detector and inside the observed quantum system under observation was firstly emphasised by D. Aerts in [15], who established clearly the distinction between hidden measurement variables and hidden state variables (see also [16,17,18,19,20,21]). The aim of the model and experiment that we shall describe in the present work is, among others, to investigate experimentally the relevance of some of the concepts developed in such axiomatic approaches of quantum mechanics. The experimental proposal [20,21] that ultimately led to the present work is a natural continuation of these ideas[e].

To conceive an experiment which allows us to check the presence of hidden variables inside the detector (hidden measurement variables), it is sufficient to reconsider Papaliolios experiment, and to permute in it the roles played by

[e]Note that the irreducible role played by the measuring apparatus was also outlined by Bohr from the beginning of quantum theory [22,23]. The classical nature of the macroscopic objects (observers, measuring devices) was often invoked by him in order to solve the measurement problem. Formally, no-go theorems related to the concepts of non-locality or contextuality [12,24,25,20,26,27], also emphasise the role played by the measuring apparatus.

the system under measurement and the measuring apparatus. We must thus consider the probability of, for instance, measuring successively, inside the same polariser, two photons initially prepared in a same polarization state.

The assumption of the existence of hidden measurement variables characterized by a measurable memory time leads to the prediction of results which will differ from standard predictions [20,21]. Roughly speaking, *we expect, if hidden measurement variables exist and vary continuously in time, that successive measurements performed rapidly enough on systems prepared in the same superposition state will exhibit a tendency to provide successive identical results which is higher than what we could expect in the case of independent, non-correlated measurements.*

This is entirely analog to so called bunching experiments of quantum optics in which, because of their bosonic nature, photons were shown to have a tendency to arrive together. In this case, the distribution of the arrival times was shown to be a surPoissonian distribution [28,29]*. In our case, photons are replaced by atoms. The standard distribution of arrival times of the atoms being Poissonian, we also expect to observe deviations from the Poissonian distribution.* The model aimed at deriving such effects is developed in the coming section.

3.2 A Simple Model

Let us develop a model in which hidden variables are assumed to characterize the "hidden state" of the detector, and in which the randomization time of these variables is not negligibly small. We shall consider the most simple case in which the quantum system is a two-level system, the state of which belongs to a 2-dimensional Hilbert space. This is the case for the spin of a spin 1/2 particle or for the polarization of a photon.

Let us assume for convenience that we want to describe the measurement of a spin 1/2 polarization realised inside a Stern-Gerlach device. As it is well known, the pure spin state in which we prepare the system can, in the case of pure spin states, be represented by a point A on the normalised sphere, the components of which are equal to the averaged Pauli matrices. Similarly, we can represent the direction of the magnet inside the Stern-Gerlach device by a point B on the same sphere. Standard quantum computations show that the quantum probability for getting spin up (spin down), along the direction B at the outcome of the Stern-Gerlach device is equal to $cos^2(\frac{\theta_{AB}}{2})$ $(sin^2(\frac{\theta_{AB}}{2}))$, where θ_{AB} is the angle taken on the sphere between the points A and B.

Our model is the following. We associate to such an experiment a random hidden variable x, and impose the following rules.

First, we assume that the hidden variable keeps the same value (let us say x_i) during a time τ_M, but that after this period, it takes a new value (x_{i+1}) defined by the so-called tent transformation:

$x_{i+1} = 2x_i$ if $0 \leq x_i \leq 1/2$, $x_{i+1} = 2 - 2x_i$ if $1/2 \leq x_i \leq 1$

We chose this transformation law because of its well-known mixing and randomizing properties. As in Bohm-Bub's theory [8], the parameter τ_M plays the role of a memory-time as we shall see in the following.

We also assume that, initially (so to say at the beginning of the experiment or even before, which usually means many τ_M ago), the initial distribution of the variable was a sufficiently regular distribution (for instance a continuous distribution on the interval [0,1] relatively to the Lebesgue measure). *In virtue of the well known properties of the tent map, this implies that, after some iterations, and thus during the experiment (at times t_i, t_{i+1}...), the hidden variable gets homogeneously spread over [0,1]. Note that this hypothesis of non-singularity of the initial distribution of the hidden variables is similar to the coarse-graining hypothesis often made in the derivation of H-theorems, in statistical mechanics.*

In the following, we shall thus always consider that generic values x_0 are uniformly spread over [0, 1]. Furthermore, whatever the particular choice that we can do for the generic value of x_0, after a time sufficiently larger than τ_M, the distribution of x_t covers the interval [0,1] in very good approximation.

Finally, we assume that during each measurement process the value of x is fixed and that its value determines unambiguously the result of the measurement, according to the requirement that when

$$x < cos^2(\frac{\theta_{AB}}{2})$$

the result of the measurement is spin up, otherwise it is spin down.

Therefore, when we average these results over a large amount of experiments (x is then homogeneously spread over the real interval [0, 1]), we recover the standard quantum probability ($cos^2(\frac{\theta_{AB}}{2})$ for spin up, $sin^2(\frac{\theta_{AB}}{2})$ for spin down).

Let us now assume that a first particle passes through the device and is deflected upwards (downwards). If a second particle, prepared in the same spin state as the previous one passes through the device after a time shorter that τ_M, and if our model is valid, this second particle will also be deflected upwards (downwards). Effectively, the hidden variable of the detector stays frozen during the time-interval that separates both measurements, and the model predicts thus exacly the same outcome in both cases. So, *the fraction of the pairs of pulses (inside a sufficiently reduced temporal window) which are deflected upwards (downwards) is* $cos^2(\frac{\theta_{AB}}{2})$ $(sin^2(\frac{\theta_{AB}}{2}))$ *which is greater than* $cos^4(\frac{\theta_{AB}}{2})$ $(sin^4(\frac{\theta_{AB}}{2}))$*, the corresponding standard quantum mechanical prediction*[f]. If the second particle passes through the device a time t after the first one, we shall recover the standard quantum predictions when $t >> \tau_M$ because then several iterations of the transformation map occured, so that the initial distribution is now spread over the whole interval [0,1]. It is in this way that our model allows us to simulate a transient memory effect[g]: *after some iterations, the mixing is such that the initial conditions are nearly totally forgotten, and we recover the standard quantum predictions (see figure 2a).*

Note that the memory effects considered here concern individual detectors but successive particles and are assumed to have a short temporal range (τ_M)*. This kind of memory effects was not investigated during the experiments performed by Papaliolios (in which individual particles but successive detectors were considered) and Summhammer (in which the range was assumed to be long).*

4 Experimental Measurement of Temporal Correlations

4.1 Methodological Considerations

If we had at our disposal a source of polarised photons, or a source of spin 1/2 particles, and that we could arbitrarily measure and/or control their arrival times in the analysers (polariser plus detectors or Stern-Gerlach device plus detectors), we could directly check the existence of the correlations predicted in the previous section. In fact, beside the experiments of Summhammer and

[f] We also get the standard quantum result if we assume that our simple model is of application, and that the hidden variable is a property of the spin 1/2 particles themselves as in the Bohm-Bub [8] theory (the hidden variable is "carried" by the particle), because then the hidden variables in both polarization processes are no longer the same (even no longer correlated). Conversely, if we assume that our model is of application in the framework of the Papaliolios experiment, the distributions of the hidden variables inside the successive polarisers are independent and we recover the standard predictions. This confirms the intuition of Tutsch [14] according to which, if the hidden variable does not "follow" the photon in Papaliolios experiment, the statistics of the result is in accordance with the Malus law.

[g] The general question to know if the quantum probability can be simulated by Markovian (thus without memory) processes was addressed in several publications during the last decenny [30,31,32]. More recently, some interesting results were obtained relatively to non-Markovian stochastic quantum mechanics [33,34,35]. The experiment described in the present work can be considered as an attempt, in a very specific situation, to quantify the departure from the Markovian paradigm. In appendix, we prove that our model is explicitly of a non-Markovian nature.

Papaliolos, temporal correlations between quantum objects were measured with great accuracy several times in the past, not in the context of hidden variable theories, but with the aim of testing standard correlation effects. For instance, in the framework of quantum optics, correlations due to the bosonic nature of photons were measured with great accuracy more than twenty years ago [28]; the temporal correlations of various sources of light (thermal, coherent and so on) were tested with high accuracy, and provided results which confirmed standard predictions. More recently, the temporal distribution of photons emitted by an isolated, single, atom placed in a resonant laser field has been shown to differ from the Poissonian (exponential) distribution for short times [36,37]. Essentially, this is due to the fact that the atom, after having emitted a photon comes back to the fundamental, non-excited state, and does not get reexcited at once. In such experiments, *a very convenient parameter aimed at measuring the deviation towards the Poissonian statistics is provided by the normalised mean square root deviation, so to say, the ratio between the mean square root deviation and the mean value of the distribution of the time-delays between the detections of two successive photons.* If this parameter is smaller (larger) than 1, one will say that the distribution is antibunched (bunched); when the distribution is Poissonian, this parameter is equal to 1. Bosonic correlations were revealed by bunching, which expresses a tendency of the beeps to arrive in group (attractive effect), while the existence of a dead time in the emission of an atom, directly after its desexcitation, was revealed by anti-bunching. *In other words, the existence of memory effects inside the detector itself can also be revealed by the departure from Poissonian statistics in the distribution of time delays between successive "beeps" in a detector.*

The physical situation that we want to investigate, which was described in our model in the previous section, is not exactly the same as in bunching and anti-bunching experiments. Atoms are produced at the level of the source with a Poissonian distribution (this was carefully checked in the past[h] [38]). It can be shown ([39,40]) that, when the distribution of the time-delays between successive atoms at the level of the source is Poissonian, this is still true at an arbitrary distance of the source, whatever the distribution of the velocities could be, provided we can consider that successive atoms are independent. This hypothesis of independence is not fulfilled if we consider the detections

[h]This was checked with the same detectors as in the experiment aimed at revealing memory effects. Methodologically, this does not exclude the possibility of a logical loophole. Nevertheless, it is very unprobable that nature conspires in order that the distribution of the atoms at the source convoluted by the memory effects of the detector precisely provides a Poissonian distribution at the level of the detector.

realised in a non-instantaneous detector, because its memory of the past correlates the atoms. *Our strategy is thus the following: by measuring the same correlation parameter as in the quantum optical experiments (the normalised mean square root deviation), which is sensitive to the level of independence between successive detections (beeps), we shall check the (non) existence of a memory effect inside the detector.*

Remark that no significant departure from the Poissonian statistics could be observed in neutronic [9,41,42] as well as in electronic [43] interferometry. In the framework of atomic interferometry, correlations due to the bosonic nature of metastable laser-cooled atoms were measured which show an accordance with standard predictions for times comparable to 10^{-7} seconds [44]. All this indirectly confirms that the hypothesis of the existence of non-standard memory effects does not seem to be justified.

If we want to prove clearly the existence of memory effects inside the detectors, we ought to show that they appear at a time scale which is larger than the typical quantum interaction times, otherwise it would be very difficult to dissociate the standard and non-standard effects. *Untill now, the possibility of memory times in the case of atoms was not investigated for such large times (larger than 10^{-7} seconds). Our aim is thus to check experimentally the relevance of the model presented in the previous section at this "mesocopic" time scale.*

In order to check that the system was able to provide a significant response relatively to temporal disturbances the duration of which belongs to the typical domain of time that we wish to investigate, we artificially submitted it to an external periodic disturbance (see appendix). The sensitivity of the system to fastly modulated perturbations was clearly established (see figures 6a and 6b, and appendix, subsection 5), in a domain of frequencies which ranges from 50 hz up to 100 Khz when the source and the atomic beam themselves were periodically disturbed, and up to some Mhz when the detector was disturbed.

Transversal Stern-Gerlach devices and spin 1/2 atoms would furnish a direct illustration of the theoretical model of the previous section. Unfortunately, it is presently difficult to measure accurately temporal correlations inside a transversal Stern-Gerlach device because, if we let pass atoms through two successive transversal Stern-Gerlach devices (one for the preparation, one for the measurement), the error rate is quite too high and makes it impossible to obtain significant results. This is no longer the case if we consider the atomic interferometric device used by the "Laboratoire de Physique des Lasers (Paris-Nord)" [40], in which a high resolution is reached (very few "dark" counts). For sure, this device does not fit exactly with the simple model of the previous section (spin 1 atoms are involved, the spin is measured indirectly thanks to a Lamb-Retherford device). Nevertheless, we shall show in the next section how one can conceive "sur mesure" an experimental protocol which makes it possible to test the reality of the non-standard memory effects considered in the previous section.

Before we refer the experiment itself, let us firstly derive the relation which exists between the existence of a randomization time τ_M of the type considered in the hidden variable model of the previous section, and departures of the normalised mean square root deviation towards unity, in the case of a Poissonian source.

4.2 Quantitative Predictions.

4.2.1. The distribution of time-delays.

The experimental set-up will be discussed in the next section. All what we need to know presently is that the signal consists of the temporal series of "beeps" detected in a photon counter. Each beep corresponds to the induced (by an electrical field) emission of a photon by an incoming excited atom. Let us predict theoretically the distribution of time-delays between successive beeps of the detector, (A) in the absence of memory effects inside the detector, and (B), in their presence.

A. Without memory effects.

If, at the time t_0, an atom passes at the level of the photon detector, in the outcome port of the experimental device, the probability that an other atom passes during a short time-interval t_0, $t_0 + dt$ is thus, up to the first order in dt, equal to $\frac{dt}{T}$, where T is the average time-delay between two atoms. Let us now decompose the detection process as follows. If we apply the standard interpretation, when an excited atom passes at the level of the detector, it will emit a photon with a probability P_Q (P_Q depends on the quantum interference effects undergone by the atom inside the interferometer, on the intensity of the electric field at the outcome of the device and so on). This photon will then get detected with a probability P_G (in first approximation P_G is comparable to the product of the efficiency of the detector with its solid angular opening towards the atomic source divided by 4π). We have thus that the probability of no-detection $\mathcal{P}_{neg}(t_0, t_0 + n \cdot dt)$ during a time $n \cdot dt$, where n is a large integer, is equal, in first approximation, to the product of the probabilities of no-detection for all intermediate dt intervals: $\mathcal{P}_{neg}(t_0, t_0 + n{\cdot}dt) = \Pi_{i=1}^{i=n}(1 - P_Q\, P_G \cdot \frac{dt}{T})$. If we consider the limit of this expression for extremely short intervals dt, we find that $\mathcal{P}_{neg}(t_0, t_0 + n \cdot dt) = exp(-n \cdot P_Q\, P_G \cdot \frac{dt}{T})$. From this expression, it is straightforward to deduce that we have again a

Poissonian distribution $\mathcal{P}(t)$ for the time-delays. The probability $\mathcal{P}(t)$. dt that the next photon is detected inside the interval $[t_0 + t, t_0 + t + dt]$ when a first photon is detected at time t_0 is equal to the product of the probability $\mathcal{P}_{neg}(t_0, t_0 + n \cdot dt)$ that no-detection occurs between t_0 and $t_0 + t$ with $P_Q\, P_G \cdot \frac{dt}{T}$, the probability of a detection inside the interval $[t_0 + t, t_0 + t + dt]$: $\mathcal{P}(t) = \frac{1}{T} P_Q\, P_G \cdot exp(-P_Q\, P_G \cdot \frac{t}{T})$.

B. With memory effects.

If the detector obeys a hidden variable model similar to the model described in the previous section, things are different. At first sight, it is not so easy to identify from experimental data the value of the probability that corresponds to the parameter $cos^2(\frac{\theta_{AB}}{2})$ of our model. This is due to the fact that one does not detect the atoms that did not excite the (unique) detector, so that the detector is calibrated up to an arbitrary normalisation constant. For convenience, we shall fix this constant by assuming that $cos^2(\frac{\theta_{AB}}{2}) = 1$ when the intensity measured at the detector reaches the maximal value I_{max} of the interference pattern. This implies that, for an arbitrary experimental configuration, characterised by an average intensity I inside the photon detector, we can fix the corresponding phase θ_{AB} inside the interference pattern thanks to the relation $cos^2(\frac{\theta_{AB}}{2}) = \frac{I}{I_{max}}$. We shall show in the section 3.2.2. that in first approximation the observable correlation parameter that measures the departure from the Poissonian distribution is proportional to the product of the memory-time τ with the probability of desexcitation. Therefore any choice of "normalisation" for $cos^2(\frac{\theta_{AB}}{2})$ is equivalent to a renormalisation of the memory-time τ which is anyhow unknown a priori. This property authorizes us to fix the unknown normalisation factor of the detection rate at our convenience (to some reasonable extent).

Let us now consider a particular story of the detector, so to say, a particular value x_{t_0} of the hidden variable at the time t_0. Then, the detector will fire when an atom passes at the time $t_0 + t$ if $x_{t_0+t} < cos^2(\frac{\theta_{AB}}{2})$, and stay silent otherwise. Let us repeat the previous reasoning, by taking account of the fact that the probability of a detection inside the interval $[t_0 + t, t_0 + t + dt]$ is now, in virtue of our previous conventions, equal to $I_{max} \cdot dt$ if $x_{t_0+t} < cos^2(\frac{\theta_{AB}}{2})$, and to 0 otherwise. Let us denote by $f(t, x_{t_0})$ the step function which is equal to 1 when $x_{t_0+t} < cos^2(\frac{\theta_{AB}}{2})$, and to 0 otherwise. One gets the following relation: $\mathcal{P}_{neg}(t_0, t_0 + (n+1) \cdot dt) = \mathcal{P}_{neg}(t_0, t_0 + n \cdot dt) \cdot (1 - f(t_0 + (n+1) \cdot dt, x_{t_0}) \cdot I_{max} \cdot dt)$. We can integrate this relation. We obtain then: $\mathcal{P}(t) = K \cdot exp(-\int_0^t f(t, x_{t_0}) \cdot I_{max}\, dt)$ where K is a normalization constant that we shall fix later.

In order to evaluate the value of $exp(-\int_0^t f(t, x_{t_0}) \cdot I_{max}\, dt)$, let us factorise the exponential factor as follows:

$$exp(-\int_0^t f(t,x_{t_0})\cdot I_{max}\,dt) \;=\; exp\{[(lim_{t\to\infty}\frac{\int_0^t f(t,x_{t_0})\,dt}{t})\cdot I_{max}t]$$
$$-[\int_0^t f(t,x_{t_0})\cdot I_{max}\,dt]\}\cdot\; exp-(lim_{t\to\infty}\frac{\int_0^t f(t,x_{t_0})\,dt}{t})\cdot I_{max}t$$

Let us denote Δ the difference

$$[(lim_{t\to\infty}\frac{\int_0^t f(t,x_{t_0})\,dt}{t})\cdot t \;-\; \int_0^t f(t,x_{t_0})\,dt]\cdot I_{max}$$

Remember now that for generic initial values of x_{t_0}, after some iterations, the variable x_t is spread according to the equilibrium distribution associated to the hidden dynamics. As it is well known, this measure, the invariant measure of the map, is the homogeneous distribution over the interval [0,1]. Now, the function $f(t,x_{t_0})$ is in average equal to

$$\frac{I}{I_{max}}$$

when the initial distribution is spread over the interval [0,1], so to say, after a time comparable to some τ_M. Besides, $f(t,x_{t_0})$ is close to 1 for a time t taken inside the interval $[t_0,\ t_0 + \tau_M]$, because, by definition, we consider that a first photon was detected at time t_0, which means that we select the values of x_{t_0} which fulfill the condition

$$x_{t_0} \;<\; cos^2(\frac{\theta_{AB}}{2}) \;=\; \frac{I}{I_{max}}$$

During a time comparable to τ_M, the values of x will remember the initial selection. At each iteration, the mixing properties of the tent map gradually wash out this memory effect, which nearly completely disappears after a time comparable to some τ_M.

For generic initial values

$$x_{t_0},\ (lim_{t\to\infty}\frac{\int_0^t f(t,x_{t_0})\,dt}{t})$$

is thus independent on the peculiar value of x_{t_0} and equal to

$$\frac{I}{I_{max}}$$

whilst Δ is a function approximately equal to $(I-I_{max})\cdot t$ for times comparable to τ_M and decreases to 0 after some iterations of the tent-map (after some

τ_M). This implies that the absolute value of Δ is majorated by a term of the order of $\tau_M \cdot I_{max}$. Besides, *direct observations show that the memory time τ_M of the detector, if it exists, must be small, because the time-delay distribution exhibits a nearly-exponential shape (see figure 4), even for short times* [38]. *This would not be the case if the memory time of the detector was comparable to the average time between two beeps $\frac{1}{I}$.*

We shall thus investigate the presence of memory effects in a domain where τ_M, the typical memory time is quite smaller than $\frac{1}{I}$, the average time between two detections. In this domain, it is coherent to evaluate the two first orders of the development of $\mathcal{P}(t)$ in the variable $\tau_M \cdot I$ which can be considered as a small perturbation parameter. The zeroth order term corresponds to the limit in which τ_M is negligibly small[i] (vanishing memory time). If we limit ourselves to the "zeroth order term in $\tau_M \cdot I$", we can then neglect Δ so that

$$\begin{aligned} \mathcal{P}(t) &= K \cdot exp - (lim_{t\to\infty} \frac{\int_0^t f(t, x_{t_0})\, dt}{t}) \cdot I_{max} \cdot t \\ &= K \cdot exp - \frac{I}{I_{max}} \cdot I_{max} \cdot t \ = \ I \cdot exp(-I \cdot t) \end{aligned}$$

We recover then the same expression as in the situation without memory effect:

$$\mathcal{P}(t) = \frac{1}{T} P_Q\, P_G \cdot exp(-P_Q\, P_G \frac{t}{T})$$

because

$$I = \frac{P_Q\, P_G}{T}$$

Let us now consider the first order effects in $\tau_M \cdot I$.

We noted that $|\Delta|$ is comparable to and majorated by a function equal to $(I_{max} - I) \cdot t$ for times comparable to τ_M and that it fastly "decays" to zero for larger times, whilst

$$(lim_{t\to\infty} \frac{\int_0^t f(t, x_{t_0})\, dt}{t})$$

is independent of x_{t_0} and equal to

$$\frac{I}{I_{max}}$$

[i]It is no worth considering the limit in which I vanishes because, experimentally, the visibility of the pattern is never equal to 100 % (typically 30 % in the experiment considered)

Now, $\mathcal{P}(t)$ is, in average, equal to the product of the zeroth order term $I \cdot exp(-I \cdot t)$ with the average over x_{t_0} of the exponential of the function Δ. We can consistently, at the first order in $\tau_M\, I$ replace the average of $exp\Delta$ over x_{t_0} by the exponential of the average of Δ over x_{t_0}. We can then approximate the exponential of the average of Δ by the sum of a constant close to unity and of a fast-decay distribution (exponential function of lifetime τ_M[j]), weighted by a term comparable to $\tau_M \cdot I$. Inside the interval of time for which the fastly decaying function Δ is not negligibly small, the slowly decaying function $exp - I \cdot t$ is close to 1.

We can thus express $\mathcal{P}(t)$, in good approximation, up to the first order in $\tau_M \cdot I$, as the sum of a fastly decaying distribution (exponential function of lifetime comparable to τ_M), with weight proportional to $\tau_M \cdot I$, and of a slowly decaying distribution (exponential function of lifetime $\frac{1}{I}$).

In the figures 2a-b, we present a numerical expression based on the exact expression of $\mathcal{P}(t)$. *In appendix, we also present an exact, analytical, expression obtained on the basis of a model slightly different from the model that we presented in the previous section. In both cases, we obtain for* $\mathcal{P}(t)$ *an expression of the type:*

$$w_1 \cdot \tau_M \cdot exp(-\frac{t}{\tau_M}) \;+\; (1 \;-\; w_1) \cdot I \cdot exp(-I \cdot t)$$

where w_1 *is a small positive parameter proportional to* $\tau_M \cdot I$ *(which validates the above approximations). It is the existence of this parameter that we shall try to reveal experimentally, thanks to the measurement of the normalised mean root square deviation.*

4.2.2 The theoretically predicted correlation parameter g

On the basis of the expression of the distribution of time-delays, it is straightforward to compute, up to the first order in $\tau_M \cdot I$, the value of the correlation parameter, the normalised mean square root deviation (the ratio between the mean square root deviation and the mean value of the distribution of the time-delays between the detections of two successive photons), that we shall from now on denote g. We obtain for instance with our fully solvable model (see appendix) that

$$g \approx 1 + w_1\,(sin^2(\frac{\theta_{AB}}{2}))\cdot \tau_M \cdot I$$

[j] In fact, this is an oversimplificating approximation because for certain values of θ, a periodically modulated exponential decay provides a better fit to the numerical estimations of the exact distribution that can be deduced from our hidden variable model. We shall study this alternative in a forthcoming paper.

In fact we generalised the computations made in appendix and the results show that at this order of approximation, the shape of the function which expresses the loss of memory is rather irrelevant: exponential functions, but also step functions, can be shown to yield the same first order correction to g. If we consider a realistic situation, we ought to be less categorical about the value of g for several reasons: (A) As we discussed it already in the previous section, the normalization of the observed intensities at the level of the detector is not strictly determined, and some arbitrariness remains thus always present in our choice of normalization. (B) Strictly speaking, it is not correct to consider that the phase of an atom inside the interference pattern is a well peaked function, because the phase is velocity-dependent (this dependence is a consequence of the de Broglie relations) and, in our experiment, the velocities are spread between, roughly, one half and three halves of the average velocity. (C) When one evaluates g by averaging the numerically integrated dynamics of the hidden variable (figure 3a, 3b) for a broad sample of initial conditions (figures 1, 2a, 2b), no analytical expression for g can be deduced from our model unless we make some approximations (see appendix). This is not astonishing, because the dynamics itself is not continuous, but discretised in time from the beginning. (D) the intensity of the source drifts a little bit during the experiment (fluctuations of the order of one percent were observed).

We can, formally, incorporate these corrections by introducing a new parameter of order 1, α, that it would cumbersome and useless to compute in detail and rewrite the expression of g as follows: $g = 1 + \alpha \cdot \tau_M \cdot I$. At this level, it is sufficient to consider that α is an adjustable parameter comparable to unity that eventually will be fitted to the experiment.

Note that if the detector has a non-negligible dead-time, which means that after a detection it is saturated for a while and insensitive during a typical dead time τ_D, we get the same expression with a minus factor, and τ_D instead of τ_M. This is analogous to the antibunching effect (sub-Poissonian distribution, $g < 1$) observed in certain quantum optical experiments [36,37]. In our case, the hidden variable model implies that, after that the detector clicks, it is more "receptive" during a time comparable to τ_M, so that we are in a situation analogous to the one met during bunching experiments ($g > 1$). Remark that even if the hypothesis of hidden variables is illusory and false, our approach makes it possible to obtain a better determination of the real dead-time of the detector.

Now that we modellized the effect that we expect to occur, let us describe the experimental conditions.

4.3 Experimental Results

4.3.1 The experimental device

The experimental set-up is described in the figure 7. It consists of a source S of metastable H-atoms, which enter a preparation zone P, an interferometer B, an analyser A and a detector D in which the excited states which survived the previous steps are desexcited and emit a photon that is amplified in a channeltron. The beeps that occur in the channeltron are collected in a multi-channel analyzer MCA coupled to a clock. This is how we establish the temporal distribution of the beeps. By varying the intensity I inside the interferometer we vary the amplitude of interference at the level of the detector (this is actually a longitudinal Stern-Gerlach interferometer, and it is surrounded by Majorana transition zones). It is not our aim here to describe in detail these experiments. *All what we need to know is that the detection of a photon emitted by an atom corresponds in the idealized gedanken-experiment presented in the previous section (transversal Stern-Gerlach experiment) to a "spin-up" detection. Note that this gedanken-experiment involved two-level states and thus two detectors and not only one (the "up" and "down" detectors). In the real experiment, we have only one detector and thus no analog of, say, the "down" detection. The role of the "up" detector is played in our case by the counter which detects light emitted by the atoms which decayed at the level of the final electric field.* The case of two detections of which one only is observed was not explicitly taken into account in the ideal model of the previous section. This is why, as we mentioned already in the section 3.2.1 and 3.2.2., when we adapt the model to the set up of the laboratory, we obtain probabilities up to a normalization constant, the unknown calibration factor of the detector P_G. In comparison to the ideal two-detectors case considered in the model, the experimental situation that we considered lacks thus somewhat of accuracy. Nevertheless, the advantage of such an experimental set-up is that the amount of spurious detections inside the photodetectors can be shown to be very low, so that the observed signal has a very high quality. Furthermore, as we already mentioned, all the unaccuracies that characterise the passage from our idealised model to the real experimental situation can be absorbed and compensated into a renormalisation of the parameter τ_M that we do not know a priori.

4.3.2 Experimental determination of the correlation parameter: the law of large numbers

In order to exhibit the existence of a memory time inside the detector, it is thus sufficient to measure with great accuracy the correlation parameter g.

The accuracy on this measurement can be evaluated by the use of the law of large numbers and by assuming that the distribution is perfectly Poissonian (this is consistent at the order of approximation considered here and with observations [38]).

Remark that, even if $\tau_M \cdot I$ is an extremely small parameter, its existence can be revealed provided we have a large enough statistics at our disposal. This situation is somewhat similar to the situation encountered in [39,41] in which postselections in time delays (thus study of a subpart of the whole temporal distribution) had as a consequence an improvement of the visibility of the interference pattern: a judicious use of the information available in the details of the temporal statistics makes it possible, thanks to the law of large numbers, to obtain very accurate informations about the system under study.

Let us now apply the law of large numbers in order to evaluate the typical error on the measurement of g.

4.3.3 Application of the law of large numbers

Before we make use of the law of large numbers, let us recall an elementary, self-consistent, derivation of its weak form, Chebyshev's inequality. Let us assume that a random, real and positive variable t obeys a normalised distribution function $P(t)$ (we assume that P is sufficiently regular so that all the integrals introduced in the following treatment uniformly converge). Let us denote $< t >$ the average value of the variable t: $< t > = \int_o^\infty dtP(t) \cdot t$. Obviously,

$$< t > \geq \int_\epsilon^\infty dtP(t) \cdot t \geq \epsilon \cdot \int_\epsilon^\infty dtP(t) = \epsilon \cdot Proba(t \geq \epsilon)$$

Let us consider the variable $(t - < t >)^2$. By the same reasoning, we get that:

$$Proba(|t - < t > | \geq \epsilon) \leq \frac{< |t - < t > |^2 >}{\epsilon^2} = \frac{\sigma^2}{\epsilon^2}$$

where σ is the mean square root deviation of the distribution.

Let us denote by $t_{\overline{N}}$ the variable that we obtain by an averaging process after having measured N times the value of t, then

$$t_{\overline{N}} = \frac{t_1 + t_2 + ... + t_N}{N}$$

Obviously $< t_{\overline{N}} > = < t >$. Besides

$$\sigma^2_{\overline{N}} = < (\frac{1}{N}\Sigma_{i=1}^N t_i)^2 > - < t_{\overline{N}} >^2$$

$$= \Sigma_{i=1}^{N} \frac{< t_i^2 >}{N^2} + \Sigma_{i \neq j} \frac{< t_i \cdot t_j >}{N^2} - < t >^2$$

$$= \frac{< t^2 >}{N} + \frac{(N-1) < t >^2}{N} - < t >^2 = \frac{< t^2 > - < t >^2}{N} = \frac{\sigma^2}{N}$$

Note that we implicitly made use of the fact that different measurements are independent when we replaced $< t_i \cdot t_j >$ by $< t >^2$ $(i \neq j)$. This is not absolutely true if hidden variables are present, because then memory effects correlate successive measurements. Nevertheless, one can consider that this effect rapidly disappears and that if we perform many successive measurements the different measurements are in average not correlated.

For instance, in our case, it is reasonable to assume that at most $N\tau_M \cdot I$ pairs of successive measurements are correlated (where $\tau_M \cdot I$ is a very small parameter in our case), we obtain by the same treatment that

$$\sigma_{\overline{N}}^2 \leq (1 + \tau_M \cdot I) \cdot \frac{\sigma^2}{N}$$

this shows that we must still expect the scaling law in $\frac{1}{N}$ in our case, which is an essential ingredient of the law of large numbers. If we replace t by $t_{\overline{N}}$ in the inequality

$$Proba(|t - < t > | \geq \epsilon) \leq \frac{\sigma^2}{\epsilon^2}$$

we get that:

$$Proba(|t_{\overline{N}} - < t_{\overline{N}} > | \geq \epsilon) \leq \frac{\sigma_{\overline{N}}^2}{\epsilon^2} = \frac{|t - < t > |^2}{\epsilon^2 N}$$

When we perform a large number of experiments, N goes to infinity, and the probability that the averaged value of t over N experiments differs from $< t >$ goes to zero. This is the essence of the law of large numbers: even if the result of each individual experiment is not predetermined with great accuracy because of the dispersion of the distribution, the averaged values behave nearly deterministically. For instance, if we measure t N times, and that we desire that our estimation $t_{\overline{N}}$ of $< t_{\overline{N}} >$ "fits" the average value $< t_{\overline{N}} >$ in the sense that

$$Proba(|t_{\overline{N}} - < t_{\overline{N}} > | \leq 0,05) \geq 0,95$$

this imposes that N must be large enough:

$$N \geq \frac{\sigma^2}{0,05^2 \cdot 0,05}$$

This can be advantageously exploited when one tries to evaluate g because g is obtained after averaging over a large amount of experiments.

Note that the Chebyshev inequality can be improved if one takes into account the fact that for large N the distribution of $t_{\overline{N}}$ will be nearly Gaussian. One gets then

$$N \geq \frac{1,96 \cdot \sigma^2}{0,05^2}$$

Let us now apply these results in order to estimate the experimental error on the measurement of g. To determine g, we measure $(t^2)_{\overline{N}}$ and $t_{\overline{N}}$, and compute g thanks to the following equality:

$$g^2_{\overline{N}} = \frac{(t^2)_{\overline{N}}}{t^2_{\overline{N}}} - 1$$

Experimental data show that the distribution is Poissonian at least at the lowest order. We can thus consistently, at this order of the perturbative development, evaluate $t_{\overline{N}}$ and $(t^2)_{\overline{N}}$ as if the distribution was exactly Poissonian. Then, standard computations show that the mean square root deviations of $t_{\overline{N}}$ and $(t^2)_{\overline{N}}$ (denoted $\sigma(t_{\overline{N}})$ and $\sigma((t^2)_{\overline{N}})$ respectively) are equal to $\frac{<t>}{\sqrt{N}}$ and

$$\frac{\sqrt{8}\cdot < t >^2}{\sqrt{N}}$$

We can majorate $\sigma(g^2_{\overline{N}})$ the mean square root deviation of $g^2_{\overline{N}}$ as follows:

$$\sigma(g^2_{\overline{N}}) = \sigma(\frac{(t^2)_{\overline{N}}}{t^2_{\overline{N}}}) \leq \frac{\sigma((t^2)_{\overline{N}}) < t_{\overline{N}} >^2 + 2 < t_{\overline{N}} > \sigma(t_{\overline{N}})(t^2)_{\overline{N}}}{< t_{\overline{N}} >^4}$$

$$= \frac{\sqrt{8}+4}{\sqrt{N}} \simeq \frac{6.83}{\sqrt{N}}$$

4.3.4 Experimental data

The experimental data are summarised on the figures 4-5. They correspond to 20 experimental runs. Each of them consists of the measurement of $1.3 \cdot 10^5$ time-delays between two successive beeps in the detector. The mean square root deviation $\sigma_{\overline{N}}$, denoted σ, with

$$\sigma^2_{\overline{N}} = < (\frac{1}{N}\Sigma_{i=1}^N \, t_i)^2 > - < t_{\overline{N}} >^2$$

and

$$t_{\overline{N}} = \frac{t_1 + t_2 + ... + t_N}{N}$$

is plotted in function of the mean $t_{\overline{N}}$, denoted μ, for each of them (this corresponds to the points 1 to 20 on the figure 5). The global average which gathers the contributions of these 20 experiments (and thus $2.6 \cdot 10^6$ time delays) is plotted as the "ensemble" point (the central point of the figure 5). The dashed line represents the perfectly Poissonian situation ($\sigma = \mu$). Obviously, the departures from the Poisson distribution are minimal in the "ensemble" point for which we get an average delay-time μ (so to say $\frac{t_{tot}}{2.6 \cdot 10^6}$) equal to $9.99945 \cdot 10^{-1} \cdot \tau$ and a mean square root deviation σ equal to $1.00005 \cdot \tau$, where τ is a reference time close to the average time between two beeps μ (approximately 0,5 ms). We obtain a reasonable fit of all the points inside the margin predicted by the law of large numbers. For instance, the mean square root deviation of μ ($t_{\overline{N}}$) is predicted in a Poissonian situation to be equal to

$$\frac{<t>}{\sqrt{N}}$$

which is approximately equal to $3 \cdot 10^{-3} \cdot \tau$ when $N = 1.3 \cdot 10^5$, and to $6 \cdot 10^{-4} \cdot \tau$ when $N = 2.6 \cdot 10^6$. The dispersion of the horizontal coordinates of the points plotted in the figure 5 is compatible with these estimations (taking account a drift of intensity at the level of the source of the order of one percent). Let us consider the interval of values of g centered around the average, central, value which ought to be occupied in 95 % of the cases. The extent of this interval is equal to

$$\epsilon = \sqrt{\frac{1,96}{N}} \cdot \sigma = \sqrt{\frac{1,96}{N}} \cdot \frac{6,83}{2}$$

so to say, to $1.33 \cdot 10^{-2}$ when $N = 1.3 \cdot 10^5$, and to $2.96 \cdot 10^{-3}$ when $N = 2.6 \cdot 10^6$. One can check that for all the 20 experimental points (figure 5), as well as the central point, g belongs to the predicted interval. As we mentioned already, for μ, the situation is less ideal, and it seems that fluctuations of the intensity of the source must be taken into account. Nevertheless, g, which is a normalised dimensionless parameter, is more robust against such fluctuations, and its dispersion is in very good agreement with the Poisson statistics. *The conclusion of this analysis is that, even if some memory time exists, such a time may not exceed* $2.96 \cdot 10^{-3} \cdot 0.5$ *ms so to say* $\sim 1.5 \cdot 10^{-6}$ *seconds*[k]. *As*

---

[k]Even if the central value of g is in accordance with the Poisson statistics, one could "cheat" and try to obtain an upper bound for τ_M by extrapolating directly from the data. This can be done for instance if one assumes that Poissonian fluctuations exactly compensate each other (a non-justified assumption, but useful in order to get an estimation of what would be the value of a systematic deviation if it exists). Then, the values observed during the experiment ($\mu = 9.99945 \cdot 10^{-1}$ and $\sigma = 1.00005$) can be fitted into the expression

we mentioned already, if hidden variable correlation times were shorter than the time measured by Shimizu ($5 \cdot 10^{-7}$ seconds), the hidden variable theory would be non-falsifiable (see also footnote 6) because standard effects would allow us to simulate the correlations predicted by the hidden variable model. We can thus conclude from our experiment that the experimental answer provided to the question of hidden variables is in our case negative. By repeating 10 times the same experiment, one could definitively rule out the possibility that a hidden memory time exists (so to say a time inside the interval $[5 \cdot 10^{-7} sec, +\infty[$).

5 Conclusion.

The experiment described in the previous section does not only help to bring on the field of experience the old polemics between partisans and detractors of hidden variable theories. Beyond the technicalities inherent to the theoretical considerations that we developed here, they also make it possible to ask directly to mother nature a very general question, that can be formulated as follows:

"Do measurable correlation times exist inside a quantum signal ?"

This kind of question is not so crazy if we think of the mysterious process of building of a quantum interferometric pattern. All particles seem to arrive at random on the screen, but, after some time, a structure emerges, the interferometric pattern. It is natural, for a physicist who sees order emerging from chaos, to try to put into evidence the existence of a "guiding force". Considered so, the existence of memory effects inside the elaboration of the quantum signal is a possible answer to the naive question "how, by Jove, do the particles know that they must build this pattern ?".

Note that some"sulfurous" theories, as the non-ergodic interpretation of Buonomano [11], and the shape wave theory of Sheldrake [45] contain a quite similar concept: they assume that maybe the particles (living beings in Sheldrake's theory) are in some way informed of the contribution of the previous

given in appendix where it is shown that, up to first order terms in $\tau_M \cdot I$, $\mu = (1 - w_1)$ and $\sigma = (1 + w_1)$, where w_1 is of the order of $\tau_M \cdot I$. By doing so, one obtains that $\tau_M \cdot I \simeq 5 \cdot 10^{-5}$. Now, $\frac{1}{I} \simeq 5 \cdot 10^{-4}$ seconds, so that we obtain that $\tau_M \simeq 2,5 \cdot 10^{-8}$ seconds. It is interesting to compare this result with the correlation time that was measured in the atomic interferometric experiments of Shimizu [44], which revealed a bunching effect of memory time $5 \cdot 10^{-7}$ seconds, due to coherent bosonic correlations between laser cooled atoms. It could be that we measured by our method such a standard effect but a careful and systematic treatment of this idea is not the object of the present paper.

particles to the elaboration of the pattern (to the past history of the world in Sheldrake's theory).

Surely, to paraphrase a famous sentence of Laplace, we do not need such an hypothesis, but, in last resort, experiments decide. *In our case, once more, facts confirm well that, at a fundamental level, nature does not remember but "lives in the present". This is in accordance with the generally accepted description of the quantum measurement in terms of an instantaneous collapse process.*

Beside the old and often academical polemics about hidden variables, our approach, in which we consider the quantum signal as the output of a black box [46], and in which we apply some techniques of treatment of data fruitfully developed in other domains (electronics [43], physical biology [47], quantum optics [28,36,37], neutronic interferometry [9,41]) is worth being pursued further. In last resort, a measurement is not possible without a macroscopic amplification process (a point emphasised by Bohr himself [22,23]). Non-linearities play an essential role in many amplification processes encountered in nature. This is why it would be worth, in all generality, investigating to which extent the ideas developed in the study of complex systems (chaotic dynamics, turbulence, ergodicity, mixing, decay of initial correlations, fractal dimensionality and so on) can be applied in the framework of the measurement theory in order to analyse the temporal development of a quantum interference pattern[l]. The present experiment, which was aimed at testing the validity of a model that combines contributions of hidden variable theories and of the theory of complex systems can be considered to constitute a first step of this general program. Finally, it is worth noting that recently the temporal correlations of single photons were tested in the framework of quantum cryptography. The goal of these experiments was to make use of the quantum randomness in order to create a quantum random number generator [48,49,50]. They did not reveal any "hidden" mechanism.

6 Appendix

6.1 Non-Markovian Nature of the Iteration Process

Before we evaluate numerically the value of the relative mean square root deviation g, it is useful to study some properties of the iteration map which

[l]Thanks to Simon Diner [46] for fruitful discussions and suggestions about this topic. Another justification of our approach can be found in recent attempts of formulating stochastic quantum mechanics in terms of non-Markovian processes [33,30,31,32,34,35] and will be developed in a forthcoming publication.

describes the temporal evolution -and randomization- of the hidden variables. This transformation, known as the tent map, sends, at the time, let us say, $t_i = i\,\tau_M$, the value x_i kept by the hidden variable during the time interval $[(i-1)\,\tau_M,\ i\,\tau_M,[$ on the value x_{i+1}, that it will keep during the time interval $[i\,\tau_M,\ (i+1)\,\tau_M[$, according to the following transformation: $x_{i+1} = 2\,x_i$ if $0 \leq x_i \leq 1/2$, $x_{i+1} = 2 - 2\,x_i$ if $1/2 \leq x_i \leq 1$. Obviously, this iteration is neither reversible, nor continuous in time, what is not so astonishing if we remember that the aim of our model is to provide a simulation of the quantum measurement and thus of the collapse process. A specific feature of the model, which differentiates it from most "Monte-Carlo-like simulations" of stochastic processes applied in the same framework, is that the averaged transition probabilities associated to the tent transformation are not Markovian. To show this, let us consider that the result of the Stern-Gerlach measurement performed at the time t_i is spin up, so that

$$0 \leq x_i \ < \ cos^2(\frac{\theta_{AB}}{2})$$

For generic values of x_0, when $1 << i$, x_i (see section 2) is distributed at random inside $[0,\ cos^2(\frac{\theta_{AB}}{2})]$. Now, let us consider the probabilities $P_{+,+}$ ($P_{+,-}$) that, at time t_{i+1}, another Stern-Gerlach measurement will yield the result spin up (down). Obviously, $P_{+,+} = 1 - P_{+,-}$. It is easy to show that, if

$$cos^2(\frac{\theta_{AB}}{2}) \geq \frac{2}{3}$$

$P_{+,+}$ is equal to

$$\frac{2cos^2(\frac{\theta_{AB}}{2}) - 1}{cos^2(\frac{\theta_{AB}}{2})}$$

whilst, if

$$cos^2(\frac{\theta_{AB}}{2}) \leq \frac{2}{3}$$

$P_{+,+}$ is equal to $\frac{1}{2}$. Similarly, one can show that the probability $P_{-,+}$ of having an outcome spin up at time t_{i+1} after that we get the outcome spin down at time t_i is equal to

$$\frac{cos^2(\frac{\theta_{AB}}{2})}{2 \cdot (1 - cos^2(\frac{\theta_{AB}}{2}))}$$

if

$$cos^2(\frac{\theta_{AB}}{2}) \leq \frac{2}{3}$$

and to 1 if

$$cos^2(\frac{\theta_{AB}}{2}) \geq \frac{2}{3}$$

One can also show that the probability of getting spin down at time t_i and also at time t_{i+2} is equal to $\frac{1}{2}$ when

$$\frac{1}{2} \leq cos^2(\frac{\theta_{AB}}{2}) \leq \frac{2}{3}$$

This value ($\frac{1}{2}$) is different from the value that one would expect if the randomization process was Markovian, in which case we would find a probability equal to $P_{-,+} \cdot P_{+,-} + P_{-,-} \cdot P_{-,-}$ (where $P_{-,-} = 1 - P_{-,+}$). Obviously, this last expression is equal, when

$$\frac{1}{2} \leq cos^2(\frac{\theta_{AB}}{2}) \leq \frac{2}{3}$$

to

$$\frac{cos^2(\frac{\theta_{AB}}{2})}{4 \cdot (1 - cos^2(\frac{\theta_{AB}}{2}))} + (\frac{2 - 3cos^2(\frac{\theta_{AB}}{2})}{4 \cdot (1 - cos^2(\frac{\theta_{AB}}{2}))})^2$$

an expression which differs from $\frac{1}{2}$ inside the interval of values of θ_{AB} under consideration.

6.2 Numerical Estimations

The results of our numerical estimations are summarised in the figures 1, 2, and 3. In the figure 1, we show the probability (in function of time) of getting the result "spin up" for a particular, randomly chosen, value of the hidden variable at time zero. We took $cos^2(\frac{\theta_{AB}}{2}) = \frac{1}{2}$ here. It can be checked that, in average over time, the probability is close to $\frac{1}{2}$. In the figures 2 a and b, we show how, when one averages the probability of particular histories over a large set of initial hidden variables taken at random inside the interval [0, 1], the system quickly forgets the past and the average probability converges to the "equilibrium", quantum probability, after some iterations of the tent map. In the figures 3 a and b, we show how the distribution of the time-delays, obtained after averaging over a large sample of numerically simulated histories, exhibits bunching (over-Poissonian behaviour).

6.3 A Fully Solvable Approached Model

A model for which one can find analytical solutions is often more appealing than a model for which one has nothing more than numerical estimations.

Therefore, we will introduce in this subsection a new method of deduction of the distribution of time delays, thanks to a slightly different modellization of the memory effect. Let us assume that one atom passes at the level of the detector at the time zero, emits a photon, and that this photon is detected. Let us now decompose the detection process of the following photon (at time t) in an infinity of elementary processes during which N atoms passed at the level of the detector ($0 \leq N \leq \infty$), but none of them caused a beep, either because the atom did not emit a photon, or because the detector "refused to see the photon". Usually, even when one assumes that the detector remembers previous detections, for instance when its dead time is not negligibly small, one considers that such a remanent effect is independent of the quantum properties of the system under measurement,*which is not the case here.* Let us assume that the phase inside the interference pattern is equal to θ_{AB}, and let us denote $P_{\theta_{AB}}(t)$ the corresponding distribution of time delays ($0 \leq t \leq \infty$). Let us now evaluate the time dependent sensibility of the detector $P_{D,\ \theta_{AB}}(t)$. We must take into account the quantum emission process but also the sensibility of the detector, which was assumed to fire at the time $t = 0$. In the present approach, $P_{D,\ \theta_{AB}}(t)$ is an averaged expression where individual histories are globally considered, and not followed in detail as in the numerical estimations of the previous subsection. Let us then represent $P_{D,\ \theta_{AB}}(t)$ as a fastly decaying, exponential, function, of life time τ_M, equal to 1 for $t = 0$, and asymptotically converging towards $cos^2(\frac{\theta_{AB}}{2})$ for large times:

$$P_{D,\ \theta_{AB}}(t) = (1 - cos^2(\frac{\theta_{AB}}{2})) \cdot exp(\frac{-t}{\tau_M}) + cos^2(\frac{\theta_{AB}}{2})$$

In accordance with the choice that we made relatively to the calibration of the detector (this point was discussed already in the sections 3.2.1, 3.2.2 and 3.3.1), the maximum of the interference pattern is identified with the value $\theta_{AB} = 0$ in our model presented in the section 2. Note that in accordance with our hidden variable model, no memory effect occurs when $\theta_{AB} = 0$ because the probability of detection is then independent of the value of the hidden variable. We have thus that $P_{D,\ \theta_{AB} = 0}(t) = 1$ a situation in which the detector is assumed to be perfectly efficient. This implies that $P_{\theta_{AB}=0}(t)$ represents the temporal density of probability that a photon emitted by a (maximally) excited atom will arrive in the detector. This allows us to evaluate the contribution of the Nth elementary process to the distribution function $P_{\theta_{AB}}(t)$, so to say in the case where N photons passed at the level of the detector without being detected. This is formally equal to the following expression:

$$\int_0^t dt_1 \cdot \int_0^{t_1} dt_2 \cdot \int_0^{t_2} dt_3 ... \int_0^{t_{N-2}} dt_{N-1} \cdot \int_0^{t_{N-1}} dt_N \ P_{D,\ \theta_{AB}}(t - t_1)$$

$$\cdot P_{\theta_{AB}=0}(t - t_1) \cdot (1 - P_{D,\, \theta_{AB}}(t_1 - t_2)) \cdot P_{\theta_{AB}=0}(t_1 - t_2)$$
$$\cdot (1 - P_{D,\, \theta_{AB}}(t_2 - t_3)) \cdot P_{\theta_{AB}=0}(t_2 - t_3) \cdots$$
$$\cdots (1 - P_{D,\, \theta_{AB}}(t_{N-2} - t_{N-1})) \cdot P_{\theta_{AB}=0}(t_{N-2} - t_{N-1})$$
$$\cdot (1 - P_{D,\, \theta_{AB}}(t_{N-1} - t_N)) \cdot P_{\theta_{AB}=0}(t_{N-1} - t_N)$$

Inverting the order of the indices, we can rewrite this expression as follows:

$$\int_0^{t_2} dt_1 \cdot \int_0^{t_3} dt_2 \cdot \int_0^{t_4} dt_3 \ldots \int_0^{t_{N-1}} dt_{N-2}$$
$$\cdot \int_0^{t_N} dt_{N-1} \cdot \int_0^{t} dt_N \; P_{\theta_{AB}=0}(t_1)$$
$$\cdot (1 - P_{D,\, \theta_{AB}}(t_1)) \cdot P_{\theta_{AB}=0}(t_2 - t_1) \cdot (1 - P_{D,\, \theta_{AB}}(t_2 - t_1))$$
$$\ldots$$
$$P_{\theta_{AB}=0}(t_{N-1} - t_{N-2}) \cdot (1 - P_{D,\, \theta_{AB}}(t_{N-1} - t_{N-2}))$$
$$\cdot P_{\theta_{AB}=0}(t_N - t_{N-1}) \cdot (1 - P_{D,\, \theta_{AB}}(t_N - t_{N-1}))$$
$$\cdot P_{\theta_{AB}=0}(t - t_N) \cdot P_{D,\, \theta_{AB}}(t - t_N)$$

This expression is clearly a rough estimation of the predictions that can be made in the framework of the original model presented in the section 2, because the sensibility of the detector, even taken in average, ought to depend on the whole history that it underwent in the past. The requirement that a particular history will happen ought to induce a selection on the initial values of the hidden variables which are consistent with this history. In the previous expression, we implicitly consider that these initial values do not depend on the history and fulfill the relation $0 \leq x_i < cos^2(\frac{\theta_{AB}}{2})$ $(i = 1 ... N)$. This is rather crude, because by doing so we overevaluate the sensibility of the detector at all the times $(t, t_1, t_2, ...)$ excepted at the time t_1. The results that we obtain by this method constitute thus an overevaluation of the populations for long times, which tends to render the distribution more overPoissonian (bunched) than it is in reality. Nevertheless, at the first order in τ_M, the dominating contributions precisely come from the lowest values of N, so that, at this order of the perturbation series, our approximation is valid.

By taking the Laplace transform $L(P_{\theta_{AB}}(t))$ of the distribution $P_{\theta_{AB}}(t)$ and summing over all the elementary processes $(0 \leq N \leq \infty)$, we can replace the convolutions of the previous expression by products, so that we obtain the following relation:

$$L(P_{\theta_{AB}}) = L(P_{D,\, \theta_{AB}} \cdot P_{\theta_{AB}=0}) + L(P_{D,\, \theta_{AB}} \cdot P_{\theta_{AB}=0})$$
$$\cdot (L(P_{\theta_{AB}=0}) - L(P_{D,\, \theta_{AB}} \cdot P_{\theta_{AB}=0})) + L(P_{D,\, \theta_{AB}} \cdot P_{\theta_{AB}=0})$$

$$\cdot(L(P_{\theta_{AB}=0}) - L(P_{D,\,\theta_{AB}} \cdot P_{\theta_{AB}=0}))^2 + ...$$
$$+L(P_{D,\,\theta_{AB}} \cdot P_{\theta_{AB}=0}) \cdot (L(P_{\theta_{AB}=0}) - L(P_{D,\,\theta_{AB}} \cdot P_{\theta_{AB}=0}))^N + ...$$

We can resume this series, which provides the final result:

$$L(P_{\theta_{AB}}) = \frac{L(P_{D,\,\theta_{AB}} \cdot P_{\theta_{AB}=0})}{(1 - L((1 - P_{D,\,\theta_{AB}}) \cdot P_{\theta_{AB}=0}))}$$

Now, $P_{\theta_{AB}=0}$ was accurately measured and it turns out that it is a Poissonian distribution:

$$P_{\theta_{AB}=0}(t) = \frac{1}{T} exp(\frac{-t}{T})$$

where T is inversely proportional to the maximal intensity I_{max} of the interference pattern taken at the phase $\theta_{AB} = 0$: $\frac{1}{I_{max}} = T$. We estimated $P_{D,\,\theta_{AB}}$ to be equal to

$$(1 - cos^2(\frac{\theta_{AB}}{2})) \cdot exp(\frac{-t}{\tau_M}) + cos^2(\frac{\theta_{AB}}{2})$$

so that we can now compute the value of $L(P_{\theta_{AB}})$. We obtain that

$$L(P_{\theta_{AB}}) = \frac{\frac{p}{T} + \frac{1}{T} \cdot (\frac{p_1}{\tau_M} + \frac{1}{T})}{p^2 + p \cdot (\frac{1}{\tau_M} + \frac{2}{T}) + \frac{1}{T} \cdot (\frac{p_1}{\tau_M} + \frac{1}{T})}$$

The inverse Laplace transform of this expression can be obtained explicitly, and is equal to the sum of a shortly decaying and of a fastly decaying exponential functions. This is a rather intricate expression, but, if we limit ourselves to the zeroth and first order terms of the analytical development in $\tau_M \cdot I_{max}$ of the coefficients which appear in it, we obtain:

$$P_{\theta_{AB}}(t) = \{1 + \tau_M \cdot I_{max} \cdot (cos^2(\frac{\theta_{AB}}{2}) - 1)\} \cdot \gamma exp(-\gamma t)$$
$$+ \{\tau_M \cdot I_{max} \cdot (1 - cos^2(\frac{\theta_{AB}}{2}))\} \cdot \delta exp(-\delta t)$$

where

$$\gamma = \frac{1}{T} \cdot \{cos^2(\frac{\theta_{AB}}{2}) + (1 - cos^2(\frac{\theta_{AB}}{2}))^2 \cdot \tau_M \cdot I_{max}\}$$

and

$$\delta = \frac{1}{T \cdot \tau_M \cdot I_{max}} \cdot \{1 + (2 - cos^2(\frac{\theta_{AB}}{2})) \cdot \tau_M \cdot I_{max}\}$$

This expression is similar to the one that we deduced from an intuitive reasoning in the section 2, in which we took all histories into account.

Note that, once again, in the limit of vanishing memory times ($\tau_M = 0$), we obtain that

$$P_{\theta_{AB}}(t) = \gamma exp(-\gamma t) = \frac{1}{T}\cdot cos^2(\frac{\theta_{AB}}{2})exp(-\frac{1}{T}\cdot cos^2(\frac{\theta_{AB}}{2})t) = Iexp(-It)$$

which shows that in absence of memory effects, the distribution is Poissonian everywhere inside the interference pattern.

6.4 Theoretical Estimation of the Correlation Factor g

In order to compute the correlation factor g associated to the distribution of time-delays $P_{\theta_{AB}}(t)$, let us write the distribution function as the sum of two normalised exponential functions of life times α_1 and α_2, with weights (1 - w_1) and w_1, where the lowest power in the development of w_1 and $\frac{\alpha_2}{\alpha_1}$ in terms of $\tau_M \cdot I_{max}$ is equal to one, as it is the case in the framework of our model. Let us evaluate g by limiting ourselves to the zeroth and first order terms in w_1 and $\frac{\alpha_2}{\alpha_1}$ which appear in its development. This corresponds to the first order perturbation term, in a series of powers of $\tau_M \cdot I$.

Let us express g as $\frac{\sigma}{\mu}$ where σ and μ are the mean square root deviation and the average of the distribution $P_{\theta_{AB}}(t)$:

$$\mu = w_1 \cdot \alpha_1^{-1} + (1 - w_1) \cdot \alpha_2^{-1}$$

and

$$\sigma^2 = 2(w_1 \cdot \alpha_1^{-2} + (1 - w_1) \cdot \alpha_2^{-2}) - \mu^2$$

So

$$\begin{aligned} g^2 &= 2(\frac{w_1 \cdot \alpha_1^{-2} + (1 - w_1) \cdot \alpha_2^{-2}}{(w_1 \cdot \alpha_1^{-1} + (1 - w_1) \cdot \alpha_2^{-1})^2}) - 1 \\ &= 2(\frac{w_1 \cdot (\frac{\alpha_2}{\alpha_1})^2 + (1 - w_1)}{w_1^2(\frac{\alpha_2}{\alpha_1})^2 + 2w_1(1 - w_1)(\frac{\alpha_2}{\alpha_1}) + (1 - w_1)^2}) - 1 \end{aligned}$$

g^2 is then, up to first order terms in $\tau_M \cdot I_{max}$, equal to

$$\frac{2(1 - w_1)}{(1 - 2w_1)} - 1$$

so to say, to $1 + 2w_1$. Equivalently, at this order, $g = 1 + w_1$. From the previous section, where we evaluated the value of w_1, we have that:

$$g \approx 1 + \tau_M \cdot I_{max} \cdot (1 - cos^2(\frac{\theta_{AB}}{2}))$$

Remark that we obtain exactly the same expression for g if, instead of an exponential average memory time of the detector, which means

$$P_{D,\ \theta_{AB}}(t) = (1 - cos^2(\frac{\theta_{AB}}{2})) \cdot exp(\frac{-t}{\tau_M}) \ + \ cos^2(\frac{\theta_{AB}}{2})$$

we assume that the memory time is a step function, which means

$$P_{D,\ \theta_{AB}}(t) = 1$$

when $0 \leq t \leq \tau_M$ and

$$P_{D,\ \theta_{AB}}(t) \ = \ cos^2(\frac{\theta_{AB}}{2})$$

when $\tau_M \ < t \ < \infty$, and that we repeat the computations of $P_{\theta_{AB}}(t)$ and g up to the same order in $\tau_M \cdot I_{max}$. Remark also that the value of the ratio $\frac{\alpha_2}{\alpha_1}$ does not influence the value of g at the order considered here, provided the lowest power in its development in terms of $\tau_M \cdot I_{max}$ is equal to one. It is the value of the weight w_1 of the fastly decaying distribution that matters. This confirms that the value of g is somehow independent of the shape of the average memory time function of the detector.

6.5 *Modulation of the Poissonian Signal by an External Periodical Perturbating Field*

It would be difficult to justify the validity of our experimental results if the device was not sensitive enough to short perturbations. In order to check the response of the device to such perturbations, we measured the perturbation brought to the distribution of time-delays by external periodical disturbances the period of which is comparable to the duration of the transient effects that we expect to occur if the hypothesis of memory-times is valid. These disturbances simply consisted of a periodical modulation of the external electrical field undergone by the atoms inside the device (normally, this field is equal to zero) obtained thanks to an external AC voltage generator of modulable frequency. The resulting distribution of time-delays is represented in the figures 6 a and b. We checked the response of the system in the case of a modulation at the level of the source, of the detector and during the time of flight. The results are in excellent agreement with the theoretical predictions that we shall deduce now. We showed in the third section that when successive atoms undergo independent histories, the probability $\mathcal{P}(t)$. dt that the next photon is detected inside the interval $[t_0 + t,\ t_0 + t + dt]$ when a first photon is detected at time t_0 is equal to

$$\frac{P_Q\, P_G}{T} \cdot exp(-\frac{P_Q\, P_G}{T}t) \cdot dt$$

This hypothesis of independence is no longer true if the interactions undergone by the atoms inside the interferometer are time-dependent, because this dependence makes it possible to differentiate successive atoms. We have now that the external modulation "kills" a part of the atoms so that the intensity of the atomic beam is periodically modulated. For simplicity, let us consider a cosinusoidal modulation, the generalisation to arbitrary periodical pertubations is straightforward. The probability $\mathcal{P}_{neg}(t_0, t_0 + n \cdot dt)$ that no detection occurs between t_0 and $t_0 + t$, where $t = n \cdot dt$ is now equal in first approximation [51] to

$$\Pi_{i=1}^{i=n}\left[(1 - \frac{P_Q P_G}{T} \cdot dt) \cdot (1 - \alpha^2 cos^2(\omega t))\right]$$

where α is proportional to the modulation depth and ω is equal to 2π times its frequency. By taking the limit of infinitesimally short time intervals, we obtain an integral form of this relation:

$$\mathcal{P}_{neg}(t_0, t_0 + t) = K \cdot exp(-\int_{t_0}^{t_0 + t} \frac{P_Q P_G}{T} \cdot (1 - \alpha^2 cos^2(\omega t)) \cdot dt)$$

where K is a normalization constant. The probability to detect the first atom between t_0 and $t_0 + dt_0$ and the second one between $t_0 + t$ and $t_0 + t + dt$ is proportional to

$$(1 - \alpha^2 cos^2(\omega t_0)) \cdot dt_0 \cdot \mathcal{P}_{neg}(t_0, t_0 + t) \cdot (1 - \alpha^2 cos^2(\omega t_0 + t)) \cdot dt$$

If we average this relation on the detection times of the initial photon, we get that the probability $\mathcal{P}(t)$. dt that the next photon is detected inside the interval $[t_0 + t,\ t_0 + t + dt]$ when a first photon is detected at time t_0 is, up to a normalization constant, equal to

$$\int_0^{2\pi} dt_0\ (1 - \alpha^2 cos^2(\omega t_0))(1 - \alpha^2 cos^2(\omega(t_0 + t))$$

$$\cdot exp(-\int_{t_0}^{t_0 + t'} \frac{P_Q P_G}{T} \cdot (1 - \alpha^2 cos^2(\omega t')) \cdot dt').dt$$

This formula corresponds very well to the observations (see figures 1 and 2). A straightforward generalisation of this expression was also successfully tested for different shapes of periodical perturbations (triangle-shaped or square-shaped for instance), in a wide spectrum of frequencies.

Acknowledgements

One of the authors (T.D.) is presently a post-doctoral fellow of the Flemish Fund for Scientific Research (FWO). During the realisation of a part of this

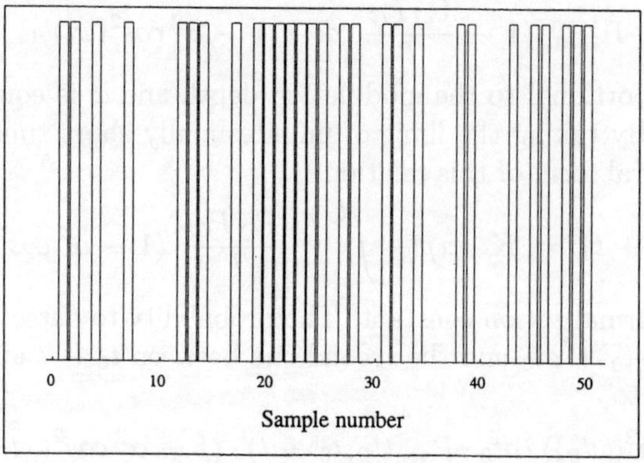

Figure 1: $f(t, x_{t_0})$ with $x_{t_0} \in [0, 0.5], t = 50\,\tau$ and $\cos^2(\frac{\theta_{AB}}{2}) = 0.5$

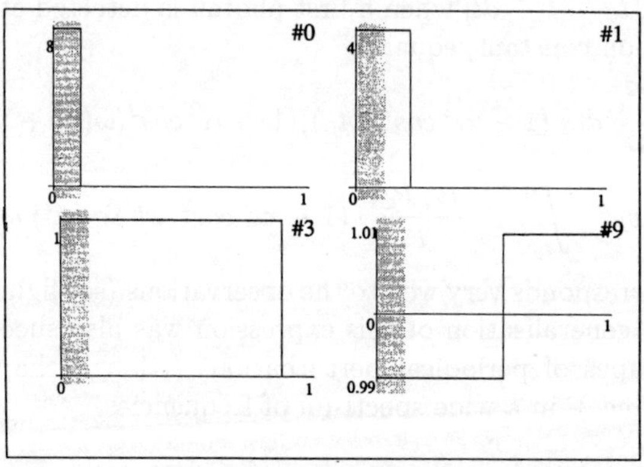

Figure 2a: distribution of x_i when $\cos^2(\frac{\theta_{AB}}{2}) = \frac{1}{8}$, and $i = 0, 1, 3, 9$.

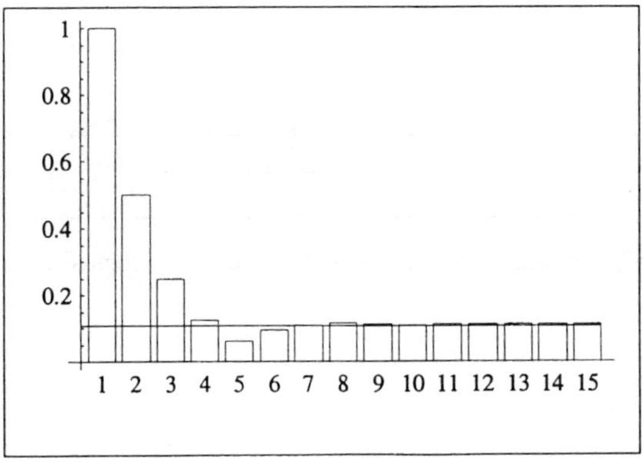

Figure 2b: $f(x_i, t_0)$ averaged over x_{t_0} with $0 \leq x_{t_0} \leq \cos^2(\frac{\theta_{AB}}{2})$ as a function of i. $i : 0 \to 15$

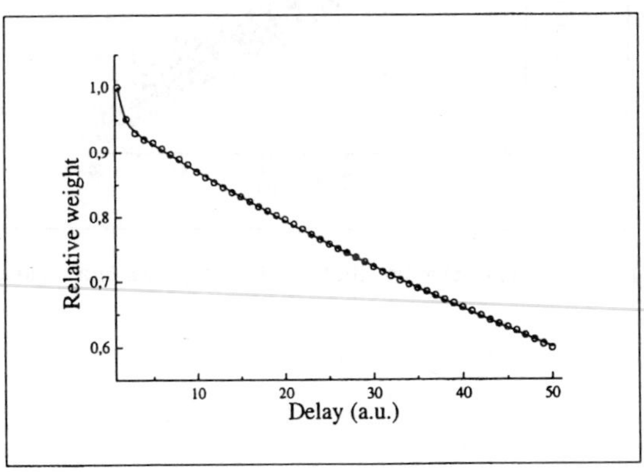

Figure 3a: Numerical expectation of $P(t)$ as a function of time, with $t \in [0, 50\,\tau]$, in logarithmic scale, with $\tau \ll T$.

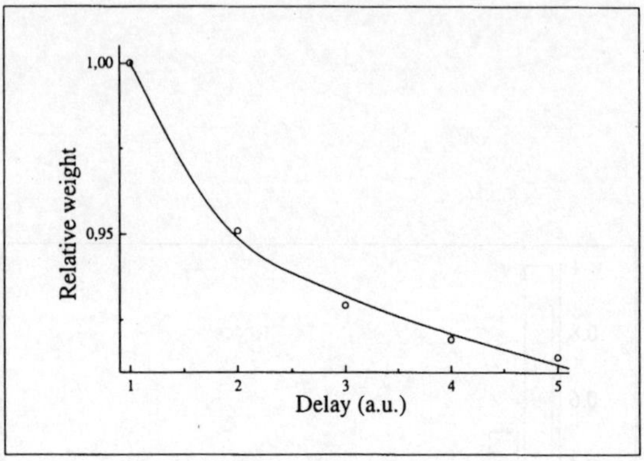

Figure 3b: Bunching contribution of the curve represented in figure 3a

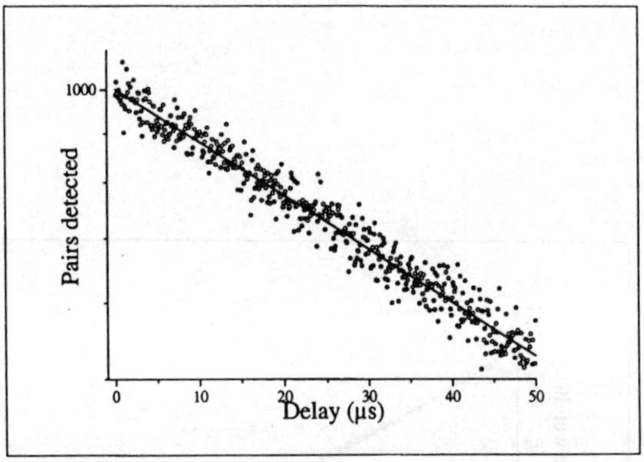

Figure 4: Histogram of the distribution of time-delays in a logaritmic scale

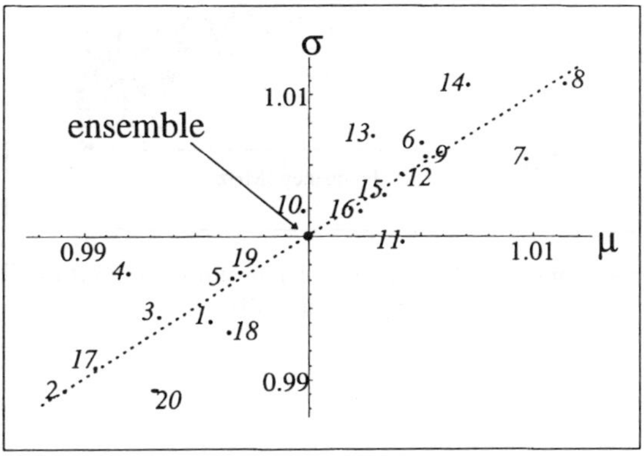

Figure 5: σ_N versus t_N for 20 experimental runs of $1.3\,10^5$ time-delays each.

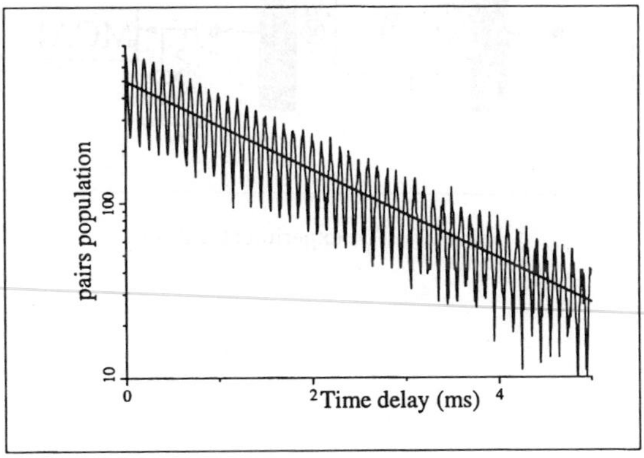

Figure 6a: Distribution of the time-delays in the presence of an external disturbance at a frequency of 10 kHz.

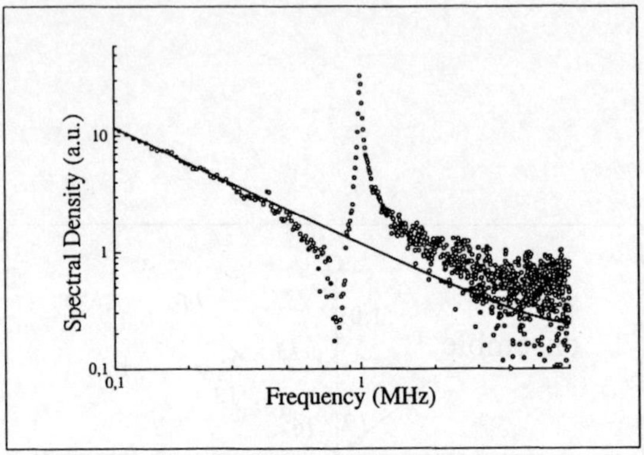

Figure 6b: Fourier-transform of the distribution of time-delays with a disturbance at a frequency of 1 MHz.

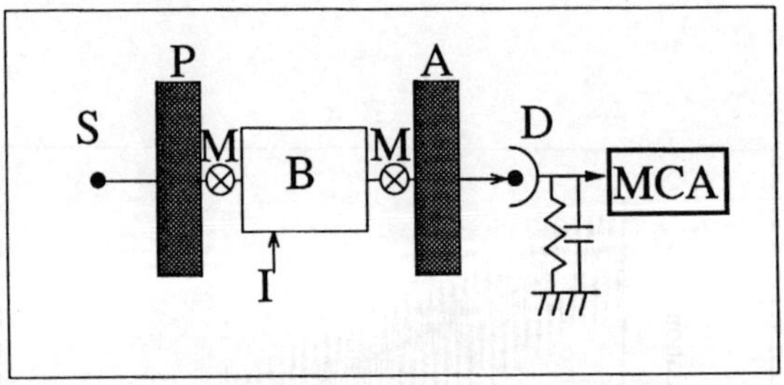

Figure 7: The experimental device

work, he enjoyed the support of the Flemish-Polish Scientific Collaboration Program No. 007 entitled "Probing the structure of Quantum Mechanics: New probability models for new experiments on quantum particles", and of the FWO project entitled "Study of the effects of fluctuations during the interactions between measuring apparatus and system in the framework of new experiments on individual quantum and mesoscopic systems".

References

1. E. Schrödinger, "Die Gegenwärtige Situation in Der Quantenmechanik", *Naturwissenschaften*, **23**, 807 (1935), translated by J. Trimmer under the title "The present situation of quantum mechanics: a translation of Schrödinger's 'cat paradox' paper", in *Proc. Amer. Phil. Soc.*, **124**, 323 (1980).
2. D. Bohm, "A suggested interpretation of quantum theory in terms of hidden variables", *Phys. Rev.*, **85**, 166 (1952).
3. D. Bohm and J. P. Vigier, "Model of the causal interpretation of quantum theory in terms of a fluid with irregular fluctuations", *Phys. Rev.*, **96**, 208 (1954).
4. J. S. Bell, "On the EPR paradox", *Physics*, **1**, 195 (1964).
5. See for instance G. Weihs, T. Jennewein, C. Simon, H. Weinfurter and A. Zeilinger, *Phys. Rev. Lett.*,**81**, 5039 (1998), and references therein.
6. H. Zbinden, J. Brendel, W. Tittel and N. Gisin, "Experimental test of relativistic quantum state collapse with moving reference frames", *quant-ph0002031* (2000).
7. C. Papaliolos, "Experimental test of a hidden variable quantum theory", *Phys. Rev. Letters*, **18**, 622 (1967).
8. D. Bohm and J. Bub, "A proposed solution of the measurement problem in quantum mechanics by a hidden variable theory", *Rev. Mod. Phys.*, **18**, 453 (1966).
9. J. Summhammer, "Neutron interferometric test of the nonergodic interpretation of quantum mechanics", *Nuov. Cim.* **103 B**, 265 (1985).
10. V. Buonomano, "Summhammer's experimental test of the non-ergodic interpretation of quantum mechanics", *Founds. Phys. Lett.*, **2**, 565 (1989).
11. V. Buonomano and F. Bartmann, "Testing the ergodic assumption in the low-intensity interference experiments", *Nuov. Cim.*, bf 95 B, 99 (1986).
12. F. J. Belinfante, *A Survey of Hidden Variable Theories*, Pergamon, Oxford (1973).
13. G. Cerofolini, "On the formal equivalence between a reformulation of Bohm's and Bub's hidden-variable theory and subquantum Mechanics",

Lettere al Nuov. Cim., **35**, 457 (1982).
14. J. H. Tutsch, "Simultaneous measurement in the Bohm-Bub hidden variable theory", *Phys. Rev.*, **183**, 1116 (1969).
15. D. Aerts, "A possible explanation for the probabilities of quantum mechanics", *J. Math. Phys.*, **27**, 203 (1986).
16. L. Accardi and A. Fedullo, "On the statistical meaning of the complex numbers in quantum mechanics", *Nuov. Cim.*, **34**,161 (1982).
17. L. Accardi, "The probabilistic roots of the quantum mechanical paradoxes", in *The Wave-Particle Dualism*, eds. S. Diner *et al.*, Kluwer Academic, Dordrecht (1984).
18. D. Aerts, S. Aerts, T. Durt and O. Leveque, "Classical and quantum probability in the ϵ-model", *Int. J. Theor. Phys.*, **38**, 407 (1999).
19. M. Czachor, "On classical models of spin", *Found. Phys. Lett.*,**5**, 249 (1992).
20. T. Durt, *From quantum to classical, a toy model*, Doctoral thesis, Brussels Free University, Brussels (1996).
21. T. Durt, "Do dice remember?", *Int. J. Theor. Phys.*, **38**, 457 (1999).
22. N. Bohr, "Can quantum mechanical description of physical reality be considered complete?", *Phys. Rev.*, **48**, 696 (1935).
23. N. Bohr, "Discussion with Einstein on epistemological problems in atomic physics", in *Albert Einstein: Philosopher-Scientist*, ed. P. A. Schilpp, The Library of Living Philosophers, Evanston, 200 (1949).
24. J. S. Bell, "On the problem of hidden variables in quantum mechanics", *Rev. Mod. Phys.*, **38**, 447 (1966).
25. H. R. Brown, "Bell's other theorem and its connection with non-locality, Part 1", in *Bell's Theorem and the Foundations of Modern Physics*, eds. A. van der Merwe *et al*, World Scientific, Singapore (1992).
26. T. Durt, "Three interpretations of the violation of Bells inequalities", *Found. Phys.*, **27**, 415 (1997).
27. S. Kochen and E. Specker, "The problem of hidden variables in quantum mechanics", *J. Math. Mech.*, **17**, 1967, 59 (1967).
28. R. Hanbury-Brown and R. Q. Twiss, "Interferometry of the intensity fluctuations in light I and II *Proc. Roy. Soc.* (London) **A 242**, 300 and **A 243**, 291 (1957 and 1958).
29. L. Mandel and E. Wolf, *Optical Coherence and Quantum Optics*, Cambridge University Press (1995).
30. D. T. Gillespie, "Untenability of simple ensemble interpretations of quantum measurement probabilities", *Am. J. Phys.* **54**, 889 (1986).
31. D. T. Gillespie, "Why quantum mechanics cannot be formulated as a Markov process", *Phys. Rev. A* **49**, 1607 (1994).

32. G. Skorobogatov and R. Svertilov, "Quantum mechanics can be formulated as a non-Markovian process", *Phys. Rev. A*, **58**, 3426 (1998).
33. L. Diosi, N. Gisin and W. T. Strunz, "Non-markovian quantum state diffusion", *Phys. Rev. A*, **58**, 1699 (1998).
34. W. T. Strunz, L. Diosi and N. Gisin, "Open system dynamics with Non-Markovian quantum trajectories", *Phys. Rev. Lett.*, 1801 (1999).
35. T. Yu, L. Diosi, N. Gisin and W. T. Strunz, "Post-Markov master equation for the dynamics of open quantum systems", *quant-ph/9905006*, 1-6 (1999).
36. F. Dietrich and H. Walther, "Non-classical radiation of a single stored ion", *Phys. Rev. Lett.*, **58**, 203 (1987).
37. L. Mandel and R. Short, "Observation of sub-Poissonian photon statistics", *Phys. Rev. Lett.*, **51**, 384 (1983).
38. J. Lawson-Daku, *Effets Transverses et Temporels en Interferometrie Atomique Stern-Gerlach*, PhD Thesis, Universite de Paris-Nord (1997).
39. B. J. Lawson-Daku, R. Asimov, S. Nic Chormaic, O. Gorceix, Ch. Miniatura, J. Robert and J. Baudon, "Time selection in atomic Stern-Gerlach interferometry", *Phys. Rev. A.*, **52**, 1457 (1995).
40. J. Baudon, J. Robert, Ch. Miniatura, O. Gorceix, B. J. Lawson-Daku, K. Brodsky, R. Mathevet and F. Perales, "Interferometry with atoms", *Comments At. Mol. Phys.*, **34**, 161 (1999).
41. M. Zawisky, H. Rauch and Y. Hasegawa, "Contrast enhancement by time-selection in neutron-interferometry", *Phys. Rev. A*, **50**, 5000 (1994).
42. E. Jericha, H. Rauch, J. Summhammer and M. Zawisky, "Low contrast and low-counting-rate measurements in neutron-interferometry", *Phys. Rev. A*, **50**, 5000 (1994).
43. A. Tonomura, J. Endo, T. Matsuda and T. Kawasaki, *Am. J. Phys.*, **57**, 117 (1989).
44. F. Shimizu, "Observation of two-atom correlation of an ultracold neon atomic beam", *Phys. Rev. Lett.*, **77**, 3090 (1996).
45. R. Sheldrake, *Une Nouvelle Science de la Vie*, eds. du Rocher, Monaco (1985).
46. S. Diner, *Private communications and discussions* at the Symposium *Murmures quantiques*, Peyresq, France, July 1997.
47. J. R. Brochon, F. Merola and A. K. Livesey, "Time-Resolved fluorescence study of dynamics parameters in biosystem, in *Proceedings of the 48th International Meeting of Physical Chemistry*, France 1991, eds. A. Beswick *et al*, Conf. Proc. n°258, American Institute of Physics (1992).
48. N. Gisin, G. Ribordy, W. Tittel and H. Zbinden, "Quantum cryptography", quant-ph/0101098, submitted to the Review of Modern Physics

(2001).
49. T. Jennewein, U. Achleitnert, G. Weihs, H. Weinfurter and A. Zeilinger, "A fast and compact quantum random number generator", quant-ph/9912118 (1999).
50. A. Stefanov, N. Gisin, O. Guinnard, L. Guinnard and H. Zbinden, "Optical quantum random nuber generators", *quant-ph/9907006* (1999).
51. J. P. Dowling, C. P. Williams and J. D. Franson, "Maxwell duality, Lorentz invariance and topological phase", *Phys. Rev. Letters*, **83**, 2486 (1999).

Note that the references 7, 8, 9, 12, 13 and 40 are reproduced in the book *Quantum theory and measurement* eds. J. A. Wheeler and W. H. Zurek, Princeton, N-J (1983).

REALITY AND PROBABILITY: INTRODUCING A NEW TYPE OF PROBABILITY CALCULUS

DIEDERIK AERTS
Center Leo Apostel (CLEA) and Department of Mathematics (FUND), Brussels Free University, Krijgskundestraat 33, 1160 Brussels, Belgium
E-mail: diraerts@vub.ac.be

We consider a conception of reality that is the following: An object is 'real' if we know that if we would try to test whether this object is present, this test would give us the answer 'yes' with certainty. The knowledge about this certainty we gather from our overall experience with the world. If we consider a conception of reality where probability plays a fundamental role, which we should do if we want to incorporate the microworld into our reality, it can be shown that standard probability theory is not well suited to substitute 'certainty' by means of 'probability equal to 1'. We analyze the different problems that arise when one tries to push standard probability to deliver a conception of reality as the one we advocate. The analysis of these problems lead us to propose a new type of probability theory that is a generalization of standard probability theory. This new type of probability theory is a function to the set of all subsets of the interval $[0, 1]$ instead of to the interval $[0, 1]$ itself, and hence its evaluation happens by means of a subset instead of a number. This subset corresponds to the different limits of sequences of relative frequency that can arise when an intrinsic lack of knowledge about the context and how it influences the state of the physical entity under study in the process of experimentation is taken into account. The new probability theory makes it possible to define probability on the whole set of experiments within the Geneva-Brussels approach to quantum mechanics, which was not possible with standard probability theory. We introduce the formal mathematical structure of a 'state experiment probability system', by using this new type of probability theory, as a general description of a physical entity by means of its states, experiments and probability. We derive the state property system as a special case of this structure, when we only consider the 'certain' aspects of the world. The category **SEP** of state experiment probability systems and their morphisms is linked with the category **SP** of state property systems and their morphisms, that has been studied in earlier articles in detail.

1 Introducing the Problem

We first introduce the conceptual tools that we consider for our problem. We suppose that we have at our disposal a physical entity S that can be in different states $p, q, r, \ldots \in \Sigma$, where Σ is the set of all relevant states of the entity S. We also have at our disposal different experiments $\alpha, \beta, \gamma, \ldots \in Q$ that can be performed on the entity S. We suppose that each experiments has only two possible outcomes, that we label 'yes' and 'no'. Q is the set of all relevant yes/no-experiments.

1.1 *What is Real*

Let us consider an example. The physical entity S that we consider is a piece of wood[a]. We consider two yes/no-experiments α and β that can be performed on this piece of wood and that test respectively whether the piece of wood 'burns well' and whether the piece of wood 'floats on water'. The experiment α consists, for example, in setting the piece of wood on fire and observing whether it keeps burning for a sufficiently long time. We give the outcome 'yes' if this is the case, and the outcome 'no' if this is not the case. The experiment β consists of putting the piece of wood in water and observing whether it floats. We give the outcome 'yes' if this is the case and the outcome 'no' if this is not the case[b].

We can say that experiment α tests the property of 'burning well' of the piece of wood, let us call this property a, while experiment β tests the property of 'floating on water' of the piece of wood, and we call this property b.

First of all we have to understand clearly when we say that a piece of wood 'has' a certain property, or, in other words, when a certain property is 'actual' for a piece of wood. Let us focus on property a to analyze this. We say that a piece of wood 'has' the property of 'burning well' if its state is such that if we would perform experiment α the outcome 'yes' would come out with certainty. We remark that this is not related to the fact that we have performed the experiment, because indeed after we have put the piece of wood on fire, and after it has burned, it will not have the property of 'burning well' any longer. The experiment α destroys the property a. This is common place in the world. However not always an experiment that tests a property also destroys this property. This is for example not the case for property b. If the state of the piece of wood was such that we could predict with certainty that it would float on water in the eventuality of performing the experiment β, even after having performed experiment β, the state of the piece of wood will still be such. The important thing to notice however is that this difference, between the effect of the experiments α and β, the first destroying the property that it tests, and the second not doing so, does not play a role in the aspect of the experiments that defines the actuality of the properties that they test. If we say that a Volvo is a 'strong' car, this refers to experiments that have been carried out in the Volvo factory, consisting of

[a]This example was first introduced in [1,2].

[b]We remark that are types of wood that do not float on water. For example the African wood called 'wenge'.

hitting a brick wall with the car and measuring the amount of damage. A new Volvo however, the one of which we say that it is a strong car, has not undergone these experiments, exactly because the experiments will destroy the car. Still we all believe that this new Volvo 'has' the property of being a strong car. We all demand that the experiment to test the strength has not been carried out on the car that we buy, and still we believe that this car, the one that we buy, has this property of strength, because we believe that in the eventuality of a test, the test will give the 'yes' outcome with certainty.

It is important for the rest of this article to understand this subtle way of reality. What is real is what we believe to react in a certain way with certainty to experiments that we eventually could perform, but that in general we do not perform, because in many cases these experiments destroy what we were considering to be real. What is real is related to what we eventually could test, but in general do not test. This is also the way we conceive of what is real in our everyday world, as the example of the Volvo car makes clear.

1.2 Attributing Several Properties at Once

When do we say that a piece of wood 'has' both the properties a and b? The subtility of the nature of 'what is real' already appears in full splendor if we analyze carefully the way in which we attribute several properties at once to an entity. At first sight one could think that attributing the two properties a and b to the piece of wood has to do with performing both experiments at once, or one after the other, or ..., well at least performing the two experiments in one way or another. This is wrong. It is in fact one of the deep mistakes of classical logic, where one introduces indeed the conjunction of propositions by means of truth tables of both proposition, as if the truth of the conjunction would be defined by verifying the truth of both propositions. We have chosen the example of the piece of wood on purpose such that this mistake becomes obvious. We would indeed be in great difficulties with our common sense conception of reality if attributing both properties 'burning well' and 'floating on water' would have to do with performing both experiments α and β in one way or another. Putting a piece of wood on fire and at the same time making it float on water does indeed not seem to be a very interesting enterprise in collecting knowledge about the state of the piece of wood. Performing one experiment after the other, in whatever order, also does not seem to be very fruitful. Indeed, if we first burn the piece of wood, it does not float on water afterwards, and if we first make it float in water, it does not burn afterwards. While we all know that we can have a very simple piece of wood in a state such that it 'has' both properties at once. Most pieces of wood indeed do have both properties at once most of the time. How do we arrive at this belief in our daily conception of reality? Let us analyze this matter.

What we do is the following. Suppose that we have a dry piece of light wood. What we say to ourselves is that whether we would perform experiment α or whether we would perform experiment β, both will give the outcome 'yes' with certainty, and that is the reason that we believe that this dry piece of light wood has both properties a and b at once. What we just come to say is clear for everybody I suppose. The question is how to formalize this. Let us make an attempt. Having available the two experiments α and β we make a new experiment, that we denote $\alpha \cdot \beta$, and call the product of α and β, as follows: The experiment $\alpha \cdot \beta$ consists of choosing one of the experiments α or β, and performing this chosen experiment, and collecting the outcome, 'yes' or 'no', and considering this outcome as the outcome of the experiment $\alpha \cdot \beta$. For the piece of wood and the two considered experiments this product experiment consists of choosing whether we want to test if the piece of wood burns well 'or' whether we want to test if the piece of wood floats on water, one of the two, and then to perform this chosen experiment, and collect the outcome. It is clear that $\alpha \cdot \beta$ will give with certainty the outcome 'yes' if and only if both experiments α 'and' β will give with certainty the outcome 'yes'. This proves that $\alpha \cdot \beta$ is an experiments that tests whether the system 'has' both properties a and b at once.

At first sight strangely enough, our analysis shows that to test the property a 'and' b we need to use the experiment α 'or' β. The logical 'and' changes into a logical 'or' if we shift from the properties to the experiments to tests them. That is the reason that it makes sense to attribute a lot of properties at once, or the conjunction of all these properties, to an entity, even if these properties cannot be tested at once.

If the reader reflects on how we conceive ordinary reality around us, he will find that this is exactly what we do. We say that a Volvo is a strong car 'and' is x meters long, for example, because we know that if we would test one of the two properties, this one would give us an outcome 'yes' with certainty. Although testing the strength of the Volvo, by hitting it against a wall, would definitely change its length.

1.3 How Infinity Comes In

Remark that the definition that we have given for the product experiment works for any number of properties. Suppose that we have a set of properties $(a_i)_i$ of any size, and for each property an experiment α_i to test it, then the product experiment $\Pi_i \alpha_i$, consisting of choosing one of the experiments α_j

and performing it, and collecting the outcome, tests all of the properties a_i at once. The 'product operation' does not pose any problems with infinity. If it is physically relevant to consider an infinite set of properties for an entity, then the product experiment tests the conjunction of this infinite set of properties.

1.4 What About Probability?

What we have explained so far is old stuff, and can be found in the earlier articles that have been written on the subject[1,2,3,4,5,6,7]. From here on we want to start to introduce new things, more specifically how probability should be introduced and understood. We have considered only situations so far where the state of the entity was such that the considered experiments would give the outcome 'yes' with 'certainty', and the concept 'certainty' has been used somewhat loosely. Would it be possible to introduce the concept 'probability' such that what we have called 'certainty' in the foregoing coincides with the notion of probability equal to 1?

Let us explain first why this is not a trivial matter, and why it even seems impossible at first sight. We will see in the following sections that indeed we have to introduce probability in a completely different manner than usually is done to be able to see clear in the situation and to solve the problem. Consider for now that we have introduced probability in the conventional way.

This means that for a state p of the entity and an experiment α, we consider a number $\mu(\alpha, p)$ between 0 and 1, such that $\mu(\alpha, p)$ is the probability that if the entity is in state p the experiment α gives the outcome 'yes'. This means that we introduce a function

$$\mu : \Sigma \times Q \to [0,1] \tag{1}$$

$$(\alpha, p) \mapsto \mu(\alpha, p) \tag{2}$$

where Σ is the set of relevant state of the entity S and Q is the set of relevant yes/no-experiments. As we said already, $\mu(\alpha, p)$ is the probability that the experiment α gives the outcome 'yes' if the entity is in state p.

For standard probability theory $\mu(\alpha, p)$ is an element of the unit interval $[0, 1]$ of the set of real numbers $\mathbb{R}$. If we want to express the situation 'the state p of the entity S is such that experiment α would give outcomes 'yes' with certainty', by '$\mu(\alpha, p) = 1$', we have to make certain that our analysis of the product experiments remains valid.

1.5 *Probability of Product Experiments*

Let us consider a set of experiments $(\alpha_i)_i$, and suppose that for a certain state p of the entity S, all the probabilities $\mu(\alpha_i, p)$ are given. Consider now the product experiment $\Pi_i\alpha_i$. What could be the meaning and definition of $\mu(\Pi_i\alpha_i, p)$? Let us analyze this.

The product experiments $\Pi_i\alpha_i$ consists of choosing one of the experiments of the set $(\alpha_i)_i$ and performing this chosen experiment. We can express the 'act of choice' between the different elements of the set $(\alpha_i)_i$ by using the notion of probability itself. Indeed, let us suppose that there is probability x_i that we would choose the experiment α_i. This means that our act of choice is described by a set $(x_i)_i$ of real numbers, such that $x_j \in [0,1]$, and such that $\sum_i x_i = 1$. With this act of choice, the probability of the product experiments, the entity being in state p, is given by

$$\mu(\Pi_i\alpha_i, p) = \sum_i x_i\mu(\alpha_i, p) \tag{3}$$

Let us express the situation 'the state of the entity is such that the product experiment gives with certainty the outcome yes' by the formula '$\mu(\Pi_i\alpha_i, p) = 1$'. The question is now whether this implies that 'the state of the entity is such that all of the experiments α_i give with certainty the outcome yes'. This means that from $\mu(\Pi_i\alpha_i, p) = 1$ should follow $\mu(\alpha_i, p) = 1$ for all i. This is only true if all of the x_i are strictly greater than 0. Let us prove this.

Proposition 1 *Consider a product experiment for which the act of choice is defined by means of a set of real numbers $(x_i)_i \in [0,1]$ such that $\sum_i x_i = 1$, and hence $\mu(\Pi_i\alpha_i, p) = \sum_i x_i\mu(\alpha_i, p)$. We have:*

$$\{\mu(\Pi_i\alpha_i, p) = 1 \Rightarrow \mu(\alpha_j, p) = 1\ \forall j\} \Leftrightarrow \{0 < x_j\ \forall j\} \tag{4}$$

Proof: Suppose that $x_j = 0$ then $\sum_i x_i\mu(\alpha_i, p) = \sum_{i\neq j} x_i\mu(\alpha_i, p)$. This means that $\sum_i x_i\mu(\alpha_i, p) = 1 \Leftrightarrow \sum_{i\neq j} x_i\mu(\alpha_i, p) = 1$. This can however never imply that $\mu(\alpha_j, p) = 1\ \forall j$, since $\mu(\alpha_j, p)$ can be arbitrary without the sum changing its value. Take now $0 < x_j\ \forall j$ and suppose that there is a k such that $\mu(\alpha_k, p) < 1$. We have $\mu(\Pi_i\alpha_i, p) = \sum_i x_i\mu(\alpha_i, p) = x_k\mu(\alpha_k, p) + \sum_{i\neq k} x_i\mu(\alpha_i, p)$. Since $\mu(\alpha_i, p) \in [0,1]\ \forall i$ we have $\sum_{i\neq k} x_i\mu(\alpha_i, p) \leq \sum_{i\neq k} x_i$. And since $\sum_i x_i = 1$ we have $\sum_{i\neq k} x_i = 1 - x_k$. From this follows that $\mu(\Pi_i\alpha_i, p) \leq x_k\mu(\alpha_k, p) + (1 - x_k) = 1 - x_k(1 - \mu(\alpha_k))$. Because $\mu(\alpha_k, p) < 1$ we have $0 < 1 - \mu(\alpha_k, p)$, and because $0 < x_k$ also $0 < x_k(1 - \mu(\alpha_k))$. From this follows that $1 - x_k(1 - \mu(\alpha_k)) < 1$. So we have proven that $\mu(\Pi_i\alpha_i, p) < 1$. □

Suppose that the set of experiments $(\alpha_i)_i$ that we consider for the product experiments $\Pi_i\alpha_i$ is finite, and that we describe the act of choice corresponding to this product experiment by means of the set of numbers $(x_i)_i$, $x_j \in [0,1]$ with $\sum_i x_i = 1$. In this case it would make sense to take out all the $\mu(\alpha_j, p)$ for which $x_j = 0$, because anyhow the probability to choose such a α_j during the act of choice is 0, and redefine a new set of experiments $(\alpha_k)_k$ for the product experiment that only retain those ones for which $0 < x_k$. From proposition 1 follows that for this new set the property that we want is satisfied: the outcome 'yes' is certain for the product experiment if and only if the outcome 'yes' is certain for all the component experiments, where 'certainty' is defined as 'probability equals 1'.

For infinite sets this however does not work in an obvious way. Suppose for example that we have a set of experiments coordinated by an interval I of the set of real numbers $\mathbb{R}$. Hence $(\alpha_i)_{i \in I}$ is our set. The probability that one of the α_i is chosen will in general be equal to 0 then, because the requirement $\sum_i x_i = 1$ forces the individual x_i's to be 0. It is common to express everything by means of integrals instead of sums in this situation. For example $\sum_i x_i = 1$ will be replaced by $\int_I \rho(x)dx = 1$, where ρ is a function from I to $[0,1]$. For an interval $[j,k] \subset I$ we interpret $\int_{[j,k]} \rho(x)dx$ as the probability that an element α_i is chosen with $i \in [j,k]$.

There is an aspect that is very uncomfortable from the physical point of view with this standard description of the infinite case. The probability to choose 'one' of the α_i's equals 0, while in reality it are the α_i's individually that are performed in the laboratory when real tests are executed. So although the probability to choose one of the α_i's equals 0, such one of the α_i's 'gets' chosen in reality. This problem is known in standard probability theory. One usually concludes by stating that probability equals 0 does not mean impossibility. That is also the reason that in this case probability equals 1 cannot be interpreted as 'certainty'.

Because of the fundamental nature of the approach that we are developing, we cannot be satisfied with this state of affairs. We will show a way to solve this problem, but it demands the introduction of a new type of probability theory that is a generalization of standard probability theory.

2 Subset Probability

In this section we introduce a new type of probability description, where a probability will not be a number of the interval $[0,1]$, but a subset of this interval. Let us tell how we arrive at this model.

2.1 *Product Experiments and Subsets*

We consider again the situation where we have at our disposal a set of yes/no-experiments $(\alpha_i)_i$, and a physical entity S in a state $p \in \Sigma$. In the foregoing sections we have described the act of choice by introducing a set of real numbers $x_i \in [0,1]$ with $\sum_i x_i = 1$. Let us reflect a little bit more to see whether this is a good way to describe the act of choice that we have in mind when we have introduced the idea of a product experiment.

Such a set of numbers x_i means that we consider the situation as if there is a 'fixed' probability for each one of the component experiments α_i to be chosen, and this fixed probability is represented by the number x_i. We have put forward this description because it seems to be the obvious one starting from standard probability. We can however also look at the situation in a different way. We can consider the act of choice to be such that we do not have fixed probabilities for each one of the α_i's, but that we can choose each time again in a different way. A description of this idea would be that we choose each time again among all the possible sets of numbers $(x_i)_i \in [0,1]$, with $\sum_i x_i = 1$, and apply then this specific chosen set for a calculation of the probability of the product experiment. This means of course that the probability of the product experiment can have different values depending on the chosen set of numbers $(x_i)_i$ that describe the act of choice. We can prove the following proposition.

Proposition 2 *Consider a physical entity S with set of states Σ and set of yes/no-experiments Q. Suppose that $(\alpha_i)_i$ is a set of experiments, and $\Pi_i\alpha_i$ the product experiment. If we allow all possible acts of choice described each time by a set of real numbers $(x_i)_i \in [0,1]$ such that $\sum_i x_i = 1$, then $\mu(\Pi_i\alpha_i, p)$ can be any number of the convex hull $V(\{\mu(\alpha_i, p)\})$ of the set $\{\mu(\alpha_i, p)\}$.*

Proof: Consider a particular act of choice represented by the set of numbers $(x_i)_i \in [0,1]$ such that $\sum_i x_i = 1$. We have $\mu(\Pi_i\alpha_i, p) = \sum_i x_i\mu(\alpha_i, p) \in V(\{\mu(\alpha_i, p)\})$. On the other hand, suppose that $x \in V(\{\mu(\alpha_i, p)\})$. This means that there exists a set of numbers $(x_i)_i \in [0,1]$ with $\sum_i x_i = 1$ such that $x = \sum_i x_i\mu(\alpha_i, p)$. This means that if we consider the act of choice described by $(x_i)_i$ for $\Pi_i\alpha_i$, then $\mu(\Pi_i\alpha_i, p) = x$. □

If we consider all possible acts of choice, the convex hull represents the subset of $[0,1]$ that described the probability involved. The general situation is however that we do not have to consider all possible acts of choice. Sometimes many different possible acts of choice are involved, but we lack knowledge about which acts of choice realize during an experiments. This situation can result again in the probability being represented by one number of the interval $[0,1]$. This means that to capture the most general possible situation we should represent the probability by a subset of the interval $[0, 1]$. The subset is the convex hull if all possible acts of choice are considered in an identifiable way, while the subset is a singleton if there is only one act of choice, or if all possible acts of choice are unidentifiable, which results again in one act of choice probabilistically distributed over several 'hidden' acts of choice.

2.2 What About Single Experiments

If we let a subset of $[0, 1]$ correspond with the probability for a product experiment, we seemingly give a special status to the product experiments as compared to single experiments. If we reflect well however we can see that in fact each experiment is a product experiment. Always when we consider an experiment we choose the location where to do the experiment, the time when to perform it, and we also choose among the possible setups, etc .... Of course, these choices are made without taking them into account because they lead to almost identical situations. In principle however there will be a little subset of the interval $[0, 1]$ centered around one point that in traditional probability theory is taken to represent the probability.

Let us reflect somewhat about the foundations of probability theory itself in this respect. A lot of different interpretations for probability have been proposed. Much of the discussion was between those defending the 'relative frequency' interpretation and those being in favor of the 'subjective' interpretation of probability. Our concern here in not about this type of issue. In fact, the interpretation of probability is rather obvious in our scheme.

If we consider a physical entity S in state p, and an experiment α, then we consider the probability $\mu(\alpha, p)$ to be an element of reality, something that 'is' there, that expresses the tendency of the experiment α to give the outcome 'yes' if α would be performed. We consider this tendency to be there also when the experiment is not performed, and in fact – think of our discussion of property – specifically to be there when the experiment is not performed. When the experiment is performed, and repeated very often on the entity prepared in an identical state p, the sequence that is formed by the relative frequency is related to this tendency, in the sense that it allows us to describe this tendency by means of a number, namely the limit of the sequence of relative frequency. This is our interpretation of standard probability. This means that there is no contradiction between the subjective and relative frequency interpretation within our approach.

As we mentioned already, even for a single experiment we always make a choice whenever we perform this single experiment. The presence of the factor choice will make that different sequences of relative frequency, although they appear from situations that we classify as repeated experiments, will in general give rise to different limits. These different limits should all be contained in a 'good' description of the probability, certainly if we want the probability to express the 'real' presence of a tendency, because such a tendency should be independent of the not to control variations on the context of the experiments. We want the probability to express a tendency of the physical entity towards contexts of experiments that we classify as equivalent, although they are perhaps not equivalent if we would be able to control them better. The situation of the product experiment is in fact a rough example of this phenomenon that however in principle is always present.

2.3 Introducing the Subset Probability

We introduce probability in the following way. Let us consider a physical entity S in a state $p \in \Sigma$ and a yes/no-experiment $\alpha \in Q$ to be performed. Because of the fluctuations on the context related to the yes/no-experiment α, the sequence of relative frequency will possibly converge to different limits, depending on the 'choices' that are made between the different possible hidden contexts.

Definition 1 (Subset Probability) *Consider a physical entity S with set of states Σ and set of yes/no-experiments Q. The probability for α to give outcome 'yes' if the entity S is in state p is a function*

$$\mu : Q \times \Sigma \rightarrow \mathcal{P}([0,1]) \tag{5}$$

$$(\alpha, p) \mapsto \mu(\alpha, p) \tag{6}$$

where $\mathcal{P}([0,1])$ is the set of all subsets of $[0,1]$. The subset $\mu(\alpha,p)$ is the collection of the limits of relative frequency for outcome 'yes' of repeated application of experiment α on the entity S in state p. When $\mu(\alpha,p) = \emptyset$ this means that the experiment α cannot be performed on the entity S in state p. We call μ a 'subset probability'.

Standard probability theory is retrieved as a special idealized case of our probability theory if all the subsets that are images of the subset probability are singletons.

2.4 Inverse Experiments

We have considered always the probability for a yes/no-experiment to give the outcome 'yes'. We introduce the probability for outcome 'no' by means of the inverse experiment.

Definition 2 (Inverse Experiment) *Consider a physical entity S with set of states Σ, set of yes/no-experiments Q and subset probability μ. The inverse experiment $\widetilde{\alpha}$ of $\alpha \in Q$ is the same experiment where 'yes' and 'no' are interchanged. We suppose that if $\alpha \in Q$ then also $\widetilde{\alpha} \in Q$.*

To be able to express the relation between $\mu(\alpha, p)$ and $\mu(\widetilde{\alpha}, p)$ we introduce an additional structure on $\mathcal{P}([0,1])$.

Definition 3 *For a subset $V \in \mathcal{P}([0,1])$ we define $1 - V = \{x \,|x \in [0,1], 1 - x \in V\}$.*

Proposition 3 *For $V, W, (V_i)_i \in \mathcal{V}$ we have:*

$$V \subset W \Rightarrow 1 - V \subset 1 - W \tag{7}$$
$$1 - \cap_i V_i = \cap_i (1 - V_i) \tag{8}$$
$$1 - (1 - V) = V \tag{9}$$
$$1 - [0,1] = [0,1] \tag{10}$$
$$1 - \emptyset = \emptyset \tag{11}$$

Proof: Suppose that $V \subset W$, and consider $x \in 1 - V$. This means that $1 - x \in V$ and hence $1 - x \in W$. As a consequence we have $x \in 1 - W$. This proves 7. We have $\cap_i V_i \subset V_j \ \forall j$. From 7 follows then that $1 - \cap_i V_i \subset 1 - V_j \ \forall j$. As a consequence we have $1 - \cap_i V_i \subset \cap_i (1 - V_i)$. Consider $x \in \cap_i (1 - V_i)$. This means that $x \in 1 - V_j \ \forall j$. Hence $1 - x \in V_j \ \forall j$. From this follows that $1 - x \in \cap_i V_i$, and hence $x \in 1 - \cap_i V_i$. This shows that $\cap_i (1 - V_i) \subset 1 - \cap_i V_i$. This proves 8. We have $x \in V \Leftrightarrow 1 - x \in 1 - V \Leftrightarrow 1 - (1 - x) = x \in 1 - (1 - V))$. This proves 9. □

Proposition 4 *Consider a physical entity S with set of states Σ, set of yes/no-experiments Q and subset probability μ. For $\alpha \in Q$ and $p \in \Sigma$ we have:*

$$\mu(\widetilde{\alpha}, p) = 1 - \mu(\alpha, p) \tag{12}$$

Proof: Consider $x \in \mu(\widetilde{\alpha}, p)$. This means that there is a sequence $(x_i)_i$ of relative frequency for outcome 'yes' of $\widetilde{\alpha}$ such that $\lim x_i \in \mu(\widetilde{\alpha}, p)$. From definition 2 follows that $(x_i)_i$ is a sequence of relative frequency for outcome 'no' for the experiment α. Then $(1 - x_i)_i$ is a sequence of relative frequency for outcome 'yes' for α. If $\lim x_i = x$ then $\lim(1 - x_i) = 1 - x$. This shows that $1 - x \in \mu(\alpha, p)$, and hence $x \in 1 - \mu(\alpha, p)$. So we have proven that $\mu(\widetilde{\alpha}, p) \subset 1 - \mu(\alpha, p)$. Consider $y \in 1 - \mu(\alpha, p)$, then $1 - y \in \mu(\alpha, p)$. This means that there is a sequence of relative frequency $(x_i)_i$ for outcome 'yes' for α such that $\lim x_i = 1 - y \in \mu(\alpha, p)$. The sequence of relative frequency for outcome 'no' for α, and hence the sequence of relative frequency for outcome 'yes' for $\widetilde{\alpha}$, is then given by $(1-x_i)_i$. This means that $\lim(1-x_i) = 1-(1-y) = y \in \mu(\tilde{\alpha}, p)$. So we have proven that $1 - \mu(\alpha, p) \subset \mu(\tilde{\alpha}, p)$. This proves 12. □

2.5 Subset Probability and Product Experiments

In section 2.1 we have analyzed how probability behaves in relation with the product experiment. In proposition 2 it is shown that $\mu(\Pi_i \alpha_i, p)$ can be any number of the convex hull $V(\{\mu(\alpha_i, p)\})$. This is due to the fact that we consider all acts of choice to be possible identifiable acts of choice. There is a subtle matter involved that we will identify first.

Consider again a set of yes/no-experiments $(\alpha_i)_i$. If we define $\Pi_i \alpha_i$ as the experiment that consists of choosing one of the α_i and then performing this experiment, we implicitly suppose that we know how to choose consciously between all of the α_i. And if we know how to choose consciously between all of the α_i, it also means that we can identify all of the α_i, *i.e.* we know exactly which one of the α_i is performed. The set of all these possible conscious acts of choice where each of the α_i is individually identifiable leads to the convex hull $V(\{\mu(\alpha_i, p)\}$, as we have shown in section 2.1. For single experiments we have remarked in section 2.2 that there is also always a process of choice involved. It is however always the case that during the actual experimental process for a single experiment a process of 'unconscious' choice takes place. That is one of the reasons that a single experiment executed repeatedly on a physical entity in the same state can give rise to different outcomes. We have proven in other work that the indeterminism of quantum mechanics can even be fully explained in this way, an approach that we have called the hidden measurement approach [9,10,11,12,13,14,15,16,17,18,19,20].

This means that the presence of choice does not necessarily lead to the probability being defined on a subset rather than on a singleton, as it is the case for standard probability theory. If the 'choice' is randomized again, because it is an unconscious choice, that cannot be consciously manipulated, this process of randomization will again lead to a probability defined on a singleton. The other extreme case is when the process of choice is completely conscious and can be manipulated at will. In this case the probability will give rise to a convex set. The general situation is however the one that is neither a singleton, which is one extreme, nor a convex set, which is the other extreme.

For the case of the product experiment $\Pi_i \alpha_i$ of a set of experiments $(\alpha_i)_i$, we will make the hypothesis that at least each of the experiments individually can be chosen as a single repeated experiment. This means that the procedure that gives rise to the sequence of relative frequency for outcome 'yes', the physical entity being in state p, is the following: one of the experiments α_j is chosen, and this experiment α_j is repeated to give rise to a sequence of relative frequency. If this is the procedure that we introduce for the product experiment, we can show that the probability of the product experiment equals the set theoretical union of the probabilities of the single experiments.

Definition 4 (Product Experiment Procedure) *Consider a physical entity S with set of states Σ, set of yes/no-experiments Q and subset probability μ. For a set of yes/no-experiments $(\alpha_i)_i \in Q$ we define the procedure of repeated experiments for $\Pi_i\alpha_i$ as follows. One of the experiments α_j is chosen, and this experiment is repeatedly executed on the entity S prepared in the same state $p \in \Sigma$ to give rise to a sequence of relative frequency for outcome 'yes' of α_j. This sequence is one of the sequences to define $\mu(\Pi_i\alpha_i, p)$.*

Proposition 5 *Consider a physical entity S with set of states Σ, set of yes/no-experiments Q and subset probability μ. For a set of yes/no-experiments $(\alpha_i)_i \in Q$, and for $p \in \Sigma$ we have:*

$$\mu(\Pi_i\alpha_i, p) = \cup_i \mu(\alpha_i, p) \tag{13}$$

Proof: If we execute the procedure for the product experiment $\Pi_i\alpha_i$ as in definition 4 we find a sequence of relative frequency with a limit that is an element $\mu(\alpha_j, p)$ for one of the α_j chosen from the set $(\alpha_i)_i$. This proves 13. □

3 Reality as a Special Case of Probability

After a large detour into mathematical subtleties related to probability theory, we come back now to the origin of our proposal. We want to show that with the subset probability we can express consistently the concept of 'certainty' as 'probability equals 1'. Of course, since probability is now considered to be a subset, the statement 'probability equals 1' from standard probability theory has to be replaced by the statement 'probability is a subset of $\{1\}$'.

3.1 Certainty for Several Experiments

Let us prove that with the subset probability the certainty for several experiments is expressed well by identifying certainty with probability contained in the singleton $\{1\}$.

Definition 5 (Certainty) *Consider an entity S with set of states Σ, set of yes/no-experiments Q, equipped with a subset probability $\mu : Q \times \Sigma \rightarrow \mathcal{P}([0,1])$.*

For $\alpha \in Q$ we say

$$\alpha \text{ gives with certainty yes for } S \text{ in state } p \Leftrightarrow \mu(\alpha, p) \subset \{1\} \quad (14)$$

Proposition 6 *Consider an entity S with set of states Σ, set of yes/no-experiments Q, equipped with a subset probability $\mu : Q \times \Sigma \rightarrow \mathcal{P}([0,1])$. For $(\alpha_i)_i \in Q$ and $p \in \Sigma$ we have:*

$$\mu(\Pi_i \alpha_i) \subset \{1\} \Leftrightarrow \mu(\alpha_i, p) \subset \{1\} \ \forall i \quad (15)$$

Proof: From 13 we know that $\mu(\Pi_i \alpha_i, p) = \cup_i \mu(\alpha_i, p)$. We have $\cup_i \mu(\alpha_i, p) \subset \{1\} \Leftrightarrow \mu(\alpha_i, p) \subset \{1\} \ \forall i$. □

3.2 Back to the Original Problem

Let us come back to what we discussed in section 1.1. We claimed there that we believe that a piece of wood has both properties a and b at once, the property a of 'burning well', and the property b of 'floating on water'. This is because if we would perform one of the tests of these properties, so trying out whether it burns well, which we called test α, or trying out whether it floats, which we called test β, this would in any case deliver us a positive outcome with certainty. Introducing the subset probability as we have done in the last few sections makes it possible to replace the concept of certainty by that of probability contained in the singleton $\{1\}$. Following proposition 6 we indeed know that $\mu(\alpha \cdot \beta, p) \subset \{1\} \Leftrightarrow \mu(\alpha, p) \subset \{1\}$ and $\mu(\beta, p) \subset \{1\}$, because the subset probability that we have introduced describes well the idea that certainty is independent of whether we choose to perform α or β. This is effectively what we have in mind when we introduce certainty speaking about things that really exist.

At first sight one might think that we have perhaps attempted too strong to describe certainty by means of probability. Indeed, could it not be claimed that the standard probabilistic approach has just the advantage of making the concept of certainty somewhat weaker and replacing it by probability equals 1, which does not really means certainty, but something like 'very very close to certainty'. And is such a concept of 'very very close to certainty' not more realistic than absolute certainty? We totally agree with this. Often it are cases of 'very very close' to certainty that appear in the world around us. In fact, it is this remark that will make it possible for us to reveal much more profoundly the motivation of all what we are doing here. We do not want so much to concentrate on how to describe 'complete certainty', because indeed such 'complete certainty' should be a kind of extreme case. More profound to our motivation is that the level of certainty, whether this level is the extreme case of complete certainty, or whether it is very very close to certainty, should be in equilibrium, in the sense that this level should be attained independent of whether we consider one property or whether we consider several properties at once. This is what standard probability does not accomplish and what we do accomplish by means of our subset probability. As a bonus it would be nice to have within the mathematical model a way available to express also the ideal situation of complete certainty. Also in the mathematical model this should be a extreme case. Also this is not accomplished by standard probability theory and it is by our subset probability. Let us make this more clear in the next section when we formally consider situations close to certainty.

3.3 Situations Close to Certainty

Suppose that we consider the situation where we want to express that the outcomes of an experiment α is very close to certainty. We can do this in an obvious way by demanding that $\mu(\alpha, p) \subset [1-\epsilon, 1]$ where ϵ is a small real number.

Proposition 7 *Consider a physical entity S with set of states Σ and set of yes/no-experiments Q, equipped with a subset probability $\mu : Q \times \Sigma \rightarrow \mathcal{P}([0,1])$. For a set of experiments $(\alpha_i)_i$ and $p \in \Sigma$ we have:*

$$\mu(\Pi_i \alpha_i, p) \subset [1-\epsilon, 1] \Leftrightarrow \mu(\alpha_i, p) \subset [1-\epsilon, 1] \ \forall i \tag{16}$$

Proof: From 13 we know that $\mu(\Pi_i \alpha_i, p) = \cup_i \mu(\alpha_i, p)$. We have $\cup_i \mu(\alpha_i, p) \subset [1-\epsilon, 1] \Leftrightarrow \mu(\alpha_i, p) \subset [1-\epsilon, 1] \ \forall i$. □

This proposition proves that the level of uncertainty remains the same whether we consider one of the experiments α_j or whether we consider the product experiment $\Pi_i \alpha_i$. And this should be the case. This is exactly what expresses what we intuitively have in mind when we think of the procedure of testing several properties at once. Our level of uncertainty should not depend on the number of properties tested.

We cannot formulate an equivalent proposition within standard probability theory, because of the problems that we have mentioned already in the foregoing sections. Indeed, in standard probability theory we are forced to define the probability related to $\Pi_i \alpha_i$ by means of a convex combination of the probabilities of the α_i, for example expressed by a set of numbers $(x_i)_i$, such that $x_i \in [0,1]$ and $\sum_i x_i = 1$, and the demand that this convex combination probability, expressed by $\sum_i x_i \mu(\alpha_i, p)$, is contained in an interval $[1-\epsilon, 1]$ does not imply that the single probabilities $\mu(\alpha_j, p)$ have to be contained in this same interval. Except when we would demand that $0 < x_i \ \forall i$

(see proposition 1). But if we demand $0 < x_i \ \forall i$ we come into problems with infinite sets of properties and related experiments.

3.4 Transfer of Other Statements

It is clear from proposition 3.3 that the equivalence between statements expressed about a collection of properties, or their tests, and the individual properties, or test, is not limited to statements about closeness to certainty. The interval $[1-\epsilon, 1]$ in formula 16 can be replaced by an arbitrary subset of the interval $[0, 1]$, and remains valid.

Proposition 8 *Consider a physical entity S with set of states Σ and set of yes/no-experiments Q, equipped with a subset probability $\mu : Q \times \Sigma \to \mathcal{P}([0,1])$. For a set of experiments $(\alpha_i)_i$, $p \in \Sigma$ and $A \subset [0,1]$ we have:*

$$\mu(\Pi_i \alpha_i, p) \subset A \Leftrightarrow \mu(\alpha_i, p) \subset A \ \forall i \tag{17}$$

Proof: From 13 we know that $\mu(\Pi_i \alpha_i, p) = \cup_i \mu(\alpha_i, p)$. We have $\cup_i \mu(\alpha_i, p) \subset A \Leftrightarrow \mu(\alpha_i, p) \subset A \ \forall i$. □

So the probability of the product test is contained in an arbitrary subset of $[0, 1]$ iff the probabilities of all the component tests are contained in this subset. A special case of this is that the probability of the product test is a singleton, and hence corresponds to one converging series of relative frequency, iff the probability of all the component tests equal this same singleton and hence also correspond to this one convergent series of relative frequency.

Proposition 9 *Consider a physical entity S with set of states Σ and set of yes/no-experiments Q, equipped with a subset probability $\mu : Q \times \Sigma \to \mathcal{P}([0,1])$. For a set of experiments $(\alpha_i)_i$, $p \in \Sigma$ and $c \in [0,1]$ we have:*

$$\mu(\Pi_i \alpha_i, p) \subset \{c\} \Leftrightarrow \mu(\alpha_i, p) \subset \{c\} \ \forall i \tag{18}$$

Proof: From 13 we know that $\mu(\Pi_i \alpha_i, p) = \cup_i \mu(\alpha_i, p)$. We have $\cup_i \mu(\alpha_i, p) \subset \{c\} \Leftrightarrow \mu(\alpha_i, p) \subset \{c\} \ \forall i$. □

So we see that by means of the subset probability we can not only express the idealized situations of complete certainty, as the probability contained in singleton $\{1\}$, but also the idealized situation of a series of relative frequency that converges to one specific limit, as the probability contained in an arbitrary singleton.

4 SEP: The Category of State Experiment Probability Systems

In the foregoing section we have introduced the subset probability for a physical entity S. In this section we introduce the mathematical structures involved, independent of whether they are used to describe a physical entity. The reason to do so, is that in this way these structures can be studied independent of their physical meaning. They can then be used as a mathematical model for the description of a physical entity. This section is self-contained from a mathematical point of view. We introduce specifically the mathematics that is needed for a mathematical model of the physical situation that we have considered in the foregoing sections. We introduce immediately the categorical structure.

4.1 State Experiment Probability Systems

Let us first introduce the mathematical structure of a state experiment probability system.

Definition 6 (State Experiment Probability System) *A state experiment probability system or SEP* $(\Sigma, Q, \Pi, \tilde{}, \mu)$ *or shorter* (Σ, Q, μ) *consists of two sets* Σ *and* Q *and a function*

$$\mu : Q \times \Sigma \rightarrow \mathcal{P}([0,1]) \tag{19}$$

On Q *there exists a product, that associates with each family* $(\alpha_i)_i \in Q$ *an element* $\Pi_i \alpha_i \in Q$ *such that for* $p \in \Sigma$

$$\mu(\Pi_i \alpha_i, p) = \cup_i \mu(\alpha_i, p) \tag{20}$$

There also exists an inverse operation on Q*, which is a function* $\tilde{}: Q \rightarrow Q$ *such that for* $\alpha, (\alpha_i)_i \in Q$ *and* $p \in \Sigma$ *we have:*

$$\tilde{\tilde{\alpha}} = \alpha \tag{21}$$

$$\widetilde{\Pi \alpha_i} = \Pi_i \tilde{\alpha}_i \tag{22}$$

$$\mu(\tilde{\alpha}, p) = \widetilde{\mu(\alpha, p)} \tag{23}$$

where $\widetilde{\mu(\alpha, p)} = \{1 - x \mid x \in \mu(\alpha, p)\}$*. There exists a unit element* $\tau \in Q$ *such that for* $p \in \Sigma$*:*

$$\mu(\tau, p) = \{1\} \tag{24}$$

4.2 The Morphisms of State Experiment Probability Systems

Consider two state experiment probability systems (Σ, Q, μ) and (Σ', Q', μ'). These state experiment probability systems respectively describe entities S and S'. We will arrive at the notion of a morphism by analyzing the situation where the entity S is a subentity of the entity S'. In that case, the following three natural requirements should be satisfied:

i) If the entity S' is in a state p' then the state $m(p')$ of S is determined. This defines a function m from the set of states of S' to the set of states of S;
ii) If we consider an experiment α on the entity S, then to α corresponds an experiment $l(a)$ on the "bigger" entity S'. This defines a function l from the set of experiments of S to the set of experiments of S'.
iii) We want α and $l(\alpha)$ to be two descriptions of the 'same' experiment of S, once considered as an entity in itself, once as a subentity of S'. In other words we want α and $l(\alpha)$ to relate to the states $m(p')$ and p' with the same probabilities. This means that for a state p' of S' (and a corresponding state $m(p')$ of S) we want the following 'covariance principle' to hold:

$$\mu(\beta, m(p')) = \mu'(l(\beta), p') \quad \beta \in P, p' \in \Sigma' \tag{25}$$

Furthermore, since the inverse operation is just a switching of 'yes' and 'no', this switching should have the same result whether we consider the description of the experiment, namely α, on the entity S itself, or the description of the experiment, namely $l(\alpha)$, on the entity as a subentity. This is expressed by

$$\widetilde{l(\alpha)} = l(\widetilde{\alpha}) \tag{26}$$

The covariance principle applied for the situation of a set of experiments $(\alpha_i)_i \in Q$ produces in a similar way the following requirement for the product operation:

$$l(\Pi_i \alpha_i) = \Pi_i l(\alpha_i) \tag{27}$$

We are now ready to present a formal definition of a morphism of state experiment systems.

Definition 7 *Consider two state experiment probability systems* (Σ, Q, μ) *and* (Σ', Q', μ'). *We say that*

$$(m, l) : (\Sigma', Q', \mu') \longrightarrow (\Sigma, Q, \mu) \tag{28}$$

is a 'morphism' (of state experiment probability systems) if m *is a function:*

$$m : \Sigma' \to \Sigma \tag{29}$$

and l *is a function:*

$$l : Q \to Q' \tag{30}$$

such that for $\alpha, (\alpha_i)_i \in Q$ *and* $p' \in \Sigma'$ *the following holds:*

$$\mu(\alpha, m(p')) = \mu'(l(\alpha), p') \tag{31}$$

$$l(\Pi_i \alpha_i) = \Pi_i l(\alpha_i) \tag{32}$$

$$\widetilde{l(\alpha)} = l(\widetilde{\alpha}) \tag{33}$$

We introduce the category with objects state experiment probability structures and morphisms the ones that we come to introduce and denote this category **SEP**.

5 The Categories SEP and SP

We can find back the structure of a state property system that we have studied in great detail in earlier articles [22,23,21,24,25,26,27,27,28,29,30,31] related to a state experiment probability structure, and describing the properties that are operationally defined by the experiments of the state experiment probability system. Also the category of state property systems and its morphism, called **SP**, appears as a substructure of the category **SEP**.

5.1 The Related State Property System

Let us repeat the definition of a state property system.

Definition 8 (State Property System) *We say that* $(\Sigma, <, \mathcal{L}, <, \wedge, \vee, \xi)$ *is a state property system if* $\Sigma, <$ *is a pre-ordered set,* $\mathcal{L}, <, \wedge, \vee$ *is a complete lattice, and* ξ *is a function:*

$$\xi : \Sigma \rightarrow \mathcal{P}(\mathcal{L}) \tag{34}$$

such that for $p \in \Sigma$*,* I *the maximal element of* $\mathcal{L}$*, and* 0 *the minimal element of* $\mathcal{L}$*, we have:*

$$I \in \xi(p) \tag{35}$$

$$0 \notin \xi(p) \tag{36}$$

$$a_i \in \xi(p)\ \forall i \Leftrightarrow \wedge_i a_i \in \xi(p) \tag{37}$$

and for $p, q \in \Sigma$ *and* $a, b \in \mathcal{L}$ *we have:*

$$p < q \Leftrightarrow \xi(q) \subset \xi(p) \tag{38}$$

$$a < b \Leftrightarrow \forall r \in \Sigma : a \in \xi(r) \text{ then } b \in \xi(r) \tag{39}$$

Proposition 10 *Consider a state experiment probability system* (Σ, Q, μ)*. For* $p, q \in \Sigma$ *and* $\alpha, \beta \in Q$ *we define:*

$$p < q \Leftrightarrow \forall \gamma \in Q : \mu(\gamma, q) \subset \{1\} \text{ then } \mu(\gamma, p) \subset \{1\} \tag{40}$$

$$\alpha < \beta \Leftrightarrow \forall r \in \Sigma : \mu(\alpha, r) \subset \{1\} \text{ then } \mu(\beta, r) \subset \{1\} \tag{41}$$

$$\alpha \approx \beta \Leftrightarrow \alpha < \beta \text{ and } \beta < \alpha \tag{42}$$

then $\Sigma, <$ *and* $Q, <$ *are pre-ordered sets, and* $\approx$ *is an equivalence relation on* Q*.*

Proof: Straightforward verification. □

Definition 9 *Consider a state experiment probability system* (Σ, Q, μ)*. We define* $\mathcal{L}$ *to be the set of equivalence classes of* Q *for the equivalence relation* $\approx$ *defined in 42, and for* $a, b \in \mathcal{L}$ *we define* $a < b$ *iff* $\alpha \in a$ *and* $\beta \in b$ *such that* $\alpha < \beta$*. We denote* $I \in \mathcal{L}$ *the equivalence class of* τ *as defined in 24.*

Proposition 11 $\mathcal{L}$ *of definition 9 is a complete lattice for the partial order relation* $<$*, where the infimum of a set of elements* $(a_i)_i \in \mathcal{L}$ *is given by the equivalence class of* $\Pi_i \alpha_i$*, with* $\alpha_j \in a_j \ \forall j$*. We denote this infimum of the set* $(a_i)_i$ *by* $\wedge_i a_i$*. The minimal element of* $\mathcal{L}$ *is given by* $0 = \wedge_{a \in \mathcal{L}} a$*,* I *is the maximal element of* $\mathcal{L}$*, and the supremum of a set of elements* $(a_i)_i$ *is given by* $\vee_i a_i = \wedge_{a_j < b \forall j, b \in \mathcal{L}} b$*.*

Proof: Since $<$ is a pre-order relation on Q, we have that $<$ is a partial order relation on $\mathcal{L}$. Let us prove that $\wedge_i a_i$ is an infimum for the set $(a_i)_i$. First we want to show that $\wedge_i a_i$ is a lower bound, hence that $\wedge_i a_i < a_j \ \forall j$. We know that $\wedge_i a_i$ is the equivalence class of $\Pi_i \alpha_i$ for $\alpha_j \in a_j \ \forall j$. Consider an arbitrary element $\gamma \in Q$ such that $\gamma \in \wedge_i a_i$, and $r \in \Sigma$ such that $\mu(\gamma, r) \subset \{1\}$. Since $\gamma \approx \Pi_i \alpha_i$ and hence $\gamma < \Pi_i \alpha_i$, from 41 follows that $\mu(\Pi_i \alpha_i, r) \subset \{1\}$. We know that $\mu(\Pi_i \alpha_i, r) = \cup_i \mu(\alpha_i, r)$, and hence we have $\cup_i \mu(\alpha_i, r) \subset \{1\}$. From this follows that $\mu(\alpha_j, r) \subset \{1\} \ \forall j$. Using again 41 this shows that $\gamma < \alpha_j \ \forall j$. Since $\gamma \in \wedge_i a_i$ and $\alpha_j \in a_j \ \forall j$ we have proven that $\wedge_i a_i < a_j \ \forall j$. Let us prove now that $\wedge_i a_i$ is the greatest lower bound of the set $(a_i)_i$. Consider an arbitrary $b \in \mathcal{L}$ such that b is a lower bound of the set $(a_i)_i$. We have to show that $b < \wedge_i a_i$. Take $\gamma \in b$. Since b is a lower bound we have $\gamma < \alpha_j \ \forall j$. Consider $r \in \Sigma$ such that $\mu(\gamma, r) \subset \{1\}$. From 41 follows that $\mu(\alpha_j, r) \subset \{1\} \ \forall j$. Then $\cup_i \mu(\alpha_i, r) = \mu(\Pi_i \alpha_i, p) \subset \{1\}$. Again using 41 we have shown that $\gamma < \Pi_i \alpha_i$. And this proves that $b < \wedge_i a_i$. For any partially ordered set $\mathcal{L}$ where the infimum of any set of elements exists, the minimal element is given by $\wedge_{a \in \mathcal{L}} a$. Let us show that I is the maximal element of $\mathcal{L}$. Consider any $a \in \mathcal{L}$, and $\gamma \in a$. Since $\mu(\tau, p) \subset \{1\} \ \forall p \in \Sigma$ we have $\gamma < \tau$. This proves that $a < I$. For any partially ordered set $\mathcal{L}$ with a maximal element and where each set $(a_i)_i$ has an infimum, each set $(a_i)_i$ has also a supremum that is given by $\vee_i a_i = \wedge_{a_j < b \forall j, b \in \mathcal{L}} b$. This proves that $\mathcal{L}$ is a complete lattice. □

Definition 10 *Consider a state experiment probability system* (Σ, Q, μ) *and the complete lattice* $\mathcal{L}$ *of definition 9. For* $p \in \Sigma$ *we define:*

$$\xi : \Sigma \to \mathcal{P}(\mathcal{L}) \tag{43}$$

$$p \mapsto \xi(p) \tag{44}$$

where

$$\xi(p) = \{a \ |a \in \mathcal{L}, \mu(\alpha, p) \subset \{1\}, \alpha \in a\} \tag{45}$$

Proposition 12 *Consider a state experiment probability system* (Σ, Q, μ). *Consider* $\Sigma, <$, *with* $<$ *defined as in 40,* $\mathcal{L}, <, \wedge, \vee$, *as in definition 9, and* ξ *as in definition 10. Then* $(\Sigma, <, \mathcal{L}, <, \wedge, \vee, \xi)$ *is a state property system.*

Proof: Consider $p \in \Sigma$, then $\mu(\tau, p) \subset \{1\}$. Hence $I \in \xi(p)$, which proves 35. Let us consider $\widetilde{\tau} \in Q$. We have $\mu(\widetilde{\tau}, p) \subset \{0\}\ \forall p \in \Sigma$. Let us denote the equivalence class of $\widetilde{\tau}$ by a. Then $0 < a$. Take an arbitrary element $\gamma \in Q$ such that $\gamma \in 0$. The following is satisfied: $\forall r \in \Sigma$ such that $\mu(\widetilde{\tau}, r) \subset \{1\}$ we have $\mu(\gamma, r) \subset \{1\}$. This shows that $a < 0$, and hence $a = 0$. As a consequence $\widetilde{\tau} \in 0$. Since $\mu(\widetilde{\tau}, p) \subset \{0\}\ \forall p \in \Sigma$ we have $0 \notin \xi(p)\ \forall p \in \Sigma$, which proves 36. Consider $(a_i)_i \in \mathcal{L}$ and suppose that $\wedge_i a_i \in \xi(p)$. This means that for $\alpha_i \in a_i$ we have $\mu(\Pi_i \alpha_i, p) = \cup_i \mu(\alpha_i, p) \subset \{1\}$. Hence $\mu(\alpha_j, p) \subset \{1\}\ \forall j$, which proves that $a_j \in \xi(p)\ \forall j$, which proves one of the implications of 37. Consider again $(a_i)_i \in \mathcal{L}$ such that $a_j \in \xi(p)\ \forall j$. With $\alpha_i \in a_i$ it follows that $\mu(\alpha_j, p) \subset \{1\}\ \forall j$. Hence $\cup_i \mu(\alpha_i, p) = \mu(\Pi_i \alpha_i, p) \subset \{1\}$. This shows that $\wedge_i a_i \in \xi(p)$ which proves that other implication of 37. Consider $p, q \in \Sigma$, such that $p < q$ and consider $a \in \xi(q)$. This means that for $\alpha \in a$ we have $\mu(\alpha, q) \subset \{1\}$. From 40 and $p < q$ follows that $\mu(\alpha, p) \subset \{1\}$, and hence $a \in \xi(p)$. This shows that $\xi(q) \subset \xi(p)$, and so we have proven one of the implications of 38. Suppose now that $\xi(q) \subset \xi(p)$, and consider $\gamma \in Q$ such that $\mu(\gamma, q) \subset \{1\}$. If a is the equivalence class of γ we have $a \in \xi(q)$, but then also $a \in \xi(p)$. As a consequence we have $\mu(\gamma, p) \subset \{1\}$. Using 40 we have shown that $p < q$. This proves that other implication of 38. Consider $a, b \in \mathcal{L}$ such that $a < b$, and $r \in \Sigma$ such that $a \in \xi(r)$. Take $\alpha \in a$ and $\beta \in b$. We have $\mu(\alpha, r) \subset \{1\}$, and since $a < b$ it follows from 41 that $\mu(\beta, r) \subset \{1\}$. As a consequence we have $b \in \xi(r)$. This proves one of the implications of 39. Suppose that $\forall r \in \Sigma$ we have that $a \in \xi(r)$ implies that $b \in \xi(r)$. Take $\alpha \in a$, $\beta \in b$ and consider $r \in \Sigma$ such that $\mu(\alpha, r) \subset \{1\}$. This means that $a \in \xi(r)$, and hence $b \in \xi(r)$. As a consequence we have $\mu(\beta, r) \subset \{1\}$. So we have shown that $\alpha < \beta$ and hence $a < b$. This proves that other implication of 39. □

Definition 11 (Related State Property System) *We call* $(\Sigma, \mathcal{L}, \xi)$, *as defined in proposition 12, the state property system related to the state experiment probability system* (Σ, Q, μ).

5.2 The Related Morphisms

We will prove that a morphism of **SEP** gives rise to a morphism of **SP**, the category of state property systems as defined in [22,23]. Let us put forward the definition of a morphism for a state property system as in definition 11 of [23].

Definition 12 (Morphism of State Property Systems) *Consider two state property systems $(\Sigma, \mathcal{L}, \xi)$ and $(\Sigma', \mathcal{L}', \xi')$. We say that*

$$(m, n) : (\Sigma', \mathcal{L}', \xi') \to (\Sigma, \mathcal{L}, \xi) \tag{46}$$

is a morphism of state property systems, if m is a function:

$$m : \Sigma' \to \Sigma \tag{47}$$

and n is a function:

$$n : \mathcal{L} \to \mathcal{L}' \tag{48}$$

such that for $a \in \mathcal{L}$ and $p' \in \Sigma'$ the following holds:

$$a \in \xi(m(p')) \Leftrightarrow n(a) \in \xi'(p') \tag{49}$$

Proposition 13 *Consider two state experiment probability systems (Σ, Q, μ) and (Σ', Q', μ'), and the corresponding state property systems $(\Sigma, \mathcal{L}, \xi)$ and $(\Sigma', \mathcal{L}', \xi')$. For a morphism (m, l) between (Σ, Q, μ) and (Σ', Q', μ') we keep m and define $n : \mathcal{L} \to \mathcal{L}'$ such that $n(a)$ is the equivalence class in Q' of $l(\alpha)$ with $\alpha \in a$. The couple (m, n) is a morphism between the state property systems $(\Sigma, \mathcal{L}, \xi)$ and $(\Sigma', \mathcal{L}', \xi')$.*

Proof: For $p' \in \Sigma'$ consider $a \in \xi(m(p'))$ and $\alpha \in a$. This means that $\mu(\alpha, m(p')) \subset \{1\}$. Form 31 we have $\mu(\alpha, m(p')) = \mu'(l(\alpha), p')$, and hence $\mu'(l(\alpha), p') \subset \{1\}$. This shows that $n(a) \in \xi'(p')$ because it is the equivalence class of $l(\alpha)$. This proves one of the implications of 49. Suppose that $n(a) \in \xi'(p')$ for $p' \in \Sigma$. This means that the equivalence class of $l(\alpha)$ for $\alpha \in a$ belongs to $\xi'(p')$. Hence $\mu'(l(\alpha, p')) \subset \{1\}$. Again using 31 this implies that $\mu(\alpha, m(p')) \subset \{1\}$, and hence $a \in \xi(m(p'))$, which proves the other implication of 49. □

Definition 13 (Related Morphism) *Let us consider the state property system $(\Sigma, \mathcal{L}, \xi)$ related to the state experiment probability system (Σ, Q, μ). We call (m, n), as defined in proposition 13, the morphism of $(\Sigma, \mathcal{L}, \xi)$ related to the morphism (m, l) of (Σ, Q, μ).*

It is easy to see that for any subset $A \in \mathcal{P}([0, 1])$ we can link a state property system with the state experiment probability system that we consider. Indeed, for the propositions that we have proved in section 5 and the definitions that are given there, we can readily replace the singleton $\{1\}$ by the subset A of the interval $[0,1]$, and everything can be proven and defined in an analogous way. Of course, it is the state property system and the morphisms that we introduced in the foregoing section that describe the state and properties of the physical entity under study. For a further elaboration of the structure of the state experiment probability system and its relation with the different state property systems that we can define in this way, it would be interesting to study also these other state property systems, and how they are interrelated. We will leave this however for future research.

We end this article with a last remark. Relative frequencies are rational numbers. This means that if we consider limits of sequences of relative frequency, it does not have to be a priori so that we have to consider these limits within the completion of $\mathbb{R}$ of $\mathbb{Q}$. There is, to the best of our knowledge, no operational reason why the completion $\mathbb{R}$ of $\mathbb{Q}$ would be superior. Other completions of $\mathbb{Q}$ might be a better choice [32]. There are even operational reasons to believe that other completions of $\mathbb{Q}$ might indeed be a better choice. If we think of the completion used in nonstandard analysis, where infinitesimals are possible, it would be possible there to explore the problem that we have considered here. Indeed, it might well be possible that infinite convex of numbers between 0 and 1 can be introduced there, such that all the elements of the convex combination can be taken to be different from zero while the sum itself remains in the interval $[0,1]$. This would be another possibility to explore that could lead to a solution of the problem of how to express 'certainty' by means of probability. Since the subset probability works on the level of $\mathcal{P}([0,1])$ and not on the level of $[0,1]$ itself, it might even be that the structure that we have investigated in this article can more readily be transposed to the situation where another completion of $\mathbb{Q}$ than $\mathbb{R}$ is considered.

References

1. D. Aerts, *The One and the Many: Towards a Unification of the Quantum and the Classical Description of One and Many Physical Entities,* Doctoral Thesis, Brussels Free University (1981).
2. D. Aerts, "Description of many physical entities without the paradoxes encountered in quantum mechanics", *Found. Phys.* **12**, 1131-1170 (1982).
3. C. Piron, *Foundations of Quantum Physics,* Reading, Mass., W. A. Benjamin (1976).
4. C. Piron, "Recent developments in quantum mechanics", *Helv. Phys. Acta* **62**, 82 (1989).
5. C. Piron, *Mècanique Quantique: Bases et Applications,*, Press Polytechnique de Lausanne (1990).

6. D. Aerts, D. "Classical theories and nonclassical theories as a special case of a more general theory", *J. Math. Phys.* **24**, 2441-2453 (1983).
7. D. Aerts, "The description of one and many physical systems", in *Foundations of Quantum Mechanics*, eds. C. Gruber, A.V.C.P. Lausanne, 63 (1983).
8. D. Aerts and B. D'Hooghe, "The subexperiment problem in quantum mechanics", in *Probing the Structure of Quantum Mechanics: Nonlocality, Computation and Axiomatics*, World Scientific, Singapore (2002).
9. D. Aerts, "A possible explanation for the probabilities of quantum mechanics", *J. Math. Phys.* **27**, 202-210 (1986).
10. D. Aerts, "The origin of the non-classical character of the quantum probability model", in *Information, Complexity, and Control in Quantum Physics*, eds. Blanquiere, A., Diner, S. and Lochak, G., Springer-Verlag, Wien-New York, 77-100 (1987).
11. D. Aerts, "The description of separated systems and quantum mechanics and a possible explanation for the probabilities of quantum mechanics", in *Micro-physical Reality and Quantum Formalism*, eds. van der Merwe, A., et al., Kluwer Academic Publishers, 97-115 (1988).
12. D. Aerts, "Quantum structures due to fluctuations of the measurement situations", *Int. J. Theor. Phys.* **32**, 2207-2220 (1993).
13. D. Aerts,"Quantum structures, separated physical entities and probability", *Found. Phys.* **24**, 1227-1259 (1994).
14. D. Aerts and S. Aerts, "The hidden measurement formalism: quantum mechanics as a consequence of fluctuations on the measurement", in *Fundamental Problems in Quantum Physics II*, eds. Ferrero, M. and van der Merwe, A., Kluwer Academic, Dordrecht (1997).
15. D. Aerts, S. Aerts, B. Coecke, B. D'Hooghe, T. Durt and F. Valckenborgh, "A model with varying fluctuations in the measurement context", in *Fundamental Problems in Quantum Physics II*, eds. Ferrero, M. and van der Merwe, A., Kluwer Academic, Dordrecht (1997).
16. D. Aerts, "The hidden measurement formalism: what can be explained and where paradoxes remain", *Int. J. Theor. Phys.* **37**, 291-304 (1998)
17. S. Aerts, "Interactive probability models: inverse problems on the sphere, *Int. J. Theor. Phys.* **37**, 1 (1998).
18. D. Aerts, S. Aerts, T. Durt and O. Lévêque, "Quantum and classical probability and the epsilon-model, *Int. J. Theor. Phys.* **38**, 407-429 (1999).
19. D. Aerts, B. Coecke and S. Smets, "On the origin of probabilities in quantum mechanics: creative and contextual aspects", in *Metadebates on Science*, eds. Cornelis, G., Smets, S., and Van Bendegem, J.P., Kluwer

Academic, Dordrecht (1999).

20. S. Aerts, "Hidden measurements from contextual axiomatics", this volume.
21. D. Aerts, "Quantum mechanics: structures, axioms and paradoxes", in *Quantum Mechanics and the Nature of Reality*, eds. Aerts, D. and Pykacz, J., Kluwer Academic, Dordrecht (1999).
22. D. Aerts, "Foundations of quantum physics: a general realistic and operational approach", *Int. J. Theor. Phys.* **38**, 289 - 358 (1999).
23. D. Aerts, E. Colebunders, A. Van der Voorde and B. Van Steirteghem, "State property systems and closure spaces: a study of categorical equivalence", *Int. J. Theor. Phys.* **38**, 359-385 (1999).
24. D. Aerts, E. Colebunders, A. Van der Voorde and B. Van Steirteghem, "The construct of closure spaces as the amnestic modification of the physical theory of state property systems", to appear in *Applied Categorical Structures*.
25. B. Van Steirteghem, *Quantum Axiomatics: Investigation of the Structure of the Category of Physical Entities and Solér's Theorem*, Dissertation for the dregee of Bachelor in Science, Brussels Free University (1998).
26. B. Van Steirteghem, "T_0 separation in axiomatic quantum mechanics, *Int. J. Theor. Phys.* **39**, 955 (2000).
27. A. Van der Voorde, "A categorical approach to T_1 separation and the product of state property systems", *Int. J. Theor. Phys.* **39**, 947-953 (2000).
28. A. Van der Voorde, *Separation Axioms in Extension Theory for Closure Spaces and Their Relevance to State Property Systems*, Doctoral Thesis, Brussels Free University (2001).
29. D. Aerts, A Van der Voorde and D. Deses, "Connectedness applied to closure spaces and state property systems", *Journal of Electrical Engineering*, **52**, 18-21 (2001).
30. D. Aerts, A Van der Voorde and D. Deses, "Classicality and connectedness for state property systems and closure spaces", submitted to *International Journal of Theoretical Physics.*
31. D. Aerts and D. Deses, "State property systems and closure spaces: extracting the classical en nonclassical parts", this volume.
32. A. Khrennikov, "Interpretations of Probability", VSP, Utrecht, The Netherlands (1999).

QUANTUM COMPUTATION: TOWARDS THE CONSTRUCTION OF A 'BETWEEN QUANTUM AND CLASSICAL' COMPUTER

DIEDERIK AERTS
Center Leo Apostel (CLEA) and
Foundation of the Exact Sciences (FUND),
Brussels Free University, Krijgskundestraat 33,
1160 Brussels, Belgium
E-mail: diraerts@vub.ac.be

BART D'HOOGHE
Foundations of the Exact Sciences (FUND), Department of
Mathematics, Brussels Free University,
Pleinlaan 2, 1050 Brussel, Belgium
E-mail: bdhooghe@vub.ac.be

Using the 'between quantum and classical' models that have been constructed explicitly within the hidden measurement approach of quantum mechanics we investigate the possibility to construct a 'between quantum and classical' computer. In this view, the pure quantum computer and the classical Turing machine can be seen as two special cases of our general computer. We have shown in earlier research that the intermediate 'between quantum and classical' systems cannot be described within standard quantum theory. We argue that the general categoral approach of state property systems might provide a unified framework for the study of these 'between quantum and classical' models, and hence also for the study of classical and quantum computers as special cases.

1 Introduction

The theory of quantum computation (*i.e.*, calculations performed on a computer which has a quantum system as register [1,2]) has gained importance after the discovery of quantum algorithms which allow to solve problems much faster than with the known (classical) algorithms written for a classical, digital computer. The most famous quantum algorithm is a factorization algorithm by P. Shor [3,4] which allows to factorize integers in a number of steps polynomial in the size of the input. This would imply that if quantum computers could be built in practice, then the most important of modern cryptography systems could be broken easily, since they are based on the assumption that polynomial factorization algorithms can not be found for classical, digital computers [4,5]. From a more theoretical and philosophical point of view quantum computing is also interesting since it contributes to the study of fundamental aspects of physics and information science: *e.g.* by the definition of a universal quantum Turing-machine [6,7]; *e.g.* by putting the many world interpretation of quantum mechanics into a new perspective [7,8].

In the first section we give a brief overview of how a quantum computer works. The quantum analogue of a classical bit is a so-called qubit, which has very different properties in comparison with a classical bit since in a sense it can be in any superposition of the two classical bit-values. The register of the quantum computer is given by a system of N spin-$\frac{1}{2}$ particles, such that the state of each qubit is encoded with the state of the corresponding spin-$\frac{1}{2}$ particle. Therefore, to study a quantum computer is to study its register, *i.e.*, study a system of N spin-$\frac{1}{2}$ particles.

In the second section we discuss the hidden measurement approach of quantum mechanics, which assumes that random fluctuations in the measurement context lead to a probability distribution over the set of outcomes, which coincides with the quantum probabilities if the random fluctuations are uniformly distributed. Not only is it possible to show that for any quantum entity one can define a hidden measurement model, also explicit hidden measurement models for quantum entities have been put forward. One of these models is a macroscopical model representing the spin properties of a spin-$\frac{1}{2}$ particle, such that each spin state is represented by a point on the unit sphere in three dimensions. Using this model, one can represent (the state of) a qubit by a point on the three-dimensional unit sphere. Two sphere models representing a spin-$\frac{1}{2}$ entity were coupled to give rise to a model that describes the situation of two coupled spin-$\frac{1}{2}$ entities by introducing so-called correlations of the first and of the second kind, and the model was realized by the concrete coupling of two ϵ-models by means of a rigid rod connecting the states of the two spins [9]. It was shown that this type of model can be generalized for the case of N coupled spin-$\frac{1}{2}$ particles [10,11]. Later [12,13] a new parameter ρ was introduced that parametrizes the coupling, in the sense that the model can evolve in a continuous way from 'rigid' coupling, making it a model for the quantum coupling of spin-$\frac{1}{2}$, hence the quantum coupling of qubits ($\rho = 1$), to 'no-coupling', making it a model for 'separated' spin-$\frac{1}{2}$ ($\rho = 0$). For the meaning of 'separated' and the problematic involved we refer to [14,15]. This makes it possible to represent a quantum register by a hidden measurement model, using the parameters ϵ and ρ, and the models that have been constructed.

In the hidden measurement approach quantum mechanical probabilities arise due to a lack of knowledge about the precise interaction between the measurement equipment and the system. This lack of knowledge about the measurement interaction is described by the parameter ϵ, that as a consequence controls this uncertainty. The introduction of ϵ makes it possible to describe a continuous transition from the (quantum) sphere model towards a (classical) deterministic spin state particle [16,17]. The parameter ρ makes it possible to describe a continuous transition from a (quantum) coupling to a completely decoupled situation, which means that both parameters allow a continuous transition from a quantum register to a classical register of N bits. The structure of the intermediate models has been studied in detail for the case of varying ϵ, and it can be proved that the model is neither quantum nor classical, since two axioms of the representation theorem of Piron [18] for quantum and classical physical systems are violated [19,20,16]. It has also been proven that the completely uncoupled situation cannot be described by standard quantum mechanics, because the same two axioms of traditional quantum axiomatics are not satisfied (see [14,15] for an overview). In forthcoming work we will study the structure of the models from an axiomatic point of view for intermediate values of ρ.

What is certain however is that one has to use a formalism more general than the quantum formalism to describe such entities and such transitions. Following the view that computation can be regarded as the evolution of a physical system – such that the initial state of the register corresponds with the input of the computation and the final state yields the output of the computation process – we can study the process of computation as the evolution of the state of the register during the computational process. We have developed a general approach where the 'between quantum and classical' models can be studied and characterized. The basic structure is the one of a state property system, where the physical entity is described by means of its states and properties [21,22]. Evolution can be described by means of the standard procedure developed in theoretical physics: *i.e.* the one parameter group of time translations is represented in the group of automorphisms of the structure. This will give rise to unitary evolution in the special case of a pure quantum computer, and allows the description of evolution for the 'between quantum and classical' computers.

2 Quantum Computation: Main Concepts

In this section we give a quick overview of how a quantum computer works (following the presentation given in the paper by Pykacz et al. [23]) and how the quantum computational process can be interpreted as the free evolution of a physical system.

2.1 Qubits Versus Classical Bits

Let us first consider a classical computer from a more physical point of view. A classical bit of information can physically be represented with any bi-stable classical physical system, such that the two possible states represent the binary digits 0 and 1. The register of the computer consists of a number N of such bi-stable physical systems. To store the input data in the register requires the preparation of the register in a particular state. A classical N-bit register can be in 2^N different states. Any such state can be denoted by $|i\rangle$ where i is a number represented by a binary word of length N. During the calculation the state of the register follows a prescribed evolution induced by means of the processor, *i.e.*, the processor forces the necessary state evolution in order to obtain a final state containing the output of the computational process.

Contrary to a classical physical system, a quantum bi-stable system (*e.g.*, the spin state of a spin-$\frac{1}{2}$ particle) can be in a superposition state of the eigenstates for the 0 and 1 digits. Therefore, each spin-$\frac{1}{2}$ particle representing a bit in the register of a quantum computer is in general in a superposition state which can be written as a linear combination of the states $|0\rangle$ and $|1\rangle$ that encode 0 and 1 with complex coefficients such that the sum of their squared moduli is 1:

$$|s\rangle = c_0|0\rangle + c_1|1\rangle, \; c_0, c_1 \in \mathbb{C}, \; |c_0|^2 + |c_1|^2 = 1. \tag{1}$$

Although any measurement of the state $|s\rangle$ necessarily yields either 0 (with probability $|c_0|^2$) or 1 (with probability $|c_1|^2$), according to standard quantum mechanics $|s\rangle$ cannot be interpreted as an unknown state that represents either 0 or 1 with respective probabilities $|c_0|^2$ and $|c_1|^2$ (see the many no-go theorems for non-contextual hidden variable theories, *e.g.* [24,25]. Because the coefficients c_0 and c_1 are complex, not real numbers, it does not represent a statistical mixture of $|0\rangle$ and $|1\rangle$. Neither it can be interpreted as representing some value "between" 0 and 1. It is an entirely new entity having no counterpart in classical physics and the unit of information it carries is customarily called qubit (= quantum bit).

2.2 A Conbit: A Contextual Bit

That a qubit is more general than a classical bit can be seen as follows. A classical bit is represented by a classical bi-stable physical device, *e.g.*, positive or negative charge, positive or negative voltage, light on or off. Therefore, no matter how one would measure the value of the bit, one would still get the same outcome, *i.e.*, a positive voltage or a negative one etc. For a qubit, the state of the entity representing a qubit can be in a superposition state, which means that for a measurement with the predefined eigenstates $|0\rangle$ and $|1\rangle$, one will only obtain a probabilistic outcome. However, if one would make a measurement such that the superposition state of the register is an eigenstate for the experiment, then one will get the outcome corresponding with this superposition state with certainty. Let us clarify this by the example of a qubit encoded in the spin state of a spin-$\frac{1}{2}$ particle. Let us consider the case where the eigenstates defining the bit value 0, resp. 1, are the states

$$|0\rangle_z = \begin{pmatrix} 0 \\ 1 \end{pmatrix}, |1\rangle_z = \begin{pmatrix} 1 \\ 0 \end{pmatrix}$$

i.e., the eigenstates of the spin down, resp. spin up outcome for a spin measurement along the z-direction (*i.e.*, the Stern-Gerlach apparatus used to measure the spin of the spin-$\frac{1}{2}$ particle is placed along the z-direction). Let us consider the case where the particle is in a superposition state such that its state is given by $s = \frac{1}{\sqrt{2}}(|0\rangle_z - |1\rangle_z) = |0\rangle_x$ then a measurement along the direction x will yield with certainty the outcome corresponding with the $|0\rangle_x$ state, *i.e.*, 'spin down' or in other words 'bit value zero along direction x'. In other words, the superposition of two eigenstates defined by a certain measurement direction, actually defines a pure state along some other direction. As such, one could interpret a qubit as a bit from which the value is determined by the highly contextual nature of how its value is measured, *i.e.*, one can regard the superposition states present in quantum computing as due to the possibility to define/apply different measurement contexts such that each defines a pure state for the qubit involved. Therefore, we could not only call such entity a qubit, emphasizing its quantum nature, but also put emphasis on its highly contextual nature by calling it a conbit (*con*textual *bit*). Indeed, depending on which context is chosen (*i.e.*, which direction is chosen for the Stern-Gerlach apparatus) different values for the bit will be found. If the Stern-Gerlach apparatus is placed along the z-direction, the superposition state

$$s = \frac{1}{\sqrt{2}}(|0\rangle_z - |1\rangle_z)$$

will yield the two bit values 0_z and 1_z with the same probability $\left(\frac{1}{2}\right)$, but if we would place the Stern-Gerlach apparatus along the x-direction, we will obtain with certainty the outcome 0_x corresponding with the spin down eigenstate $|0\rangle_x$ for the x-direction. As such, we see that according to the used measurement context, the state of the system representing a bit yields different results for the bit value. Therefore, a qubit is highly contextual and we could call it a *conbit*, referring to this contextuality.

Using the concept of conbit we can consider any physical system with a set of bi-stable states, such that the outcomes of experiments are defined by the measurement context, to represent a 'contextual' bit. Therefore, if we could define a physical system in which the contextuality could be parametrized, we could in principle cover with the concept of conbit on the one hand the qubits, which are the highly contextual conbits, and on the other hand the classical bits which can be regarded as the non-contextual limit of a conbit. Hence, in the case that the entity is a quantum system, the conbit reduces to a qubit, and in the case of a classical system, where no contextuality occurs in the measurement situation and all experiments are deterministic, the conbit reduces to a classical bit.

2.3 Quantum Processing

Let us assume now that we have a register of N qubits. The theory of quantum computation tells how to encode an input to a quantum computer in a number of qubits that form the quantum register, and how to operate on them, with the aid of a quantum processor that works according to the laws of quantum mechanics, in order to get the desired output. Let us describe this process in a more detailed way.

The Hilbert space of a collection of quantum systems is the tensor product of the Hilbert spaces of the respective subsystems. Thus, the Hilbert space of an N-qubit quantum register is the tensor product of N 2-dimensional complex Hilbert spaces, each representing a single qubit. We will abbreviate the tensor product notation, *e.g.*, $|1\rangle \otimes |0\rangle \otimes |0\rangle \otimes |1\rangle$ by the notation $|1001\rangle$ which encodes also the binary expansion of the number 9 such that we can even write $|9\rangle = |1001\rangle$. These configurations in which each qubit is in an eigenstate of bit value one or bit value zero, coincide with the states of a classical register with N classical bits. However, while a classical system can be only in a classical configuration, the corresponding quantum system can be in any linear superposition (*i.e.*, linear combination with complex coefficients with squared moduli sum up to 1) of such configurations. Thus, the possible quantum states form a subset of a complex linear space (a Hilbert space) spanned by the classical configurations. As such, the state of a quantum register of N qubits can be written in terms of the classical configurations as:

$$\sum_{i=0}^{2^N-1} c_i \,|\,i\rangle, \qquad c_i \in \mathbb{C}, \tag{2}$$

where $|i\rangle$ denotes the state of the register that encodes the binary expansion of the number i, and $\sum |c_i|^2 = 1$. These 2^N pure states of the form $|\sigma_1, \ldots, \sigma_N\rangle$ with $\sigma_k = 0, 1$, $k \in \{1, \ldots, N\}$ form a basis of the register's state space which is called 'computational basis'.

The running of the quantum computer requires the application of various state manipulations according to some quantum algorithm. These manipulations are called 'quantum logic gates' and are given by unitary transformations. During these unitary transformations induced by the logic gates, the state of the quantum register evolves continuously in time. In the case of quantum computations usually one considers unitary transformations acting only on a few (1, 2 or 3) qubits at a time, called quantum gates. It can be shown (see, *e.g.*, [26]) that this does not restrict the variety of arithmetic operations which can be performed, *i.e.*, the set of logical gates acting on few qubits at a time is a so-called universal set of gates. This means that any logic circuit can be implemented using a number of these gates. This allows us to make numerical estimations of the time it would take on a computer to run a certain algorithm. Indeed, if we could estimate the number of logic gates used in the calculation and combine this with the time it takes to apply a certain logic gate to the register, we could estimate the time it will take the physical computational device to run a certain algorithm.

Let us conclude this section by giving some examples of elementary quantum gates, and construct the unitary matrices which represent the respective state transformation induced by each quantum logic gate.

One of the frequently used quantum gates is the controlled-NOT gate that operates on two qubits and changes the second bit iff the first bit is 1:

$$C_{cnot} : \begin{array}{l} |00\rangle \to |00\rangle \\ |01\rangle \to |01\rangle \\ |10\rangle \to |11\rangle \\ |11\rangle \to |10\rangle \end{array} \tag{3}$$

The C_{cnot} gate is usually represented graphically by a following circuit:

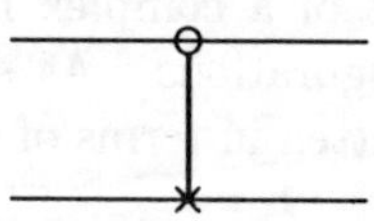

Fig. 1. Graphical representation of the controlled-NOT gate

where the circle represents the first (control) bit and the cross represents the conditional negation of the second bit. If we represent the involved bits in $\mathbb{C}^4$, *i.e.*,

$$|00\rangle = \begin{pmatrix}1\\0\\0\\0\end{pmatrix}, |01\rangle = \begin{pmatrix}0\\1\\0\\0\end{pmatrix}, |10\rangle = \begin{pmatrix}0\\0\\1\\0\end{pmatrix}, |11\rangle = \begin{pmatrix}0\\0\\0\\1\end{pmatrix}$$

then the unitary matrix $U_{C_{cnot}}$ representing the operation C_{cnot} is given by

$$U_{C_{cnot}} = \begin{pmatrix}1\,0\,0\,0\\0\,1\,0\,0\\0\,0\,0\,1\\0\,0\,1\,0\end{pmatrix}$$

The controlled-controlled-NOT gate, also called Toffoli gate, which operates on three qubits and negates the third bit iff the first two bits are 1 is represented by a circuit of the form:

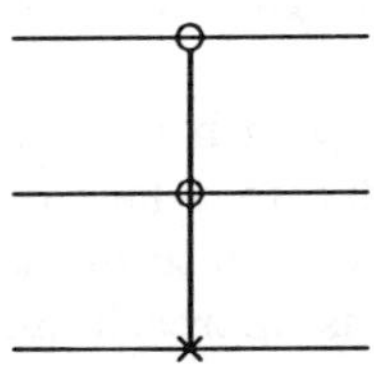

Fig. 2. Graphical representation of the Toffoli gate

This operation on the triple of qubits can be represented with the unitary matrix $U_{C_{cc-not}}$

$$U_{C_{cc-not}} = \begin{pmatrix}1\,0\,0\,0\,0\,0\,0\,0\\0\,1\,0\,0\,0\,0\,0\,0\\0\,0\,1\,0\,0\,0\,0\,0\\0\,0\,0\,1\,0\,0\,0\,0\\0\,0\,0\,0\,1\,0\,0\,0\\0\,0\,0\,0\,0\,1\,0\,0\\0\,0\,0\,0\,0\,0\,0\,1\\0\,0\,0\,0\,0\,0\,1\,0\end{pmatrix}$$

such that indeed *e.g.* $U_{C_{cc-not}} \cdot |110\rangle = |111\rangle$ etcetera.

To conclude, a quantum computer is constructed as follows. For the register one uses an N spin-$\frac{1}{2}$ particle system, such that the spin up state of a qubit corresponds with bit value +1 and spin down with bit value 0. The computation process can be regarded as the free evolution of the state of the register by running the quantum processor, *i.e.*, by applying various unitary state transformations induced by the (quantum) logic gates. However, during this unitary evolution the state of the register does not necessarily have to stay a product state of spin up and spin down eigenstates of the individual spin-$\frac{1}{2}$ entities, in general the register will be in a superposition state of such product states.

Therefore, during a quantum computation, the processor induces an evolution of the state of the quantum register along a path in Hilbert space which is not accessible for a classical device. One could expect that this larger set of available states allows a quantum computer to solve some problems faster than any classical algorithm can. And indeed, Shor's factorizing algorithm for a quantum computer allows to factorize an integer in a time polynomial in the size of the input. Despite all efforts, the best known classical algorithm still needs a time exponential in the size of the input. Whether it is actually impossible to find a classical polynomial time factorization algorithm, remains an open question.

3 Intermediate Models Between Quantum and Classical

In the hidden measurement approach to quantum mechanics the quantum probability is interpreted as due to a lack of knowledge about the precise measurement interaction which leads to indeterministic outcomes (see, *e.g.*, [16,17,19,20,27,28,29,22,30,31]). In this approach, an experiment is identified with a family of deterministic sub-measurements with a lack of knowledge about which sub-measurement actually takes place during a measurement. A concrete model has been put forward, which allows to visualize the concept of hidden measurement on a macroscopic model for a spin-$\frac{1}{2}$ measurement. Depending on the amount of uncertainty about which sub-measurement actually takes place, one obtains a continuous transition from a classical, deterministic system towards a quantum-like system in the sense that it has quantum-like state transitions induced by the measurement procedure with a quantum probability distribution over the set of outcomes. This uncertainty was modelled by a continuous real parameter, as we will discuss now in some more detail in next sections.

3.1 The Quantum Description of a Spin-$\frac{1}{2}$ Entity

In quantum theory a spin-$\frac{1}{2}$ particle is described in a two-dimensional complex Hilbert space. Pure states of the entity are represented by rays in that Hilbert space. It is well known that the unit vectors of the 2-dimensional complex Hilbert space can be represented on the surface of a unit sphere in three dimensions, usually called the Poincaré sphere. In this procedure we make use of the connection between the measurement direction u of a Stern-Gerlach experiment in three-dimensional space and the eigenstate s_u^+ for the spin up outcome corresponding with the spin observable S_u for this direction. The operator representing the spin observable along direction u is given by the Hermitian matrix S_u:

$$S_u = \frac{1}{2}\begin{pmatrix} \cos\theta & \sin\theta\, e^{-i\varphi} \\ \sin\theta\, e^{i\varphi} & -\cos\theta \end{pmatrix} \tag{4}$$

This self-adjoint spin operator has two orthogonal eigenvectors which are a basis for the Hilbert space $\mathbb{C}^2$, namely

$$s_u^+ = \begin{pmatrix} \cos\frac{\theta}{2}\, e^{-i\frac{\varphi}{2}} \\ \sin\frac{\theta}{2}\, e^{i\frac{\varphi}{2}} \end{pmatrix}, \quad s_u^- = \begin{pmatrix} -\sin\frac{\theta}{2}\, e^{-i\frac{\varphi}{2}} \\ \cos\frac{\theta}{2}\, e^{i\frac{\varphi}{2}} \end{pmatrix} \tag{5}$$

with eigenvalue $+\frac{1}{2}$ and $-\frac{1}{2}$, respectively. The physical meaning of these eigenvectors is that if the entity is in a state s_u^+ we will find with certainty the outcome $+\frac{1}{2}$ for the spin measurement along the direction defined by u. Therefore the property 'the spin entity is in a state such that spin up will be measured with certainty in a Stern-Gerlach experiment along direction u' can be represented with the eigenstate s_u^+. We now associate the measurement direction u with the eigenstate s_u^+ and simply represent the spin state s_u^+ by the point u on the Poincaré sphere. In short we let correspond a point $u \in \mathbb{R}^3$ on the surface of the Poincaré sphere with a quantum state vector $s_u^+ \in \mathbb{C}^2$, eigenvector of S_u with eigenvalue $+\frac{1}{2}$. This correspondence between the set of (pure) states of quantum spin-$\frac{1}{2}$ particles and the points on the surface of the Poincaré sphere is one to one.

A measurement on a quantum entity induces a state transition from the initial state towards an eigenstate of the observed outcome. Therefore, if the eigenvalues are not degenerated (and this is the case in spin measurements), we can identify each outcome with its eigenstate. As such, we can regard the probability for each outcome as the probability with which a state transition from the initial state towards an eigenstate of the observed outcome will occur. According to standard quantum mechanics, such a state transition happens with a probability given by the squared amplitude of the inner product of the initial and the final state. Written in spherical coordinates such that $\theta_u = 0$, the probability $P(\psi_u \mid \psi_p)$ for a state transition from initial state ψ_p towards final state ψ_u, is given by:

$$P(\psi_u \mid \psi_p) = |\langle \psi_u \mid \psi_p \rangle|^2 = \left| (1\,0) \cdot \begin{pmatrix} \cos\frac{\theta}{2}\, e^{-i\frac{\varphi}{2}} \\ \sin\frac{\theta}{2}\, e^{i\frac{\varphi}{2}} \end{pmatrix} \right|^2$$

$$= \left| \cos\frac{\theta}{2}\, e^{-i\frac{\varphi}{2}} \right|^2 = \cos^2\frac{\theta}{2}$$

3.2 The ϵ-Model

Let us now describe the hidden measurement model for a spin-$\frac{1}{2}$ entity which was first given by D. Aerts [19]. The entity consists of a point particle on the sphere. Hence, its set of states is given by the points p on the Poincaré sphere. The experiments e_u^ϵ are defined as follows. We put an elastic of length 2ϵ centered around the origin between the point u and its antipode $-u$, and attach the end points of the elastic to the points u and $-u$ with unbreakable cords. Let us denote the segment between u and $-u$ with the interval $[-u, u]$. Next, the particle falls from its position p orthogonally onto the interval $[-u, u]$ in the point p' and stays attached there. Then the elastic breaks randomly and two things can happen. If the elastic breaks between p' and $-u$, the elastic will pull the point particle towards u where it stays attached and the experiment is said to yield the outcome $+1$. If on the other hand the elastic breaks between u and p', then the elastic will pull the particle towards $-u$, where it stays attached, and the measurement is said to yield outcome -1. If the string breaks at exactly the point where the particle is attached, then we assume that in such a case the measurement always yields the outcome $+1$. However, we remark that these events are physically irrelevant since they have measure zero, but we include these situations here anyway to make the definition of the measurement complete.

Let us use the notation θ to denote the angle between the state p of the entity and the direction u of the measurement device. If $\cos\theta \geq \epsilon$, then the elastic will always pull the particle towards u, resulting in an outcome $+1$ with certainty. Analogously, if $\cos\theta \leq -\epsilon$, the measurement always yields -1. If p is such that $-\epsilon < \cos\theta < \epsilon$, the measurement yields one of the two possible outcomes $+1$ or -1. According to the definition of the experiment e_u^ϵ, the probabilities of the respective outcomes for this situation are as follows. The probability for outcome $+1$ is given by the length of the elastic between the projection point p' and the point $-\epsilon$, normalized by the total length of the elastic. This is

$$P(u \mid p) = \frac{\cos\theta + \epsilon}{2\epsilon} \tag{6}$$

Similarly we can calculate the probability for the outcome -1 as

$$P(-u \mid p) = \frac{\epsilon - \cos\theta}{2\epsilon} \tag{7}$$

Let us now consider the special cases $\epsilon = 1$ and $\epsilon = 0$. If $\epsilon = 1$ the probabilities are given by

$$P(u \mid p) = \frac{1 + \cos\theta}{2} = \cos^2\frac{\theta}{2} \tag{8}$$

$$P(-u \mid p) = \frac{1 - \cos\theta}{2} = \sin^2\frac{\theta}{2} \tag{9}$$

These probabilities coincide with the quantum probabilities for a spin measurement of a spin-$\frac{1}{2}$ particle. If $\epsilon = 0$, the experiment is deterministic, and therefore this is called the deterministic or even classical limit of the sphere model. Hence, depending on the value of the parameter ϵ controlling the lack of knowledge about the fluctuations in the measurement interaction, one obtains a physical entity varying from a quantum probabilistic spin-$\frac{1}{2}$ model towards a classical deterministic entity.

3.3 Representing a Conbit with the ϵ-Model

If we would use the ϵ-model to represent a so-called 'conbit' (*i.e.*, a bit for which the value can only be measured in a contextual way), we can define a continuous family of physical systems representing a conbit, from a quantum spin-$\frac{1}{2}$ entity, and hence a qubit, towards a classical deterministic entity, hence a classical bit.

It could be remarked that the deterministic limit of the sphere model is not a bi-stable state particle, since all points on the sphere are possible states. A true classical bit only has two possible states: either it is in a state of bit value zero or of bit value one. One way to solve this problem would be to associate the set of eigenstates of the outcome $+1$ with the bit value one, and the eigenstates of the outcome -1 with the bit value zero. As such, each bit is defined by a hemisphere containing the eigenstates of the corresponding outcome. (The set of states lying in the intersection of the two hemispheres has measure zero and as such is physically irrelevant.) In the deterministic limit of the ϵ-model, the only measurements which are considered meaningful (*i.e.*, defining a value for a bit) are the ones with a fixed measurement direction, *e.g.* along the z-direction. Hence the value of a classical bit is determined by the state of the ϵ-model entity in the deterministic limit $\epsilon = 0$, such that if the state is in the upper hemisphere the value of the conbit(classical bit) is one, and zero if the state is in the lower hemisphere.

3.4 Generalization to N Spin-$\frac{1}{2}$ Entities

It is possible to show that for any quantum entity of which the set of outcomes is in a finite dimensional space, a hidden measurement representation can be found. More specifically, in the case of an N-spin entity it was shown that a hidden measurement model exists as follows. First, the Majorana representation is used to represent the system by a system of $2N$ coupled spin-$\frac{1}{2}$ entities. Secondly, for this system a hidden measurement representation can be constructed using correlations between the so-called proper states of the $2N$ spin-$\frac{1}{2}$ entities in the system. Since for each such spin-$\frac{1}{2}$ entity there exists a concrete hidden measurement model, one can come to the conclusion that for any N-spin system a hidden measurement model can be constructed, given by $2N$ sphere models with so-called correlations of the first and the second kind. We refer to the references for a more detailed discussion of these hidden measurement models. More importantly, as a corollary, these results show how to construct a hidden measurement model for a system of N correlated spin-$\frac{1}{2}$ quantum entities, *i.e.*, a register of a quantum computer.

4 Computers 'Between Quantum and Classical'

Within the hidden measurement approach, we can represent the quantum register by N correlated sphere models. By introducing the parameter ϵ, we can construct a continuous family of physical entities with in one limit N correlated quantum-like spin-$\frac{1}{2}$ entities, and in the other limit a system consisting of N deterministic entities, representing N classical digital (stable) bits. For each value of ϵ we obtain a physical model of a computational device with a register of N conbits, such that the set of states and properties is determined by the parameter ϵ. Depending on the structure of the set of properties, one obtains different families of possible algorithms, since these depend on the nature of the physical device used to perform the computation. Let us explain this in more detail in the following subsections.

4.1 Algorithms Identified by State Transformations: Computation as Evolution of a Physical System

We consider classical and quantum computation from a physical point of view, *i.e.*, we interpret the process of computation as the evolution of the state of the physical device representing the register of the computer. Feeding the input to the computer is done by preparing the state of the computer register in a certain state. Then the processor induces some state transformations following a set of instructions encoded in a circuit of logic gates leading to a final state from which the output of the computer can be obtained. Therefore, from a physical point of view the classical and quantum computer can be treated within the same formalism. During a computation the state of the register undergoes a state transformation according to the used algorithm, and therefore one can identify an algorithm with the state transformation it induces. To get a better classification of the possible state transitions, and hence of the possible algorithms, we propose in the next subsection a general scheme to identify a physical system (in this case the register of the computer) with its set of states and the structure on its set of properties.

4.2 Between Quantum and Classical Computers and Generalized Evolution

We mentioned already that the 'between quantum and classical' models entail a structure that cannot be modelled by standard quantum mechanics. The reason is that two of the traditional axioms of standard quantum mechanics (when described axiomatically within standard quantum axiomatics) are not satisfied for the 'between quantum and classical' situations. We have investigated this aspect of the 'between quantum and classical models' in great detail [16,17,20,29,30,31], and developed a general (quantum-like) categorical formalism (of state property systems) where these models can be described [21,22,32]. Concretely this means that within this formalism we can describe quantum systems, classical systems, and 'between quantum and classical' systems. The formalism is still in full development, so we cannot call it a full fledged theory yet. However, since for quantum computation, we only need finite systems (N spins), the specific models (the ϵ-model and the ϵ, ρ-model) that we mentioned already are sufficient for our purpose, we do not in principle need the general formalism that is under development. These specific models are on the same level of concreteness as the standard quantum mechanical models. Where the fact that the specific models fit into the general theory that we are developing is important is for the description of evolution of these specific models. Indeed, the aspects of the specific models that have been studied in great details are the aspects related to their quantum nature (measurement, state transition due to measurement, entanglement, probability, etc...), but little has been investigated to their evolution. Since the unitary evolution is also an intrinsic part of the quantum computation process we will have to study in detail the evolution aspect of the 'between quantum and classical' models to be able to define a 'between quantum and classical' computer. It is in this study that we want to consider the specific models as concrete entities within this general categorical formalism that we developed [21,22,32]. There is indeed a straightforward way in theoretical physics to introduce dynamical evolution into a theory: one looks for representations of the one parameter group of time translations into the group of automorphisms of an entity in this theory. It is possible that we encounter unexpected and deep problems here, which may even make it impossible to conceive of a 'between quantum and classical' computer. For example, it might turn out that the type of evolution that we can derive for the 'between quantum and classical' models only entails the specific aspects of the quantum computation process that makes it so powerful in the limit case for a pure quantum system and unitary evolution. Even in this case however we will have learned something more about the nature of the quantum computation process and what makes it so different from the classical computation process. If, on the other hand, we can derive evolutions that allow us to also realize a 'between quantum and classical' computation process, with the same (although probably less strong) gain of power as the pure quantum computation process, we might be able to see in which way such a process could be realized in reality.

5 Conclusions

Using results obtained within the hidden measurement approach of quantum mechanics, we propose a way how to construct explicit macroscopical models for the register of a quantum computer, such that the quantum register is represented by a set of sphere models coupled by correlations of the first and second kind.

Secondly, by varying the uncertainty about the measurement interaction ϵ and the coupling ρ between the ϵ-models, one can construct a family of physical systems representing a register of N coupled contextual bits (called conbits) with a continuous transition from a quantum system with quantum entanglement (the register of the quantum computer) towards a classical system of N 'separated' bits (register of a classical digital computer). This way, quantum and classical computation can be studied in a uniform way.

Since for the intermediate sphere models, *i.e.*, for $\epsilon \in]0, 1[$, no quantum nor classical description is possible, these intermediate entities have to be described in a more general formalism, namely within the categorical setting of the state property systems. In this formalism, state transformations are described by automorphisms of the state property system. As such, the running of an algorithm (which is just a continuous state transformation from the initial state (representing the input) towards the final state (representing the output)) can be characterized, and therefore, one can study classical and quantum computation in the same formalism, namely by the automorphisms on the set of properties.

6 Acknowledgments

Part of the research for this article took place in the framework of the bilateral Flemish-Polish project 127/E-335/S/2000. B. D'Hooghe is a Postdoctoral Fellow of the Fund for Scientific Research - Flanders (Belgium)(FWO - Vlaanderen).

References

1. R. Feynman, Simulating physics with computers, *Int. J. Theor. Phys.*, **21**, 467 (1982).
2. R. Feynman, Quantum mechanical computers, *Found. Phys.*, **16**, 507 (1986).
3. P. W. Shor, Algorithms for quantum computation, discrete logarithms and factoring, *Proc. 35th Annual symposium on Foundations of Computer Science*, IEEE Computer Society Press, Los Alamitos, CA, 124 (1994).
4. P. W. Shor, Polynomial-time algorithms for integer factorization and discrete logarithms on a quantum computer, *SIAM J. Comput.*, **26**, 1484 (1997).
5. C. H. Bennett, G. Brassard, S. Breidbart and S. Wiesner, Quantum cryptography, or unforgeable subway tokens, in *Advances in Cryptology: Proceedings of Crypto '82*, eds. Chaum, D., Rivest R. L. and Sherman, A. T., Plenum Press, New York and London (1983).
6. C. H. Bennett, G. Brassard, C. Crépeau and M. Skubiszewska, Practical quantum oblivious transfer, in *Advances in Cryptology – CRYPTO '91*, ed. Feigenbaum, J., volume 576 of Lecture Notes in Computer Science, Springer-Verlag, Berlin (1992).
7. B. S. DeWitt and N. Graham, eds., *The Many-Worlds Interpretation of Quantum Mechanics*, Princeton University Press, Princeton (1973).
8. D. Deutsch, *The Fabric of Reality*, The Pengiun Press, London (1997).
9. D. Aerts, "A mechanistic classical laboratory situation violating the Bell inequalities with $2\sqrt{(2)}$, exactly 'in the same way' as its violations by the EPR experiments", *Helv. Phys. Acta* **64**, 1-23 (1991).
10. B. Coecke, Representation of a spin-1 entity as a joint system of two spin-$\frac{1}{2}$ entities on which we introduce correlations of the second kind, *Helvetica Physica Acta,* **68**, 396 (1995).
11. B. Coecke, A representation for a Spin-S entity as a compound system

in R^3 consisting of $2S$ individual spin-$\frac{1}{2}$ entities, *Foundations of Physics*, **28**, No. 8, 1347 (1998).

12. D. Aerts, S. Aerts, B. Coecke and F. Valckenborgh, "The meaning of the violation of Bell's inequalities: non-Local correlation or quantum behaviour?" preprint, Foundations of the Exact Sciences, Brussels Free University (1995).
13. D. Aerts, S. Aerts, J. Broekaert and L. Gabora, "The violation of Bell inequalities in the macroworld", *Found. Phys.* **30**, 1387-1414 (2000).
14. D. Aerts and F. Valckenborgh, "The linearity of quantum mechanics at stake: the description of separated quantum entities", this volume.
15. D. Aerts and F. Valckenborgh, "Linearity and compound physical systems: the case of two separated spin 1/2 entities", this volume.
16. B. D'Hooghe, From quantum to classical: A study of the effect of varying fluctuations in the measurement context and state transitions due to experiments, *doctoral dissertation*, VUB (2000).
17. D. Aerts, S. Aerts, B. Coecke, B. D'Hooghe, T. Durt and F. Valckenborgh, A model with varying fluctuations in the measurement context, in *New Developments in Fundamental Problems in Quantum Physics*, eds. Ferrero, M. and van der Merwe, A., Kluwer Academic, Dordrecht (1996).
18. C. Piron, *Foundations of quantum physics*, Reading, Mass (1976).
19. D. Aerts, A possible explanation for the probabilities of quantum mechanics, *J. Math. Phys.*, **27**, 202 (1986).
20. D. Aerts and T. Durt, Quantum, classical and intermediate, an illustrative example, *Found. Phys.*, **24**, 1353 (1994).
21. D. Aerts, "Foundations of quantum physics: a general realistic and operational approach", *Int. J. Theor. Phys.* **38**, 289-358 (1999), lanl archive ref: *quant-ph/0105109*.
22. D. Aerts, E. Colebunders, A. Van der Voorde and B. Van Steirteghem, State property systems and closure spaces: a study of categorical equivalence, *Int. J. Theor. Phys.*, **38**, 259 (1999), lanl archive ref: *quant-ph/0105108*.
23. J. Pykacz, B. D'Hooghe and R. R. Zapatrin, Quantum Computers as Fuzzy Computers, *Lecture notes in computer science*, **2206**, Computational intelligence: theory and applications, ed. B. Reusch, Springer-Verlag, 526 (2001).
24. J.M. Jauch and C. Piron, Can hidden variables be excluded in quantum mechanics?, *Helv. Phys. Acta*, **36**, 827 (1963).
25. S. Kochen and E.P. Specker, The problem of hidden variables in quantum mechanics, *The Logico-Algebraic Approach to Quantum Mechanics, II*, ed. C. A. Hooker, Reidel Publishing Company, 293 (1967).

26. V. Vedral, A. Barenco and A. Ekert, Quantum Networks for Elementary Arithmetic Operations, *Phys. Rev. A,* **54**, 147 (1996).
27. D. Aerts, Classical theories and non classical theories as a special case of a more general theory, *J. Math. Phys.*, **24**, 2441 (1983).
28. D. Aerts, Construction of a structure which makes it possible to describe the joint system of a classical and a quantum system, *Rep. Math. Phys.*, **20**, 421 (1984).
29. D. Aerts, T. Durt and B. Van Bogaert, Quantum probability, the classical limit and non-locality, in the proceedings of the *International Symposium on the Foundations of Modern Physics 1992, Helsinki, Finland*, ed. T. Hyvonen, World Scientific, Singapore, 35 (1993).
30. D. Aerts and T. Durt, Quantum, classical and intermediate: a measurement model, in the proceedings of the *International Symposium on the Foundations of Modern Physics 1994*, Helsinki, Finland, eds. Montonen, C. et al., Editions Frontieres, Gives Sur Yvettes, France (1994).
31. D. Aerts and B. D'Hooghe, Operator structure of a non quantum and a non classical system, *Int. J. Theor. Phys.*, **35**, 2241 (1996).
32. D. Aerts, "Being and change: foundations of a realistic operational formalism", this volume.

BUCKLEY-SILER CONNECTIVES FOR QUANTUM LOGICS OF FUZZY SETS

J. PYKACZ
Instytut Matematyki, Uniwersytet Gdański,
Wita Stwosza 57, 80-952 Gdańsk, Poland
E-mail: pykacz@delta.math.univ.gda.pl

B. D'HOOGHE
Foundations of the Exact Sciences (FUND), Department of Mathematics, Brussels Free University,
Pleinlaan 2, 1050 Brussel, Belgium
E-mail: bdhooghe@vub.ac.be

A new way of calculating intersection and union of fuzzy sets recently developed by Buckley and Siler is applied to fuzzy set models of quantum logics. The original Buckley-Siler approach is generalized to the case when one has no access to the statistical binary data that determine the membership functions of fuzzy sets.

1 Introduction

Neither the most popular Zadeh [1] operations on fuzzy sets, nor less popular but also frequently used, specially for modelling various quantum structures [2,3], Łukasiewicz [4] operations (called also *Giles, bold, truncated,* or *algebraic* operations) satisfy all identities that are satisfied by traditional set-theoretic operations on crisp sets. The main drawback of Zadeh operations is that they satisfy neither the law of excluded middle:

$$A \vee A' \neq U \tag{1}$$

nor the law of contradiction:

$$A \wedge A' \neq \emptyset \tag{2}$$

when A is a genuine fuzzy, i.e., non-crisp subset of a universe U.

Both these laws are satisfied by Łukasiewicz operations defined as follows:

$$(A \sqcup B)(x) = \min[A(x) + B(x), 1] \tag{3}$$

$$(A \sqcap B)(x) = \max[A(x) + B(x) - 1, 0] \tag{4}$$

However, on the other hand, Łukasiewicz operations are not idempotent:

$$A \sqcup A \neq A \tag{5}$$

$$A \sqcap A \neq A \tag{6}$$

whenever A is a non-crisp subset of U.

The third pair of operations on fuzzy sets that is important enough to be distinguished from the uncountable family of union-like and intersection-like operations by their own name, namely probabilistic operations:

$$(A \bigtriangledown B)(x) = A(x) + B(x) - A(x)B(x) \tag{7}$$

$$(A \bigtriangleup B)(x) = A(x)B(x) \tag{8}$$

is, in this respect, even worse since these operations satisfy neither Eq. (1),(2), nor Eq. (5),(6) when A is a genuine fuzzy set.

There are also other laws of the traditional set theory (equivalently: classical logic) that are not satisfied by Zadeh, Łukasiewicz, and probabilistic operations, but it seems that at least in the realm of quantum logics neither of them is as desirable as the laws of idempotency, excluded middle and contradiction: in a quantum logic (= orthomodular complete partially ordered set or lattice) which is an algebraic model of experimentally verifiable propositions about a physical system, conjunctions and disjunctions of propositions are traditionally modelled by meets and joins which *are* idempotent and *do satisfy* the laws of excluded middle and contradiction.

Abstract quantum logics can be modelled by families of fuzzy sets. One of such attempts was pursued by the "Bratislava school" and it is based on Zadeh connectives (see the paper by Riečan [5] and subsequent papers). The other approach of this type was developed by one of the present authors, and it is based on Łukasiewicz connectives (see the paper by Pykacz [6] and subsequent papers, esp. [2]). In both these approaches elements of a quantum logic, which represent propositions about a physical system, are modelled by suitable fuzzy sets but in view of the previously given arguments neither of these models seem to be entirely satisfactory.

At the Seventh IFSA World Congress held in Prague in June 1997, J.J. Buckley [7] proposed a new way of calculating values of membership functions of intersections and unions of fuzzy sets. The novelty of this approach consists in the fact that these values depend not only on the values $A(x)$ and $B(x)$ of membership functions of considered fuzzy sets A and B, but also on the parameter ρ that can be interpreted as the correlation coefficient between binary data a_i and b_i used to establish the values $A(x)$ and $B(x)$. The parameter ρ takes value in the interval $[-1, 1]$ and, for example, forces intersection and union to be, respectively, Zadeh, probabilistic, and Łukasiewicz operations when ρ equals 1, 0, and -1. Due to this "flexibility", for Buckley-Siler connectives more basic identities hold true than it happens in the case of any of the previously mentioned operations: in particular, Buckley-Siler connectives fulfill both idempotency laws fulfilled by Zadeh connectives (in this case necessarily $\rho = 1$) and also excluded middle and contradiction laws fulfilled by Łukasiewicz connectives (since in this case necessarily $\rho = -1$).

The objective of the present paper is to study the possibility of using Buckley-Siler connectives in fuzzy set models of quantum logics. Section 2 is devoted to the general outline of the original Buckley-Siler ideas applied to the "classical case", i.e., to studying properties of classical physical systems. In Section 3 we generalize Buckley-Siler construction to the case when an underlying sample space is unknown, i.e., to the case when one has no access to the binary data (results of tests) that determine membership functions of fuzzy sets. Finally, in Section 4 we apply the derived construction to the case of a quantum spin-$\frac{1}{2}$ particle.

2 Buckley-Siler connectives for classical observables

Buckley and Siler [8,9,7] proposed a way to evaluate the intersection and union of fuzzy sets by using a "correlation coefficient" between two fuzzy sets. In this section we will briefly describe their approach applied to the situation when one tests simultaneously two properties of a classical physical system. Let us assume (following, e.g., Pykacz [2]) that a fuzzy set A represents a property of a physical system, and that for every state x of this system we have a collection of data of n tests of this property. The value of A in a state x is determined as follows. Perform test i and check whether the property A is true or not in the state x. If it is true, we assign to the test the outcome $a_i = 1$. If it is not true, then the test yields the outcome $a_i = 0$. If we average over the n tests we get the value $A(x)$. This membership value could be viewed as representing the degree to which the system in the state x possesses the property A. In a similar way the value $B(x)$ is defined as the average over the experimental results b_i for a sequence of tests of the property B when a physical system is in the state x. We assume that a physical system obeys the law of classical physics, i.e., that in particular in every state we can simultaneously test properties A and B, so the average is over the same number of tests. From now on we will use the abbreviation $a = A(x), b = B(x)$ to denote these averages. If we define $c_i = a_i b_i$ then the average over the c_i is c, and it represents the property $C(x) = (A \text{ and } B)(x)$, which could be used as the definition of the intersection of fuzzy sets that represent properties A and B, respectively, or yet in other words: $C(x)$ represents the statement "physical system in the state x has the property A *and* the property B". This definition of $C(x)$ makes sense since we assumed that both A and B can be tested simultaneously. Once we know the outcomes a_i and b_i for every test, then also c_i are defined and hence $C(x) = (A \cap B)(x)$. In some cases, however, we might not have access to the binary data a_i and b_i of every test. If then the membership values $A(x)$ and $B(x)$ are given, we cannot draw conclusions about the value of $(A \cap B)(x) = C(x)$. Buckley and Siler showed that if we have previous information about the correlation between A and B, then we can define the value $C(x)$ of their intersection. In order to do this, they define the correlation coefficient ρ between A and B as follows:

$$\rho = \frac{\sum_{i=1}^{n}(a_i - a)(b_i - b)}{n\delta} \tag{9}$$

In this formula δ stands for the product of $\sigma_a \cdot \sigma_b$, with σ_a being the standard deviation of A in x, and σ_b being the standard deviation of B in x. The summation goes over the set $\{1, \ldots, n\}$ indicating the number of tests of the properties A and B. One could think of this correlation coefficient between A and B to be known from the past experience or, in the particular case of properties of physical systems, from the theory. Therefore, we can assume that information about the association between the values of the membership functions of the two fuzzy sets A, B in the point x is available. Since the binary data a_i and b_i can take value in $\{0, 1\}$ only, the standard deviations can be easily calculated, and one finds that $\sigma_a = \sqrt{a(1-a)}$ and $\sigma_b = \sqrt{b(1-b)}$. Beside intersection of fuzzy sets A and B which represents conjunction of properties of a physical system, we can also define union of fuzzy sets A and B which represents disjunction of respective properties. Let $d_i = a_i + b_i - c_i$ for all $i \in \{1, \ldots, n\}$. Then the average d represents the statement: "physical system has the property A *or* the property B". If one elaborates the expression for the correlation coefficient ρ, one finds that it equals

$$\rho = \frac{c - ab}{\delta}. \tag{10}$$

When a or b is zero or one, the correlation coefficient ρ is not defined but c and d are still defined (cf. [8,9]). Let us define

$$\rho_u = \frac{\min(a, b) - ab}{\delta} \tag{11}$$

and

$$\rho_l = \frac{\max(a + b - 1, 0) - ab}{\delta} \tag{12}$$

Buckley and Siler [8,9] proved that the following relations hold:

1. $\max \rho = \rho_u$,

2. $\min \rho = \rho_l$,

3. $\rho_l \leq \rho \leq \rho_u$,

where $\max \rho$ and $\min \rho$ is taken over all possible values of a_i and b_i for $1 \leq i \leq n$. From the definition of c, d, and ρ it follows immediately that $c = ab + \rho\delta$ and $d = a + b - ab - \rho\delta$. Therefore, once the average values a, b and the correlation between A and B is known, also the truth value of the conjunction of the two properties A and B, and the truth value of the disjunction of A and B when a physical system is in a state x is determined. Finally, one could easily see that:

1. If $\rho = \rho_u$, then $c = \min(a, b)$ and $d = \max(a, b)$.
2. If $\rho = 0$, then $c = ab$ and $d = a + b - ab$.
3. If $\rho = \rho_l$, then $c = \max(a + b - 1, 0)$ and $d = \min(a + b, 1)$.

These three values of ρ coincide with the case that A and B are, respectively, maximally correlated, not correlated and maximally anti-correlated. This is an important result since it shows that: (1) when A and B are maximally correlated we obtain Zadeh operations; (2) if A and B are not correlated at all we get probabilistic operations; and (3) when A and B are maximally anti-correlated, we get Łukasiewicz operations for computing intersection and union of two fuzzy sets that represent properties of a physical system. Therefore, depending on the value of ρ these connectives take different forms.

These previous results were valid for the case that one has access to all the binary data a_i and b_i, so that the expressions for c and d are somehow tautological. If we do not have access to these binary data, and only the averages a and b are available, then we cannot calculate the intersection and union of two fuzzy sets in a correct way. Indeed: if A and B have the average value $\frac{1}{2}$ in some state x, then it could be (among the other cases) that $A = B$, or that $A = B'$. In these two distinguished cases the membership value of the intersection of the two fuzzy sets that represent properties of a physical system would be expected to be, respectively, $\frac{1}{2}$ and 0. Therefore, we see that the averages alone do not allow to distinguish between these two possibilities. But if we have some knowledge of the correlation between the two fuzzy sets, then the union and intersection can be defined. Buckley and Siler generalized the expressions for c and d by introducing two functions $T(a, b)$ and $C(a, b)$ as follows:

$$T(a, b) = ab + \rho\delta \tag{13}$$

and

$$C(a, b) = a + b - ab - \rho\delta \tag{14}$$

where $-1 \leq \rho \leq 1$ and $(a,b) \in \Omega \subseteq [0,1] \times [0,1]$. The restricted domain Ω was introduced by Buckley *et al.* [8,9,7] in order to make sure that the values of a and b are compatible with the correlation coefficient ρ, i.e., that there exists a set of binary data such that the triplet $\{a, b, \rho\}$ is possible. Since we will not go into the details of their proofs we will assume from now on that ρ is always in agreement with the given averages a and b.

The interpretation of $T(a,b)$ and $C(a,b)$ is as follows: given $A(x) = a$ and $B(x) = b$, then $(A \cap B)(x) = T(a,b)$ and $(A \cup B)(x) = C(a,b)$. The correlation ρ used in the expressions for $T(a,b)$ and $C(a,b)$ is assumed to be known from the "past experience", or, in our case, from the theory. So given (a,b) in Ω and the assumed ρ we can compute $T(a,b)$ and $C(a,b)$. All the basic crisp set identities hold for the pair T and C if the parameter ρ is suitably adjusted. We write A' for the complement of A and then $\rho = -1$. This is the same as saying that in every test the outcome a_i' equals $1 - a_i$, which is the standard way of defining negation in quantum logic theory, so we have $A'(x) = 1 - A(x)$. Let us denote by X the universe whose membership function is one everywhere and which corresponds to the "always true" property and by $\emptyset$ the empty set whose membership function is everywhere zero and which corresponds to the property that is never true. The following relations were shown by Buckley and Siler [8,9] to be valid, for properly chosen ρ:

1. $A \cap A = A$, $\rho = 1$. (Idempotency Law for $\cap$).
2. $A \cap \emptyset = \emptyset$, any ρ.
3. $A \cap X = A$, any ρ.
4. $A \cap A' = \emptyset$, $\rho = -1$. (Law of Contradiction)
5. $A \cup A = A$, $\rho = 1$. (Idempotency Law for $\cup$)
6. $A \cup X = X$, any ρ.
7. $A \cup \emptyset = A$, any ρ.
8. $A \cup A' = X$, $\rho = -1$. (Law of Excluded Middle)
9. $(A \cap B)' = A' \cup B'$, any ρ. (De Morgan Law)
10. $(A \cup B)' = A' \cap B'$, any ρ. (De Morgan Law)
11. $A \cup (A \cap B) = A$, appropriate ρ. (Absorption Law)
12. $A \cap (A \cup B) = A$, appropriate ρ. (Absorption Law)

13. $A \cap (B \cup C) = (A \cap B) \cup (A \cap C)$, appropriate ρ. (Distributivity Law)

14. $A \cup (B \cap C) = (A \cup B) \cap (A \cup C)$, appropriate ρ. (Distributivity Law)

In formula (1) and (5) $\rho = 1$, and in formula (4) and (8) $\rho = -1$. This is true for the correlation coefficient of a set with itself and with its complement, respectively, and indicates the generality of this approach.

Finally we want to close the presentation of the ideas of Buckley and Siler by quoting one of their closing remarks [9]:

"One important point that has not been discussed is that if we are using $\rho = 0.7$ for evaluating $(A \cap B)(x)$ when $x = 1$, should we change ρ when computing $(A \cap B)(x)$ at $x = -1$? We believe that ... there is one value of ρ for obtaining $(A \cap B)(x)$ for all x. ... We have one value of ρ when working with A and B but another value for C and D. $T(a, b)$ and $C(a, b)$ depend on three variables a, b and ρ. The fuzzy sets determine ρ and the membership values determine the a and b."

3 Generalization of Buckley-Siler approach to the case of fuzzy sets with no underlying sample space

Buckley and Siler always started from the basic assumption that the value of the membership function of a fuzzy set in any point could be obtained by making a series of tests and that during each test the binary data could be retrieved for all fuzzy sets simultaneously. Therefore, their intersection and union of two fuzzy sets was always well defined and obeyed the properties characteristic to crisp sets. The correlation coefficient could always be thought of as being possibly unknown but at least it was knowable in principle if one would take a look at the individual tests giving rise to values of membership functions of fuzzy sets. In this section we will generalize the Buckley and Siler connectives to cases where it is a priori impossible to perform tests that could simultaneously yield the binary data for A and B, which is a typical situation encountered in quantum mechanics when measurements of non-compatible (non-commuting) observables are involved.

While Buckley and Siler proposed to use a fixed number n to denote the number of tests, we will assume that we have an infinite number of tests available. This is an idealization of the real physical experimental situation, but it will allow us to treat our generalization in a more uniform way. First we define the correlation coefficient between the two fuzzy sets A and B :

$$\rho(A, B) = \frac{\int_\Omega A(x) B(x)\, dx - \int_\Omega A(x)\, dx \int_\Omega B(x)\, dx}{\sigma_A \sigma_B} \tag{15}$$

where

$$\sigma_A = \sqrt{\int_\Omega (A(x))^2\,dx - \left(\int_\Omega A(x)\,dx\right)^2} = \sqrt{\langle A^2\rangle_\Omega - \langle A\rangle_\Omega^2} \tag{16}$$

stands for the standard deviation of A over the state space. Analogously

$$\sigma_B = \sqrt{\int_\Omega (B(x))^2\,dx - \left(\int_\Omega B(x)\,dx\right)^2} = \sqrt{\langle B^2\rangle_\Omega - \langle B\rangle_\Omega^2}, \tag{17}$$

is the standard deviation of B over the state space Ω. The integrals are taken to be over the whole state space Ω. This is a straightforward generalization of the Buckley and Siler definition of a correlation, except for the important difference that they define the correlation pointwisely, and we define it globally over the whole state space. The pointwise definition would be:

$$\rho(A,B)(x) = \frac{\int_{\mathcal{E}} A_e(x)\,B_e(x)\,de - \int_{\mathcal{E}} A_e(x)\,de \int_{\mathcal{E}} B_e(x)\,de}{\sigma_A(x)\,\sigma_B(x)} \tag{18}$$

in which the integral now goes over the set of experiments $\mathcal{E}$. In this case the standard deviations are understood to be local: they are the standard deviations for the measurements $e \in \mathcal{E}$ in one point x. As such, they are given by

$$\begin{aligned}\sigma_A(x) &= \sqrt{\int_{\mathcal{E}} (A_e(x))^2\,de - \left(\int_{\mathcal{E}} A_e(x)\,de\right)^2} \\ &= \sqrt{\int_{\mathcal{E}} A_e(x)\,de - \left(\int_{\mathcal{E}} A_e(x)\,de\right)^2} \\ &= \sqrt{A(x) - A(x)^2}\end{aligned} \tag{19}$$

since $\int_{\mathcal{E}} A_e(x)\,de = A(x)$ by definition and since $(A_e(x))^2 = A_e(x)$ because it equals either zero or one for every test e. Analogously, we find for the local standard deviation of B in x :

$$\sigma_B(x) = \sqrt{B(x) - B(x)^2} \tag{20}$$

The integral $\int_{\mathcal{E}} A_e(x)\,B_e(x)\,de$ is for some systems undefined. If A and B are compatible observables we can measure them both at the same time and the Buckley-Siler scheme can proceed as before. However, if they are not compatible, then simultaneous measurements of these two properties do not exist and we have to find another way to define connectives. In order to do this we assume, following Buckley and Siler, that the local correlation coefficient $\rho(A, B)(x)$ is the same as the global one and we put $\rho(A, B)(x) = \rho(A, B)$.

With this assumption we are able to compute the correlation coefficient if only the membership functions $A(x)$ and $B(x)$ are given. If A and B are the same or the complementary fuzzy sets, then it is easy to check that Eq. (15) yields, respectively, values $\rho(A, A) = 1$ and $\rho(A, A') = -1$, as expected, so our assumption seems reasonable. However, in these two examples properties of a physical system described by fuzzy sets A and A' are compatible which does not imply a priori that our assumption is also good for non-compatible observables. Nevertheless, we shall now apply this approach to a simple example of fuzzy sets generated by the spin measurements of a quantum spin-$\frac{1}{2}$ particle , and refer for general considerations to a previous paper [10].

4 Buckley-Siler connectives for spin-$\frac{1}{2}$ observables

Let us define a property A of a quantum spin-$\frac{1}{2}$ particle to be a spin observable in a direction u. The probability of obtaining the outcome "spin up" for the particle in a state that makes an angle θ with u is $\cos^2\left(\frac{\theta}{2}\right)$. Since we are interested only in the spin properties of the particle and all possible (spin) states of the particle are in one-to-one correspondence with the points of the interval $[0, 2\pi]$, we can identify the set of (spin) states of the particle with this interval. The membership value of the fuzzy set representing the property A is thus given by $\cos^2\left(\frac{\theta}{2}\right)$. Let another property B correspond to a Stern-Gerlach apparatus that makes an angle φ with the direction u. The membership function of B over the state space $[0, 2\pi]$ is given by $\cos^2\left(\frac{\theta-\varphi}{2}\right)$. With this we can calculate the global correlation coefficient $\rho(A, B)$. Eq. (15) yields $\rho(A, B) = \cos\varphi$. If we compare this expression with the values of the correlation coefficient for the two cases of compatible spin-$\frac{1}{2}$ observables (either $A = B$, or $A = B'$), then we see that our assumption is confirmed since we get $\rho(A, A) = \cos 0 = 1$ and $\rho(A, A') = \cos\pi = -1$. Let us now calculate intersection and union of two fuzzy sets that represent non-compatible spin-$\frac{1}{2}$ observables. First we calculate the standard deviations:

$$\begin{aligned}
\sigma_A(x) &= \sqrt{\int_{\mathcal{E}} (A_e(x))^2 \, de - \left(\int_{\mathcal{E}} A_e(x) \, de\right)^2} \\
&= \sqrt{\int_{\mathcal{E}} A_e(x) \, de - \left(\int_{\mathcal{E}} A_e(x) \, de\right)^2} \\
&= \sqrt{A(x) - A(x)^2}
\end{aligned} \tag{21}$$

and, analogously, $\sigma_B(x) = \sqrt{B(x) - B(x)^2}$. If we take $A = \cos^2\left(\frac{\theta}{2}\right)$ and $B = \cos^2\left(\frac{\theta-\varphi}{2}\right)$ then, since the correlation coefficient is given by $\cos\varphi$, we get the following formulas for the intersection and union of $A(\theta)$ and $B(\theta)$:

$$\begin{aligned}(A \cap B)(\theta) &= A(\theta) B(\theta) + \rho\delta \\ &= \cos^2\left(\frac{\theta}{2}\right)\cos^2\left(\frac{\theta-\varphi}{2}\right) + \\ &\cos\varphi\left|\cos\left(\frac{\theta}{2}\right)\sin\left(\frac{\theta}{2}\right)\cos\left(\frac{(\theta-\varphi)}{2}\right)\sin\left(\frac{(\theta-\varphi)}{2}\right)\right| \end{aligned} \tag{22}$$

$$\begin{aligned}(A \cup B)(\theta) &= A(\theta) + B(\theta) - A(\theta) B(\theta) - \rho\delta \\ &= \cos^2\left(\frac{\theta}{2}\right) + \cos^2\left(\frac{\theta-\varphi}{2}\right) \\ &- \cos^2\left(\frac{\theta}{2}\right)\cos^2\left(\frac{\theta-\varphi}{2}\right) \\ &- \cos\varphi\left|\cos\left(\frac{\theta}{2}\right)\sin\left(\frac{\theta}{2}\right)\cos\left(\frac{(\theta-\varphi)}{2}\right)\sin\left(\frac{(\theta-\varphi)}{2}\right)\right| \end{aligned} \tag{23}$$

According to our previous considerations these formulas can be thought of as representing the truth values of the conjunction: "the particle has the property A *and* the property B" and the disjunction: "the particle has the property B *or* the property B", respectively. The numerical values yielded by Eq. (22) and (23) are, in general, different from the results obtained within the standard quantum logic approach, in which case conjunction and disjunction is, respectively, represented by meet and join of elements of a logic. However, since non-compatible properties cannot be tested simultaneously, there is no direct experimental possibility to distinguish between these two models and one has to look for other, maybe indirect arguments in order to decide which model is better suited for describing properties of quantum physical systems.

Acknowledgments

The paper was written as a part of the joint Polish-Flemish Research Project 127/E-335/S/2000. J. Pykacz was also supported by the University of Gdansk Research Grant BW/5100-5-0300-1. B. D'Hooghe is a Postdoctoral Fellow of the Fund for Scientific Research - Flanders (Belgium)(F.W.O. - Vlaanderen).

References

1. L. A. Zadeh, "Fuzzy sets", *Information and Control*, **8**, 338 (1965).
2. J. Pykacz, "Fuzzy quantum logics and infinite-valued Łukasiewicz logic", *Int. J. Theor. Phys.*, **33**, 1403 (1994).
3. J. Pykacz, "Triangular norms-based quantum structures of fuzzy sets", in *Proceedings of the 7th IFSA World Congress*, Academia, Prague (1997).
4. R. Giles, "Łukasiewicz logic and fuzzy set theory", *International Journal of Man-Machine Studies*, **8**, 313 (1976).
5. B. Riečan, "A new approach to some notions of statistical quantum mechanics", *BUSFFAL*, **35**, 4 (1988).
6. J. Pykacz, "Quantum logics as families of fuzzy subsets of the set of physical states", in *Preprints of the Second IFSA World Congress, Tokyo,* **Vol. II**, 437 (1987).
7. J.J. Buckley, W. Siler and Y. Hayashi, "A new fuzzy intersection and union", in *Proceedings of the 7th IFSA World Congress*, Academia, Prague (1997).
8. J.J. Buckley and W. Siler, "L_∞ fuzzy logic", preprint, available from the authors on the request sent to: buckley@math.nab.edu.
9. J.J. Buckley and W. Siler, "A new t-norm", preprint, available from the authors on the request sent to: buckley@math.nab.edu.
10. B. D'Hooghe and J. Pykacz, "Classical limit in fuzzy set models of spin-$\frac{1}{2}$ quantum logics", *Int. J. Theor. Phys.* , **38**, 387 (1999).

SOME NOTES ON AERTS' INTERPRETATION OF THE EPR-PARADOX AND THE VIOLATION OF BELL-INEQUALITIES

WIM CHRISTIAENS

Department of Philosophy and Moral Science, University of Ghent, Blandijnberg 2, 9000 Ghent, Belgium

E-mail: wimivan@hotmail.com

I give an analysis of what happens in the case of the Einstein-Podolsky-Rosen-Bohm experiment according to the creation-discovery view. According to many physicists the correlations in the experiment indicating a violations of Bell-inequalities are cases of outcome-dependence and not parameter dependence. It is shown that we can interpret the outcome independence violating correlations as an indication that quantum entities do not always exist in space, and that the Einstein-Podolsky-Rosen-Bohm experiment involves a localization of one entity, that in the course of this localization is 'broken' into two entities.

1 Introduction

The so-called Geneva school in the foundations of physics (counting among its members J. M. Jauch, C. Piron, D. Aerts and others) is part of the field of quantum logic. I shall not comment on quantum logic itself or on the relation of this approach to other approaches in the foundations of quantum mechanics and philosophy of quantum mechanics (For recent overviews and links with other approaches to the foundations of physics see [1,2]). Within the Geneva group Diederik Aerts has studied the quantum logical description of compound quantum entities taking into account the discussion surrounding the Einstein-Podolsky-Rosen incompleteness argument and violations of Bell inequalities by quantum entities in a so-called 'entangled' state. He also proposed interpretations of the quantum paradoxes. I will focus on the Aerts' understanding of the EPR-paradox and the violations of the Bell inequality of quantum entities. I will first briefly recount the EPR-story in this introductory section following the presentation in [3].

A. Einstein, B. Podolsky and N. Rosen (I will refer to them as EPR) claimed that either quantum mechanics is incomplete or quantum mechanics predicts non-local behavior for spatially separated entities, given that one accepts a principle of locality and a principle of completeness:

> (Loc.) Elements of reality pertaining to one system cannot be affected by measurements performed at a distance on another system.

(Completeness) Every element of reality must have a counterpart in the theory.

In the Genevan approach the elements of reality correspond with properties. Properties are identified with equivalence classes of so-called questions. If α and β are questions of an entity S such that $\alpha < \beta$ and $\beta < \alpha$ (where $<$ means "when α is true, β is true" and $<$ is reflexive and transitive) then α and β are said to be equivalent: they test the same property. A property is actual iff a question testing that property is true. If a property is actual all the questions testing that property are true. A question is an experiment that can be performed on an entity. If the experiment gives the expected outcome, we say that the aswer of the test is "yes". If the experiment does not give the expected answer, we say that the answer is "no". A question on an entity S is true iff when we should decide to perform the test, the expected answer would come out with certainty (see [4,1]). A question is a yes-no measurement. So the locality principle actually says that experiments performed on one entity cannot be affected by experiments performed on another entity. EPR devised a Gedanken-experiment to prove their point [5]. David Bohm developed an experimentally realizable version of this experiment with spin-properties. Most discussions are in the context of Bohm's version of the experiment. Consider the spin-properties of quantum entities. The spin properties of one quantum entity are incompatible properties, so they cannot all have definite values at the same time. The actual experiment involves a source that produces pairs of quantum entities $S(1)$ and $S(2)$ in a superposition state of their spin properties. The entities are emitted from the source in the state described below at t_0. I refer to the compound entity as S. Perfect anti-correlation is supposed: if the spin of $S(1)$ and $S(2)$ are measured along some direction (say the Z-direction) then we will find with certainty that they are opposite. The anti-correlation-state for the Z-direction is given by:

$$| \Psi\rangle = \frac{1}{\sqrt{2}} \mid (spin - up)Z(1)\rangle \mid (spin - down)Z(2)\rangle -$$

$$\frac{1}{\sqrt{2}} \mid (spin - down)Z(1)\rangle \mid (spin - up)Z(2)\rangle$$

where $\mid (spin - up)Z(1)\rangle \mid (spin - down)Z(2)\rangle$ and $\mid (spin - down)Z(1)\rangle \mid (spin - up)Z(2)\rangle$ are states in the product state space $H(1) \bigotimes H(2)$ ($H(1)$ is the state space of $S(1)$ and $H(2)$ is the state space of $S(2)$). The state described in the formula and states like it (non-factorizable states for the joint system of two systems) are entangled states. The pairs of quantum entities $S(1)$ and $S(2)$ travel in opposite directions (the left wing and the right wing of the experimental set-up) and coincidence measurements of their spin are performed at t_1 on both quantum entities with measurement apparatuses $M(1)$ and $M(2)$. After the measurement a different state is obtained than the entangled state before the measurement. We have either

$$| (spin - up)Z(1)\rangle \mid (spin - down)Z(2)\rangle$$

or

$$| (spin - down)Z(1)\rangle \mid (spin - up)Z(2)\rangle$$

This later state does have a definite value for the variable that was measured. Therefore EPR concluded that quantum mechanics is incomplete, because the state cannot be the result of a real state transition. J. S. Bell managed to derive an inequality that any local physical theory should obey. Experimental evidence (see [6]) showed that quantum systems violate this inequality. It is generally agreed that the reason for this is that the compound system described by the singlet state violates the locality condition. Aerts proved a similar result in one form of operational quantum logic [4,1]. He constructed a lattice that describes separated quantum entities and compared this with the property lattice of a tensor product of the Hilbert spaces of the individual entities. He concluded that separated quantum entities cannot be described by the property lattice of a tensor product of the Hilbert spaces of the individual entities. He constructed a new version of his proof in response to criticism in [7]. For a simplified version of the proof see [8] (exercises 1.8.2, 1.8.3, 8.2.3). For an overview of the mathematical work of Aerts with respect to the description of compound systems in operational quantum logic, see [9].

In short, the interpretation of Aerts (and I think also Jean Reignier and Bob Coecke) goes like this. A physical entity has intrinsic monadic properties. In the case of quantum entities the properties can be in superposition states. A measurement induces a state transition. When this happens with the position variable, we have a case of "localization" in space. What are the implications for our space-time view? Space-time is usually seen as a substance that is the necessary condition for physical processes, minimally as a structure that functions as a necessary condition for the existence and causal interaction of physical entities. The experiences with quantum entities (in the specific interpretation that is given here) teach us that this space-time structure is not the necessary condition of the existence and causal interaction of all physical entities. We have to renounce the idea of a permanent location of a quantum entity. Toraldo di Francia quotes from a book by D'Espagnat and Klein: "The existence of non-separability clearly reveals at least a dissonance between quantum mechanics and the very notion of space. This is undeniably an important philosophical conclusion" [10]. Another scholar writes: "Whereas the notions of space and time figure prominently in the foundational-philosophical investigations of locality, and whereas the notions of space and time are necessarily presupposed in the very idea of locality, space and time are absent from the Bohm singlet. Moreover, space and time never enter the relevant formulae; they are served as ingredients of a verbal sauce on the mathematical pasta, which do not have counterparts in that pasta" [11].

I end this section with an overview of this paper and then a summary presentation of the model of the EPR-situation and the violation of Bell-inequalities. In section 2, I describe a stochastic model for the EPR-situation and specify the conditions such a model must satisfy to be factorizable. Factorizability can be analyzed in two more specific conditions: parameter independence and outcome independence. I then proceed to interpret the fact that one of these conditions (outcome independence) is violated. According to [12] *separability* says that spatially separate entities possess separate real states and *locality* means that the state of an entity can be changed only by local effects (effects propagated with finite subluminal speeds). (The first definition I gave of locality (i) pertains to properties and not states, (ii) says 'at a distance', which can mean 'at a spacelike separation' (Einstein locality) or 'in the absence of causal influences recognized by current physical theories' ([3]). If you define states as collection of properties ("elements of reality") with a specific mathematical structure (the structure of a Piron lattice) and fill 'at a distance' in with Einstein locality, we get the definition of Howard. The violation of outcome independence is usually interpreted as a violation of separability (or conceptually distinct: holism), rather than an instance of action-at-a-distance (which would be the case if parameter independence were violated). I look at the main reason to discard violation of OI as a causal influence and at a prima facie reason to doubt this can be done. In section 3, I look at a possible way out: we suppose that S is the common cause of the measurement outcomes. S causes the outcome in the left wing together with $M(1)$ and S causes the measurement outcome in the right wing with $M(2)$. S can do this because the causal influence of S does not propagate through space. Furthermore the causal operations of S do not factorize. But this model has been criticized, because it is still factorizable, and thus does not reproduce the probabilities predicted by quantum mechanics. In section 4 another possible interpretation of quantum entanglement is discussed: S is not a compound entity, but one entity that occupies one region of space, and gets separated into two entities by the measurement interaction. Aerts modifies this view by supposing that the non-separable entity does not occupy a region of space. In section 5 the three threads of the discussion (the non-separability, the non-spatiality of the causal influence and non-spatiality of the non-separable entity S) come together in the hidden measurement/hidden correlation representation of quantum entanglement by Aerts and Coecke. The two final sections are of a purely philosophical nature. I look at the notion of individuality in quantum mechanics (are quantum entities individuals?) in section 6, and at the nature the scientific realism of Aerts must have (given his treatment of the EPR-paradox) in section 7.

Let me sum up the idiosyncrasies of the EPR-model that is obtained in section 5:

(A) the entity S is the common cause of the measurement results in both wings (more precisely: the entity S is a partial cause acting together with $M(1)$ and $M(2)$); (B) S's causal operations (causing the measurement outcomes) are correlated in the way predicted by quantum mechanics; (C) S's causal influences do not propagate through space; (D) before any measurement interaction S is not a compound entity consisting of two entities $S(1)$ and $S(2)$ between which there exist correlations, rather it is one non-separable entity; (E) S is not localized in space before any measurement interaction, so it is not meaningful to say that the measurement results are caused by the common cause in the source since the source is a classical entity localized in space and the measurement interaction happens after S has left the source and is not localized in space anymore; (F) the localization in space of the entity S is a real state transition, more specifically, it is a "creation" of the property of position, in the course of which the entity breaks in two entities $S(1)$ and $S(2)$ and localizes them in the spatial contexts defined by the measuring apparatuses (G) during the localization, the way the measuring apparatuses $M(1)$ and $M(2)$ affect S (in the course of which S ceases to exist), is similar to the way $S(1)$ and $S(2)$ (which come into existence because of the measurement interaction) affect each other (this is particularly apparent when there are more than two compounds in S).

2 The EPR-Situation

There is a source that emits in a uniform manner compound entities in an entangled state. Because we prepared the state we know that two entities are entangled, labeled 1 and 2: $S(1)$ one the left and $S(2)$ on the right. For every entity pair there is supposed to be a state k from a set K. Hidden variables models acknowledge that an entity prepared in a given way is correctly described by the quantum mechanical state; what they add is that even this state is incomplete. (A bottle of gas in thermodynamic equilibrium is correctly described by one specific thermodynamic state and not by any other. This thermodynamic state however fails to state the microscopic constitution of the gas.) We can distinguish non-contextual or simple hidden variables models (k assigns a definite value to each variable at all times) from contextual hidden variables models (k together with some feature(s) of the context the experimental arrangement restore determinateness of all the variables). There also exist deterministic hidden variable models and stochastic hidden variables models.[a] In the case of stochastic models the complete states only determine probabilities that values will occur. In this paper I only consider stochastic models. It will turn out that there are contextual hidden variables. But for the moment I make no assumption about the states k and the state space K, except that one can define probability measures on K. There are two analyzers: one on the left, denoted $M(1)$, with controllable parameters a and a', one on the right, denoted $M(2)$, with controllable parameters b and b'. Measurement values for a will be denoted s_m where $m = 1, 2,$; measurement values for b will be denoted t_n where $n = 1, 2, 3, ...$ They can have the values 1 or 0 (spin up or spin down along the direction of measurement of a quantum entity). Each choice of k and parameters a and b determines a joint probability of values in the two wings and single probabilities in each wing. For example, the expression

$$\Pr_1(m/k\&a\&b)$$

is the probability of outcome s_m for the first entity $S(1)$ given the complete state k and the observables a and b. Similarly for t_n. The expression

$$\Pr_{1,2}(m\&n/k\&a\&b)$$

is the joint probability of outcome s_m and t_n given state k and observables a and b. One can assume either that k is the state of the compound system S at the time of emission from the source, or the state of S just before the measurements takes place. Usually the first assumption is made. Suppose $\rho(k)$ is the probability density defined over the set K. A local physical theory will per definition satisfy the following conditions, called parameter independence (PI) and outcome independence (OI):[b]

(PI) $\Pr_1(m/k\&a\&b)$ is independent of b and may be written as $\Pr_1(m/k\&a)$,

---

[a]In [13] it is pointed out that the label "deterministic" is really a misnomer. "Definite result" would have been better, because the crucial assumption is that a complete state k fixes definite results.

[b]I rely here mostly on [14]. The analysis of factorizablity was introduced by Jarret. He called PI the locality condition and OI the completeness condition.

$\mathrm{Pr}_2(n/k\&a\&b)$ is independent of a and may be written as $\mathrm{Pr}_2(n/k\&b)$.

(OI)

$$\Pr_1(m/k\&a\&b\&n) = \Pr_1(m/k\&a\&b)$$
$$\Pr_2(n/k\&a\&b\&m) = \Pr_2(n/k\&a\&b)$$

A third independence condition says that the complete state is independent of the direction of measurement:

$$\rho(k/a\&b) = \rho(k/a) = \rho(k/b) = \rho(k)$$

This means that the decision to measure from a certain direction (the decision to measure a different observable) does not affect the state of the system coming out of the source. This independency condition is different from PI and OI, but similar to PI. Factorization (denoted as (Fact.)) is the following claim:

$$\Pr_{1,2}(m\&n/k\&a\&b) = \Pr_1(m/k\&a)\Pr_2(n/k\&b)$$

The (Fact.)-condition can be derived from PI and OI and says that if you do an experiment on one entity, this does not influence the result of experiments on the other entity and vice versa. More specifically, (Fact.) says that the probability of a value of a variable stays the same (i) independently of the variable that is measured on the other entity in the compound system (PI), (ii) independently of which value appears as measurement result in the wing (OI). Condition OI means that the outcomes do not cause each other. If you measure with $M(1)$ and the result is s_m, then this will not influence the probability of the result t_n, and vice versa. A so-called Bell inequality follows from the conjunction of (Fact.) with a bunch of other conditions:

$$\text{(Fact.)}\& \mathrm{E}_1 \& \mathrm{E}_2 \&...\& \mathrm{E}_n \supset \text{Bell-inequality}$$

The conditions E_1 & E_2 & ... E_n contain: practical requirements, metaphysical assumptions and epistemological assumptions. (1) Practical requirements that the whole thing is a "closed" system, i.e. that the system is isolated or that interactions that take place apart from the ones we are focusing on can be effectively modeled (standard assumptions for an experimental set-up). (2) Metaphysical assumptions: determinateness and spatiality. Determinateness says that all variables of a physical entity have definite values at all times. The principle of spatiality says that all physical entities exist in space. The principle of spatiality is analyzable into two subtheses: (a) propagation through space of causal influences; (b) the localization in space of physical entities.[c] (3) Epistemological assumptions: the principle of faithful measurement (FM) which says that the measurement reveals the value the parameter had.

The Bell inequality is violated. One of the premises must be wrong. Many people think that we have violation of OI. This will also be the hypothesis in this paper. The violation is a source of puzzlement. OI says that k screens off s_m from t_n and vice versa. The state k is the common cause of a having value s_m and b having value t_n. Violation of OI means that a and b must exhibit a direct (stochastic) causal link because the correlation between a and b "can only be accounted for on the basis of stochastic links to a common cause or a direct stochastic causal link" as is mentioned in [3] on page 102. Let us look more closely at one formulation of the principle of common cause. If there is no direct causation between two events p and q (but there is a correlation between them), then there is in their common past a third event, r say, that screens them off. This is the principle of 'Past Prescribes Stochastic Independence' (PPSI).[d] PPSI is a way to show that there is a common cause that makes events independent. More precisely, events p and q are independent if

$$\Pr(p\&q) \neq \Pr(p)Pr(q)$$

and there is an r such that

$$\Pr(p\&q/r) = \Pr(p/r)\Pr(q/r)$$

But sometimes this principle fails and there is at the same time a common cause while factorizability fails:

$$\Pr(p\&q) \neq \Pr(p)\Pr(q)$$

and for any r

$$\Pr(p\&q/r) \neq \Pr(p/r)\Pr(q/r)$$

The following is an example: (i) a stationary molecule decays indeterministically into two particles; (ii) in which direction the two particles fly is a matter of chance; (iii) since momentum is conserved, the directions of the particles must be opposite to one another. Butterfield distinguishes two kinds of case: (1) the two events are spatio-temporally contiguous with r, here r has two non-independent effects; (2) p and q are not spatio-temporally contiguous with r, then in everyday and classical physical examples, there will be events

[c]Actually the spatiality of the quantum entity is a special case of determinateness, namely determinateness of the position variable.

[d]The account of PPSI and the example are taken from [13], section 7.

mediating between r and p or q. In a classical view incorporating the mediating events into r can always save PPSI. One can always take a sufficient large region of space-time as the common past of p and q. What screens off the two events is this region of space-time.

Saying that the state in the source is the common cause would mean the state acted across a gap in space-time and it did not screen off its effects. But because the common cause does not screen off, the quantum correlations seem to imply that there is a non- local influence between the events in the two wings of the experiment. There is some discussion about what kind of influence this is: does it count as a causal influence or not? Actually there are two kinds of explanation: (a) non-separability (that can be interpreted as holism); (b) some kind of action-at-distance [e]. Versions of option (a) are subscribed to by many scholars. For example Redhead who calls it "passion-at-a-distance" (an expression used by Duch and Aerts in their quantum poll (see [15]), and later also by Abner Shimony): the conditional probabilities

$$\Pr_1(m/n\&a\&b)$$
$$\Pr_2(n/m\&a\&b)$$

are candidates for inherently relational properties of the joint entity as mentioned in [3] page 107. They are relational properties of the joint entity that do not supervene on the monadic intrinsic properties of the parts. The main reason to subscribe to (a) is that the violation of OI cannot be interpreted as a causal influence between the two wings in the experiment: leaving aside nuances and technical details in case $S(1)$ and $S(2)$ really are situated in space at separated locations, they cannot interact instantaneously across the space-time interval. A causal influence must be able to travel 'the distance' between causal system and effect system: "I can't punch you in the nose unless my fist gets to where your nose is" [16]. How does a causal influence propagate? Normally we think that it cannot propagate along a space-time gap, so it must travel along a continuous path from cause to effect (In [17] this is called the At-At theory, because the causal influence is at every space-time point along the path). Typically fields of force are defined in space-time: there is a force vector in every point of space-time. If a force wants to influence something this thing must be at the location in space where the force is defined. A force becomes weaker with the distance from it and it does not work instantaneously.

[e]Butterfield defines causality with David Lewis' concept of causal dependence and gives reasons to claim that the OI-violating correlations are causal [13].

Suppose we have a model of a physical entity B. I agree with Wessels who writes [18]:

> "Suppose all the forces on B have been taken into account, and yet we find empirically that the behavior of B seems to depend on the properties of some system B' in a way not predicted by our theory. We would simply conclude that we had failed to recognize at first all of the forces on B, and suppose that B' is the source of some further force that must be added in to the total force F on B."

If we cannot account for the dependence of B on B' (no matter how minimal) with the forces that we know of, some other force (i.e. cause) must be responsible. The stochastic link is indicative of a causal dependence, albeit a very weak one.

Choosing for (a) does not seem a viable option because we end up in (b). Nevertheless I think the idea of non-separability is necessary for understanding the EPR-paradox and the violation of the Bell inequality. I will come back to non-separability in section 4 and it will play a crucial part in the model of the EPR-situation I present in section 5. First I look at another way to understand the EPR-correlations.

3 Some Causal Influences Do Not Propagate

Seemingly the dilemma can be resolved by opting for a third alternative: the original idea of looking for a common cause coming from the source. Nancy Cartwright presented a model [19] of the EPR-situation that allows for a straightforward causal interpretation, but the causal link is not between $S(1)$ and $S(2)$. With "local" she means that there is no action-at-a-distance between the two wings in the EPR-situation. She says that you can give a local, but non-factorizable common-cause model where the spatial contiguity requirement is dropped for the common cause. Non-contiguity means that there is no propagating cause connecting the common cause with its effects, i.e. the common cause acts across a gap in space-time. S is the common cause of the measurement results in both wings, and the common cause exerts its causal influence across a space-time gap.

Actually the orientation of the measuring apparatuses is not yet set at time t_0 when S leaves the source. Then it is more correct to say that the common-cause is only a partial cause; it combines with the setting of the measuring apparatuses to produce the outcomes. S is a common partial cause, operating in conjunction with states of each of the apparatuses to produce the measurement results. Cartwright's model is local in the following senses (I quote from [20]):

- No Spacelike Causation: There is no direct causal connection between spacelike separated events, i.e. S in conjunction with the operation of $M(1)$ is a complete cause of a measurement result in the left wing, and S in conjunction with the operation of $M(2)$ is a complete cause of the measurement result in the right wing.
- Localization of Partial Causes: The partial causes of the measurement results in the left wing and the right wing (S and the operation of respectively $M(1)$ and $M(2)$) are well localized in space and time.
- Localization of Operations at the Source: The operations of S, which occur at the source at t_0, are independent of the states the apparatuses take on somewhere else at a later time, say just before t_1.

I will now look at Cartwright's model of the EPR-situation in more detail. To describe what happens in the Cartwright-model a specific notation is necessary.

(1) The situation where $X_1, ..., X_n$ are deterministic causes of E is written as:

$$E \equiv A_1 \& X_1 \vee A_2 \& X_2 \vee ... \vee A_n \& X_n$$

X_i is a partial cause of E and A_i is a helping factor. "&" is logical conjunction (the conjunction is true iff all the conjuncts are true), "$\equiv$" is logical equivalence (two propositions are equivalent when they have the same truth value) and "$\vee$" is classical disjunction (one of the disjuncts must be true).

(2) A triangular array is a set of equations:

$$\begin{aligned} x_1 &= u_1 \\ x_2 &= a_{21}x_1 + u_2 \\ &\dots \\ x_n &= a_{n1}x_1 + a_{n2}x_2 + \dots + u_n \end{aligned}$$

where the variables on the left are the dependent variables and the ones on the right are the independent variables. The x_i are ordered in time. A variable x_i is a cause of a variable x_j when x_j is a linear function of x_1 ($a_{ji} \neq 0$) and x_i precedes x_j in time. In this formalism causality is deterministic and the causal influence is permanent.

(3) The formalism used by Cartwright is a combination of the triangular array of econometrics and the formalism of deterministic causes. Cartwright

modifies the triangular array by introducing indeterministic causes and explicitly mentioning the operations of a cause. But the indeterminism of the causes is different from what is usually understood. As an example I look at the following set of equations:

$$x_1 \equiv u_1$$
$$x_2 \equiv \widehat{a}\&x_1 \vee u_2$$
$$x_3 \equiv \widehat{b}\&x_1 \vee \widehat{c}\&x_2 \vee u_3$$

The variable x_1 takes place on t_1, x_2 takes place on t_2 and x_3 takes place on t_3, t_1 precedes t_2 and t_2 precedes t_3. The symbol $\widehat{a}$ represents the operation of x_1 in producing x_2, $\widehat{b}$ represents the operation of x_1 in producing x_3, $\widehat{c}$ represents the operation of x_2 in producing x_3. The symbols $\widehat{a}$, $\widehat{b}$ and $\widehat{c}$ are yes-no propositions about whether the causes operate to produce their effect. The variable x_1 stands for a yes-no proposition about events that happen at t_1 (and similarly for the other variables). The equivalence symbol means that the variable on the left is produced by what lies on the right of the equivalence symbol. In each particular causal process a (contributing) cause either operates to produce the effect, or it does not. The variables u_1, u_2 en u_3 are again exogenous. We can say that $\widehat{a} = 0$ iff

$$\Pr(x_2\&x_3/x_1) = \Pr(\widehat{a}\&\widehat{b}/x_1)$$

and (1) x_1 does not cause $\widehat{a}$ to cause $\widehat{b}$, (2) sometimes x_1 produces x_2 without producing x_3. This condition says that x_2 plays no role in producing x_3 iff the probability of the simultaneous occurrence of x_2 and x_3 is identical to the probability of x_1 to produce them together (i.e. jointly). We are here of course leaving out the error terms. To obtain "classical" factorizability the following has to be the case:

$$\Pr(\widehat{a}\&\widehat{b}/x_1) = \Pr(\widehat{a}/x_1)\Pr(\widehat{b}/x_1)$$

In that case you can say that:

$$\Pr(x_2\&x_3/x_1) = \Pr(x_2/x_1)\Pr(x_3/x_1)$$

If the factorizability of the operation of x_1 does not obtain, then you do not have "classical" factorizability, but x_1 is still the common cause. The factorization condition depends on the joint cause acting independently.

To describe the causal model of the EPR-situation I will need some further notation. The symbol $\widehat{a}_m$ represents the operation of k when it brings about the measurement result in measuring apparatus $M(1)$. The symbol $\widehat{a}_n$ represents the operation of k when it brings about the measurement result during the measurement in $M(2)$. Both occur at t_0, the time of emission from the source. The symbol $\widehat{m}_m$ represents the operation of measuring a with measuring apparatus $M(1)$. The symbol $\widehat{m}_n$ represents the operation of measuring b with apparatus $M(2)$. Both occur at t_1. The EPR-situation has the following causal structure:

$$m = \widehat{a}_m \& k \& \widehat{m}_m \vee u(1)$$
$$n = \widehat{a}_n \& k \& \widehat{m}_n \vee u(2)$$

That the exogenous factors u_1 and u_2 can be disregarded is contained in the practical conditions among E1-E2-..En. The state k is a common partial cause and the complete cause includes the state of the apparatuses. The factorizability of the operations of k during the production of the measurement results m and n looks like this:

$$\Pr(\widehat{a}_m \& \widehat{a}_n / k) = \Pr(\widehat{a}_m / k) \Pr(\widehat{a}_n / k)$$

The operation of the common cause is not independent, so this condition is violated. If it were true we could derive factorizability:

$$\Pr(m \& n / \widehat{m}_m \& \widehat{m}_n \& k) = \Pr(m / \widehat{m}_m \& k) \Pr(n / \widehat{m}_n \& k)$$

This is called Full Factorizability in [19] on page 242. If we substitute a for $\widehat{m}_m$ and b for $\widehat{m}_n$ then we have (Fact.).

This way of presenting the EPR-situation has been criticized because it is possible to show that it does have a factorization, and therefore does not reproduce the probabilities of quantum mechanics. The problem is the modification made in the usual way of presenting causal models, i.e. the explicit mention of the operation of causes. We can define the causal past of an event variable x as the value of all the event variables that stand for its direct causes. Let us denote the direct causal past of x by $CCP(x)$, the direct causal past of y by $CCP(y)$ and the conjunction of the direct causal pasts of x and y by $CCP(x, y)$. Events x and y factorize upon $CCP(x, y)$ if

$$\Pr(x \& y / CCP(x, y)) = \Pr(x / CP(x)) \Pr(y / CP(y))$$

If you define the direct causal past of x to be the disjunction of the conjunction of each of x's actual causes and the (event of the) operation (or lack of operation) of this cause to produce x, then any two events factorize upon the conjunction of their direct causal pasts. In the EPR-case the direct causal past of outcome s_m is the conjunction of S and $M(1)$ and the (event of the) operation (or lack of operation) of these causes, the direct causal past of t_n is the conjunction of S and $M(2)$ and the (event of the) operation (or lack of operation) of these causes. One can then prove that in the EPR-case the event s_m and t_n factorize upon the conjunction of the causal pasts. For the technical details I refer the reader to [20].

In section 5, I will present the work of Aerts that avoids the problems of the Cartwright-model. Aerts ideas however are based on a particular understanding of non-separability. This idea is necessary to present the modified model. So I proceed by a further analysis of the notion of non-separability.

4 Non-Separable Physical Entities Not Situated in Space

In this section I will look closer at the notion of quantum non-separability from the perspective of the work of the Mario Bunge and Aerts. In general quantum non-separability says that you cannot attribute to $S(1)$ and to $S(2)$ by themselves definite states on the basis of the entangled state. Mario Bunge describes the EPR- situation in the following way: the source emits a single system, the state function of which occupies a region that is in one piece. He writes that one should solve the "mystery" much in the same way as a detective solves an alleged murder case by proving that there was no murder to start with (see [21], page 214, figure 2.15.):

> (1) the system was not dismantled after leaving the source, the distant components apparently traveling towards the two measurement apparatuses are still parts of the original system.
>
> (2) The original system only becomes dismantled when at least one of its original components gets integrated into another system - e.g. when it is captured or absorbed by an atom (see [21] page 215).

Point (2) is very important in the next section. I concentrate on (1) and its consequences and a possible modification. One important consequence of (1) is that if the compound entity is a single system that occupies a single region, and the compounds are traveling in opposite directions, the entity must exist in part of the region connecting the two compounds. Remark that the same observations were made by many scholars. Peres writes: "There is a paradox only because you force on this physical system a description with two separate photons. These photons exist only in your imagination. The only thing you have really prepared is a pair of photons, in a spin zero state. That pair is a single, indivisible, nonlocal object" (see [22] page 169). A similar observation was made in [19] page 263: "There is good intuitive appeal in keeping particles inside their classical dimensions. But once they are spread out, what is the sense in trying to keep one very smeared particle on the left and the other on the right?". Aerts draws attention to what he calls the property of macroscopical wholeness: for macroscopical entities, if they form a whole (hence cannot be separated entities) then they hang together through space, i.e. they cannot be localized in macroscopically separated regions without being present in the space between these regions (see [23]). Take for example the experiment of Nicolas Gisin & co, who obtained a violation of Bell inequalities with two quantum entities that are more than 10 km apart. If the entity is one whole occupying one region of space and from this follows the property of macroscopic wholeness for the entity, then it occupies a region that is more than 10 km long. Cartwright comes close to what Aerts is proposing (see [19] page 261):

> "it is the notion of a body with localized states that lies at the core of the common refusal to attribute causal structures in quantum mechanics" .

The modification we make on Bunge's view is that the one system is not part of space. It does not occupy one region. The property of macroscopic wholeness is not relevant to describe S. Probably it is the non-localization of S that Howard hints at when he writes (in a footnote): properties manifest themselves locally, but that does not mean they are localized [12].

5 The Creation-Discovery View

What I want to do now is combine (1) the idea of Cartwright of a common cause coming from the source that does not factorize, (2) the idea of non-spatiality of causal influence of the common cause, (3) the non-separability of the compound entity S and the non-spatiality of S. To do this we need one more element: Bunge's idea that the original system only becomes dismantled when at least one of its original components gets integrated into another system which is similar to Aerts' notion that the two parts of the quantum entity become located in the measurement apparatus, i.e. they become localized in regions of space defined by the measurement apparatus. Or to put it more imaginatively: they are "pulled" into space by the classical entities $M(1)$ and $M(2)$. Aerts writes (see [24] page 240):

> "The experiment consisting in finding or not finding a quantum entity in a given region of space takes place only after setting up in the laboratory the measuring apparatus used for the detection, and it requires the interaction of the quantum entity with that measuring apparatus"

So the second option at the end of the previous section has to be completed into: S is a non-localized whole that does not occupy one finite region of space but acts in two different regions of space by being localized in those two regions. If we look at the Cartwright-model this means that in Aerts' hypothesis the operations of S do not occur before the act of measurement, that is, they do not occur at the source at t_0, but at t_1 at the location of the apparatuses: $m_m\&k$ operate at t_1 to produce s_m, and $m_n\&k$ operate to produce t_n. I will now consider Aerts' representation of what happens in the EPR-experiment and later on see in what way this changes the Cartwright representation [f]. Several concepts and a mathematical representation of some aspects are introduced: as a general philosophical framework the creation-discovery view and as part of this framework hidden measurements and hidden correlations.

The first concept that is introduced is the creation-discovery view [24,25]. It follows from a characterization of quantum measurements. A quantum measurement has two aspects:

1. The measurement are not just observations but provoke a real change of state of the entity.
2. There exists a lack of knowledge on what precisely happens during a measurement process.

I begin with an explanation of the first aspect. Every measurement has a discovery-aspect and creation-aspect. When you measure a classical property, you generally discover that it has a certain value. This implies the combination of definiteness (all values have definite values independent of any measurement) and faithful measurement (the measurement actually reveals the value that was already there beforehand). But it is also possible that an interaction creates properties that were not already there before the measurement. An example from macroscopic reality illustrates what is meant with creation. Consider a uniform sphere [g]. The collection of states are the possible spatial coordinates of the center of the sphere, a state transition corresponds with a translation of the sphere. If we cut the sphere in two halve spheres, each part now has angular properties. Although the original entity had only translational properties, the two halves have both translational properties and

[f] In [20] it is shown that this new Cartwright-like model still factorizes on the basis of direct causal pasts (see section 3 supra), but violates the condition (Ind.). Since the modified Cartwright-like model I present following the creation-discovery view and the hidden measurement/hidden correlations view (see infra in this section) is quite different from the Cartwright-like model in [20] this criticism is not relevant for the discussion in the present paper.

[g] The example is discussed in [26]. It is a case of symmetry-breaking.

angular properties. Even if the two halves are kept together the presence of the cut defines angular properties that were not there before. Since an entity is defined by the set of states it has (and the structure of this state space) the creation of this new state space, means the creation of a new entity. The two new entities were potentially part of the original sphere, but the original sphere does not exist anymore, because the breaking is irreversible. When you measure a quantum variable and the entity is in an eigenstate of a value of the variable, then we have a classical measurement (value definiteness and faithful measurement). But in many cases we measure variables when the entity is in a superposition state. When the entity is in a superposition state of position, there is an element of creation in the measurement interaction. The idea that there is a creation-aspect to physical interactions with quantons is not new. Heisenberg already entertained such ideas. Popper suggests that quantons are probably "changing propensities for change" (see [27] page 159-160). The discussion that Popper engaged in is related to the interpretation of laws of nature as capacities, powers or tendencies [19,28,29] and to the interpretation of properties as dispositional [30].

The second aspect of quantum measurements is the lack of knowledge about what happens during the measurement process. It lies at the origin of the indeterministic nature of the quantum measurement process and can be formalized in the following way. (For the working physicist this may not be of great interest: it makes not much difference to say that there are hidden measurements (see below) because maybe the state transition from superposition state to one of the eigenstates in the superposition is too contextual and sensitive for the physicist to obtain any structural feedback, in the sense of a dynamical equation. It is of great interest however if we want to understand what is going on, and especially in what way and to what degree quantum phenomena challenge classical metaphysical views.)

1. With each real measurement e corresponds a collection of deterministic measurements e_λ, called "hidden measurements".

2. When a measurement is performed on an entity in a state p, then one of the hidden measurements takes place. The probability finds its origin in the lack of knowledge about which one of the hidden measurements takes place.

Let H be a collection of states (the state $p \in H$ does not depend on the parameter λ and the selection of λ is also independent of p), Λ the collection of parameters for the hidden measurements in a hidden measurement representation then we can represent the unknown content of the measurement interaction by

$$\Phi = \{\varphi_\lambda : H \to H_e \mid \lambda \in \Lambda\}$$
$$\mu : B(\Lambda) \to [0,1]$$

$B(\Lambda)$ is a σ-field of subsets of Λ. The map $\mu : B(\Lambda) \to [0,1]$ determines the relative frequency of occurrence of $\lambda \in \Lambda$. The function $\varphi_\lambda : H \to H_e$ represents a strictly classical observable. Every hidden classical measurement e_λ is determined by a probability measure $\Pr_{p,\lambda} : B(\Sigma_e) \to \{0,1\}$ such that

$$\Pr_{p,\lambda}(\{q\}) = 1 \Leftrightarrow \varphi_\lambda(p) = q$$
$$\Pr_{p,e}(q) = \mu(\{\lambda \mid P_{p,\lambda}(\{q\}) = 1\}) = \mu(\{\lambda \mid \varphi_\lambda(p) = q\})$$

The next concept we have to discuss is "hidden correlations". Correlations that were already present before a measurement and are only detected by the experiment are *correlations of the first kind.* Take an entity consisting of two material point particles moving in space and having total momentum zero. A coincidence experiment of the momenta of the individual particles will give us correlated results, but these correlations were already there before the measurement. The outcomes do not depend on each other. *Correlations of the second kind* are correlations, which were not present before the measurement but that are created during and by the measurement process. Aerts distinguishes three stages in the EPR-experiment: (i) Before the measurement after the compound entity leaves the source we have the pure state singlet state: it describes one entity. Before the measurement there are no probabilities. (ii) During the measurement the quantum entity gets separated in two entities, in the process influencing each other. The correlations of the second kind enter at this stage. (iii) The correlations are broken after the measurement (see [31] page 143). We have transition from a tensor product to a Cartesian product:

$$H(1) \otimes H(2) \to H(1) \times H(2)$$

where $H(1)$ and $H(2)$ are two-dimensional complex vector spaces. The Cartesian product is in fact $H(1) \otimes H(2)$ but with enough super selection rules so that there are no more superposition states, i.e. only factorizable states are left in the joint state space. The correlations that violate OI are correlations that are created by and during the measurement [h].

[h] Aerts does not refer to the distinction between OI and PI. Although the distinction is implicitly used in the distinction between three stages of the measurement interaction and the characterization of the hidden correlations representation.

Suppose we have a quantum entity S in an entangled state. The idea of the hidden correlations model is that S contains $S(1)$ and $S(2)$ only *counterfactually* in the basic sense of the word counterfactual (i.e. contrary to fact: the antecedent is false): at t_0 (stage (i) in the three stages Aerts distinguishes in the EPR-experiment) S is one entity, so the measurement interaction has not taken place and no state transition has taken place. Coecke gives an even stronger formulation: "one cannot speak about an initial unkown hidden measurement, but only about a formalization of the interaction that will take place when we actually decide to perform the measurement, i.e., if there is no measurement, the parameters that characterize the possible formalizations of the deterministic measurement process through hidden measurements are meaningless" [26]. But it can at that stage be represented counterfactually as a compound entity with hidden correlations, i.e., we can give a representation of S as a collection of individual entities (with their own proper states) between which there exist hidden correlations (correlations of the second kind). An individual entity is a potential separate entity in what is still one entity S. For example, the individual entities $S(1)$ and $S(2)$ in the EPR-state do not have their own spin states. The spins are created during the measurement that takes the two entities apart. But the singlet state in $H(1) \otimes H(2)$ can be represented as two individual entities with hidden correlations. Being in a singlet state means (i) being one entity in a non-separable state; (ii) having no angular properties at all. After the measurement two separated spin entities are obtained, both of them having angular properties. The singlet state does not describe two entities that already have their spin, it is a state of two possible individual entities that do not have spin properties yet. The spin is being created by the measurement experiment. So the hidden measurements/ hidden correlations representation is consistent with the non-separability of the previous section. The dependence of the outcomes of experiments in both wings depends on the fact that one non-localized object becomes two localized objects. There is no probabilistic dependence between s_m and t_n before the measurement because there were no values s_m and t_n and separate entities between which there could exist correlations. There is a state transition as a consequence of the experiments performed, but the experiments are performed on a non-localized entity. The state transition involves the localization.

Next I will look at the formalization of the hidden measurement in the case of measurements on an individual entity in a compound entity S, the hidden correlations view and especially the combination of the two in relation to a generalization of the EPR-situation. Suppose we have a collection $\{S(i)\}$ (where $i = \alpha, \beta, \gamma, ..., \kappa, \lambda, \mu, ..., \xi, \upsilon, \varsigma$) of possible individual entities in S. Every individual entity $S(\alpha)$ has a set of proper states represented by $H(\alpha)$

and a set of states H(α), which are the states the entity can have if it is separated from S. A measurement that separates $S(i)$ from S is characterized by a set $\{\phi_{\alpha,i}\}$ for $i = 1,, n$ of outcome states.

(A) The hidden measurement representation. Just like before with each real measurement on a $S(\alpha)$ in S corresponds a collection of deterministic measurements called "hidden measurements". When a measurement is performed on an entity, one of the hidden measurements takes place. There is a hidden measurement presentation for every α when:

$$\Phi_\alpha = \{\varphi_{\alpha,\lambda(\alpha)} : H_\alpha \to \{\phi_{\alpha,i}\} \mid \lambda_\alpha \in \Lambda_\alpha\}$$
$$\mu_\alpha : B(\Lambda_\alpha) \to [0,1]$$

This is the unknown but relevant context of the measurement interaction. Λ_α is the collection of parameters for the hidden measurements. $B(\Lambda_\alpha)$ is a σ-field of subsets of Λ_α. The map $\mu_\alpha : B(\Lambda_\alpha) \to [0,1]$ describes the relative frequency of occurrence of $\lambda_\alpha \in \Lambda_\alpha$. The $\mu_\alpha : B(\Lambda_\alpha) \to [0,1]$ represents a strictly classical observable.

(B) The hidden correlations representation [i]. After a measurement an entity has a definite value for the value that we were measuring, and the entity has now the state that corresponds to this outcome. Suppose we take the most general case:

> There exist *hidden correlations* between individual entities in a collection $\{S(i)\}$ (where $i = \alpha, \beta, \gamma, ..., \kappa, \lambda, \mu, ..., \xi, \upsilon, \varsigma$) if:
>
> (1) a measurement on one individual entity $S(\alpha)$ induces a change of the proper state of the other individual entities $S(\beta), S(\gamma)...$;
>
> (2) this change of proper state depends deterministically on the proper state transition of $S(\alpha)$;
>
> (3) after this measurement, the state of $S(\alpha)$ cannot be influenced anymore by measurements on other entities.

For every measurement on an individual entity $S(\alpha)$, there is a map for every other individual entity $S(\alpha)$ that characterizes the induced change of state.

$$f_{\beta\circ\alpha} : \text{H}\ (\alpha) \to H(\beta) : \phi_\alpha \mapsto \omega_{\beta\circ\alpha}$$

This map describes the state transition of $S(\alpha)$ from a proper state ω_β into a proper state $\omega_{\beta\circ\alpha}$ due to a transition of $S(\alpha)$ from ω_α into an outcome state

[i] In [26] was proven that every compound quantum system described in the tensor product of a finite number of Hilbert spaces has a unique representation as a collection of individual entities on which exist correlations of the second kind.

ϕ_α. Similarly we have that

$$f_{\gamma o\alpha}: \text{ H }(\alpha) \rightarrow H(\gamma) : \phi_\alpha \mapsto \omega_{\gamma o\alpha}$$

...

(C) Combining the hidden measurement representation and the hidden variables representation [j]. Suppose there is a hidden measurement representation for every measurement on an individual entity in S (see (A)). A measurement on S as a whole corresponds with one measurement on every individual entity in S. An outcome of such a measurement can be written as $\times_\alpha \phi_\alpha$. We can represent the proper states of the individual entities in a Cartesian product $H_S = \times_\alpha H(\alpha)$ and we can also define $\Lambda_\alpha = \times_\alpha \Lambda_\alpha$. For measurements on the individual entities according to the ordering $\alpha, \beta, \gamma, ..., \kappa, \lambda, \mu, ..., \xi, \upsilon, \varsigma$ one can define the following map:

$$\varphi_{S,\lambda} : H_S \rightarrow \times_\alpha \{\phi_{\alpha,i}\}$$

A state in H_S (which will be a n-tuple of proper states in respectively $H(\alpha)$, $H(\beta)$,) will be mapped onto the n-tuple consisting of respectively

$$\begin{aligned}
\phi_\alpha &= \varphi_{\alpha,\lambda(\alpha)}(\omega_\alpha) \\
\phi_\beta &= \varphi_{\alpha,\lambda(\beta)} \circ f_{\beta o\alpha} \circ \varphi_{\alpha,\lambda(\alpha)}(\omega_\alpha) \\
&\quad ... \\
\phi_\beta &= \varphi_{\varsigma,\lambda(\varsigma)} \circ f_{\varsigma o ... o\beta o\alpha} \circ ... \circ \varphi_{\alpha,\lambda(\beta)} \circ f_{\beta o\alpha} \circ \varphi_{\alpha,\lambda(\alpha)}(\omega_\alpha)
\end{aligned}$$

The hidden measurement presentations for the composing entities together with the hidden correlations representation result in a new hidden measurements representation for S as a whole, which takes into account the compoundness of S. The singlet state with just two individual entities is actually the simplest case.

I return now to the Cartwright-like model we looked at in the beginning of this section. What are the major ways the Cartwright-like model differs from the model of this section ? We already established at the beginning of this section that the determination of

$$\Pr(\widehat{a}_m \& \widehat{a}_n / k)$$

does not happen at the source. It happens together with the measurement interaction at t_1. Here we can point out two major changes on the basis of Bunge and especially Aerts. (i) We have an emendation of one of the principles mentioned above: *Localization of Partial Causes*, because only some of the partial causes of the measurement results in the left wing and the right wing

[j] For more mathematical theory see [26].

are well localized in space and time, namely $M(1)$ and $M(2)$. The time t_1 is the moment that the subsystems of S are spatialized. It is still S that is responsible for the OI-violating correlations. (ii) According to Cartwright there is no direct causation between the two wings of the experiment. In a sense in the modification on the basis of Aerts' interpretation, we could still say both wings influence each other: it is through the non-spatial state that the individual entities that become separate physical entities influence each other. These influences are represented by the mappings

$$f_{\beta o \alpha}, f_{\gamma o \alpha}, \dots$$

$$f_{\gamma o \beta}, \dots$$

In the EPR-situation there are two (coincident) mappings: $f_{\beta o \alpha}$ and $f_{\alpha o \beta}$. No locality condition is violated because the entities between which these interactions happen are not spatially separated. It is the measurement apparatuses that induce the state transition of the singlet state, and it is the *emerging* individuals in the singlet state that influence each other. All of these interaction happen outside of space-time.

I repeat that the non-separability idea of the previous section finds its place in the representation. The situation with individual entities between which there exist correlations of the second kind, is a *counterfactual* representation: if something like a measurement interaction would occur, then the hidden correlations would be created by the interaction. The hidden correlations representation says what the correlations would be, given that a measurement had taken place on one of the individual entities in the compound entity S. Before the measurement on one (or more) of the individual entities (and after leaving the source), there were neither individual entities nor correlations of the second kind, just the one non-separable entity S.[k]

6 Are Quantum Entities Individuals?

What are some of the classical criteria to confer individuality to an entity ? We can name three: the identity of indiscernables (Leibniz), spatio-temporal location and a purely metaphysical criterium: haeccity. (1) If properties are the basis for distinguishing entities, then we can only apply the principle of the identity of indiscernables: two entities are identical if they have the same properties. No two individuals can be absolutely indistinguishable (i.e. possess the same collection of properties). (2) If entities are impenetrable, then we can distinguish them on the basis of their spatio-temporal properties:

[k] Sometimes Aerts confuses the entangled state with the hidden correlations representation.

no two entities can have exist in the same place at the same time and must exist somewhere at all times. (3) A third criterion is haecceity or primitive thisness, which we can describe as the primitive basis of individuality, which cannot be analyzed further. Quantum mechanics is one of the big problem fields for the notion of individual. Electrons have all their intrinsic monadic properties in common. Furthermore they exist in superposition states. If we put two electron in a box, one after the other, and then we take one out, is that the first electron or the second? Paul Teller compares it to money in the bank. Suppose you deposit 5 euro in an empty bank account. Then you deposit another 5 euro in the account. Next, you withdraw 5 euro: is it the first or the second 5-euro you are withdrawing?

A possible solution is to clearly draw a distinction between distinguishability and individuality: two things can be indistinguishable without being the same entity. This seems to be the case in quantum mechanics. But this leaves us with haecceity as a means to individuate quantum entities, which is a purely metaphysical criterium.

This is not the place to review in detail the solutions that have been proposed. (Another option is to say that the idea of individuality got of on a wrong start.) I propose the following hypothesis. In the systemic view (for example in the *Treatise* of Bunge) a system always has properties or characteristics that do not supervene on the monadic properties of the components. It is a defining characteristic of systems to have non-supervenient properties. Of course most system will be spatio-temporal and have properties which individuate them (the first and second criterium I mentioned at the beginning of this section). Independently of the possibility of applying the first and second criterium, the non-supervenient properties individuate an entity for sure. So one could say that these non-supervenient relational properties are the distinguishing characteristics of individuals. This has a serious consequence for the quantum world. In general an entity is compound, and such a compound entity is an individual if the compound entities become meshed with one another, for example in the typical quantum mechanical manner. A compound entity of quantum entities with individual state spaces $\{H_i\}$ is described by a product space $\otimes H_i$. One can make millions of products of Hilbert-spaces, such a huge tensor product will never deliver a classical phase space. This may have as a consequence that we have to look at the quantum domain as one systemic whole. From the quantum perspective classical entities are not individuals. For classical entities the individuation depends for a large part on space-time and the identity of indiscernables.

7 Is The Creation-Discovery View a Scientific Realism?

Einstein's view is an expression of the metaphysical ideas that are presupposed by classical physics [32]:

> "If one asks what is characteristic of the realm of physical ideas independently of the quantum theory, then above all the following attracts our attention: the concepts of physics refer to a real external world, i.e., ideas are posited of things that claim a "real existence" independent of the perceiving subject (bodies, fields, etc.) ... Moreover, it is characteristic of these physical things that they are conceived of as being arranged in a space-time continuum. Further it appears to be essential for this arrangement of the things introduced in physics that, at a specific time, these things claim an existence independent of one another, insofar as these things "lie in different parts of space." Without such an assumption of the mutually independent existence (the "being-thus") of spatially distant things, an assumption which originated in everyday thought, physical thought in the sense familiar to us would not be possible. Nor does one see how physical laws could be formulated and tested without such a clean separation."
>
> Quoted from and translated in [33]

Since the common sense view of space-time was superseded by the mathematical space of physics (starting with Newtonian view of space-time), space-time theories have become an important mathematical toolbox of physicists to model physical phenomena. Even a relationalist or conventionalist with respect to space-time seems committed to the idea that a space-time description is absolutely necessary for every model of physical phenomena. Can a physicist that is prepared to give up space-time still call himself a scientific realist?

Epistemology can be a bad guide to metaphysics. Measurement in physics involves the creation and use of physical and cognitive instruments which instantiate a measuring scale and appear to be set of necessity within a spatio-temporal frame. The idea of space as a necessary condition for knowledge is deeply entrenched in our worldview. The influential German philosopher I. Kant compared the forms of space and time and the categories of the mind with colored glass through which we see the world. Space, time, printers, rules, and the like are impositions of our cognitive psychological framework onto physical reality. But the habits or necessities of human cognition and perception can be misleading when we meet new physical phenomena. In [34] the Kantian view is criticized (although Kant is not mentioned):

> "the structured space is like colored spectacles through which we discover, study and partly understand the world. In order to discover, to study and to understand the scenes that happen to be of the complementary color, it would be better to put the spectacles aside, or at least, to be conscious of the fact that we have them on."

I think the only conclusion is: if (a) quantum entities are indeterminate in their position properties and (b) space is a necessary condition for scientific knowledge (as Kant and many others since have thought), then knowledge of the quantum world is limited to those parts that can be spatialized. The paradoxes of quantum mechanics are the result from the attempt to impose a three-dimensional spatial framework on a physical world where it does not exist. If there is no physical space, the EPR-paradox more or less dissipates, since there is no need to treat the two sister particles as spatially separate, and thus no problem about the influence of one on the other. The paradox does not lie in the apparent non-local behavior of compound quantum entities. The paradox is rooted in the structure of the human mind, which cannot easily imagine a world where the continuum we call space would not be a convenient tool any more for describing the entities (this point of view is also mentioned in [35] page 50-51.

Aerts claims to be a scientific realist. C. A. Hooker describes a scientific realist as some one who believes that scientific theories aim at the truth and he believes in a correspondence theory of truth [36,37]. Is it possible to give up such deep-rooted metaphysical and methodological beliefs like the necessity of the space-time continuum in model building? According to the scientific realism of C. A. Hooker truth does not necessarily involve epistemically accessible truth criteria. Truth is not tied of necessity to any cognitive human construct or procedure (empirical adequacy, rational acceptance, pragmatic reliability etc.). Everything is fallible theory, from the most concrete factual claim to the highest metaphysical hypothesis. The only correct attitude towards both theory and method is fallibilism: nothing is free from questioning, of course: given good reasons for this questioning. And these good reasons are to be found in the interpretation of the EPR-paradox and the violation of the Bell inequality. Consequently, we can say that even if it were true that we cannot perceive or think without imposing a space-description, that does not mean we are forced to make it part of our metaphysics. So we can conclude that Aerts' research program is justified from the point of view of a scientific realist.

References

1. D. Aerts, "Description of many separated physical systems without the paradoxes of quantum mechanics", *Found. Phys.*, **12**, 1131 (1982).
2. B. Coecke, D. Moore and A. Wilce, "Operational quantum logic: an overview", in *Current Research in Operational Quantum Logic: Algebras, Categories, Languages*, eds. B. Coecke, D. Moore and A. Wilce, Kluwer Academic, Dordrecht (2000).
3. M. Redhead, *Incompleteness, Nonlocality and Realism. A Prolegomenon to the Philosophy of Quantum Mechanics*, Oxford University Press, Oxford (1987).
4. D. Aerts, *The One and the Many. Towards the Unification of the Quantum and the Classical Description of One and Many Physical Entities*, Doctoral Dissertation, Brussels Free University (1981).
5. A. Einstein, B. Podolsky and N. Rosen, "Can quantum-mechanical description of physical reality be considered complete?", *Phys. Rev.*, **47**, 777 (1935).
6. A. Aspect, P. Grangier and G. Roger, "Experimental realization of the Einstein-Podolsky-Rosen-Bohm Gedankenexperiment: a new violation of Bell's inequalities", *Phys. Rev. Lett.*, **49**, 91 (1982).
7. D. Aerts, "Quantum mechanics, separated physical entities and probability", *Found. Phys.*, **24**, 1127 (1994).
8. C. Piron, *Mécanique Quantique. Bases et applications*, Presses Polytechnique et universitaires romandes, Lausanne (1990), 2nd edn (1998).
9. F. Valckenborgh, "Operational axiomatics and compound systems", in *Current Research in Operational Quantum Logic: Algebras, Categories, Languages*, eds. B. Coecke, D. Moore and A. Wilce, Kluwer Academic Publishers, Dordrecht (2000).
10. See page 28-29 of Toraldo di Francia, "A World of Individual Objects?", in *Interpreting Bodies. Classical and Quantum Objects in Modern Physics*, ed. E. Castellani, Princeton University Press, Princeton (1998).
11. See page 241 of F. A. Muller, "The locality scandal of quantum mechanics", in *Language, Quantum, Music. Selected Contributed Papers of the Tenth International Congress of Logic, Methodology and Philosophy of Science*, Florence, August 1995, eds. M. L. Dalla Chiara, R. Giuntini and F. Landisa, Kluwer Academic Publishers, Dordrecht (1995).
12. D. Howard, "Einstein on locality and separability", *Stud. Hist. Phil. Sci.*, **16**, 171 (1985).
13. J. Butterfield, "Bell's theorem: what it takes", *Brit. J. Phil. Sci.*, **43**, 41, (1992).

14. A. Shimony, "An exposition of Bells inequality", in *Search for a Naturalistic World View. Volume II: Natural Science and Metaphysics*, ed A. Shimony, Cambridge University Press, Cambridge (1993).
15. W. Duch and D. Aerts, "Microphysical reality", *Physics Today*, **39**, 13-14 (1986).
16. See page 64 of D. Albert, *Quantum Mechanics and Experience*, Harvard University Press, Cambridge, Massachusets (1992).
17. W. Salmon, *Scientific Explanation and the Causal Structure of the World*, Princeton, Princeton University Press (1984).
18. See page 485 of L. Wessels, Locality, factorability and the Bell inequality", *Nous*, textbf19, 481 (1985)
19. N. Cartwright, *Nature Capacities and Their Measurement*, Clarendon Press, Oxford (1989).
20. J. Berkovitz, "What econometrics cannot teach quantum mechanics", *Stud. Hist. Phil. Mod. Phys.*, **26**, 163 (1995).
21. M. Bunge, *Treatise on Basic Philosophy. Epistemology & Methodology III: Philosophy of Science and Technology. Part I: Formal and Physical Sciences*, Kluwer Academic, Dordrecht (1985).
22. A. Peres, *Quantum Theory: Concepts and Methods*, Kluwer Academic Publishers, Dordrecht (1993).
23. D. Aerts, "An attempt to imagine parts of the micro-world, in *Problems in Quantum Physics II*, eds. J. Mizerski, A. Posievnik, J. Pykacz, and M. Zukowski, World Scientific, Singapore (1990).
24. D. Aerts, "The entity and modern physics: The creation-discovery view of reality", in *Interpreting Bodies. Classical and Quantum Objects in Modern Physics*, ed. E. Castellani, Princeton University Press, Princeton (1998).
25. D. Aerts, "The game of the biomousa: A view of discovery and creation", in *Perspectives on the World*, VUB Press, Brussels (1995).
26. B. Coecke, "A representation of compound quantum systems as individual entities: hard acts of creation and hidden correlations", *Found. Phys.*, **28**, 1109 (1998).
27. K. R. Popper, *Quantum Theory and the Schism in Physics. From the Postcript to the Logic of Scientific Discovery*, Routledge, London (1982).
28. R. Harré and E. H. Madden, *Causal Powers: A Theory of Natural Necessity*, Basil Blackwell, Oxford (1975).
29. R. Bhaskar, *A Realist Theory of Science*, Harvester Press, Sussex (1978).
30. H. Sankey, (ed.), *Causation and Laws of Nature* (Dordrecht: Kluwer Academic Publishers) (1999).
31. D. Aerts, "The Description of one and many physical systems", in *Foun-*

dations of Quantum Mechanics, ed. C. Gruber, A.V.C.P., Lausanne (1983).

32. A. Einstein, Quantenmechanik und wirklichkeit", *Dialectica*, **2**, 320 (1948).
33. Howard, Don (1989), 'Holism, Separability, and the Metaphysical Implications of the Bell Experiments', in: J. T. Cushing and E. McMullin, (eds.), *Philosophical Consequences of Quantum Theory: Reflections on Bell's Theorem*, Notre Dame, University of Notre Dame Press (1989).
34. D. Aerts and J. Reignier, "On the problem of non-locality in quantum mechanics", *Helv. Phys. Acta*, **64**, 527 (1991).
35. J. O'Keefe, "Kant and the sea-horse: An essay in the neurophilosophy of space", in *Spatial Representation. Problems in Philosophy and Psychology*, eds. N. Eilan, R. McCarthy and B. Brewer, Oxford University Press, Oxford (1999).
36. C. A. Hooker, *A Realistic Theory of Science*, State University of New York Press, Albany (1987).
37. C. A. Hooker, *Reason, Regulation and Realism. Toward a Regulatory Systems Theory of Reason and Evolutionary Epistemology*, State University of New York Press, Albany (1996).

QUANTUM CRYPTOGRAPHIC ENCRYPTION IN THREE COMPLEMENTARY BASES THROUGH A MACH-ZEHNDER SET UP

THOMAS DURT

Foundations of the Exact Sciences (FUND) and Applied Physics and Photonics (TONA), Brussels Free University, Pleinlaan 2, 1050 Brussels, Belgium
E-mail: thomdurt@vub.ac.be

BOB NAGLER

Applied Physics and Photonics (TONA), Brussels Free University, Pleinlaan 2, 1050 Brussels, Belgium
E-mail: Bob.Nagler@vub.ac.be

We present in this paper a quantum cryptographic set-up in which the authorized users of the raw cryptographic key can choose between three bases of encryption instead of two bases as is usually the case. This technique generalises the coding in phase technique thanks to the introduction of an extra-degree of freedom (the internal phase-shift of a Mach-Zehnder device) which allows us to encode information through the localisation of the pulses, a local, corpuscular property, complementary relatively to phase, a non-local, ondulatory property.

1 Introduction

The essential ingredient exploited in quantum cryptography is that the signal is encoded through a quantum support, in non-commuting bases chosen at random, in such a way that, in virtue of the uncertainty principle, an eventual spy could not detect a part of the signal without causing the appearance of errors of transmission [1]. For instance, this is what happens when the cryptographic key is encoded through phase [2,3,4] in two complementary bases. We propose in this paper a simple device aimed at encoding and decoding the cryptographic raw key in three bases, which generalises the technique of phase coding in which two bases only are involved.

2 The Mach-Zehnder interferometer considered as a unitary transformation

Let us consider the Mach-Zehnder interferometer described on the figure 1; in terms of optical components such an interferometer consists of an entry $50-50$ beam splitter $BS1$, and two mirrors, that simply redirect the output

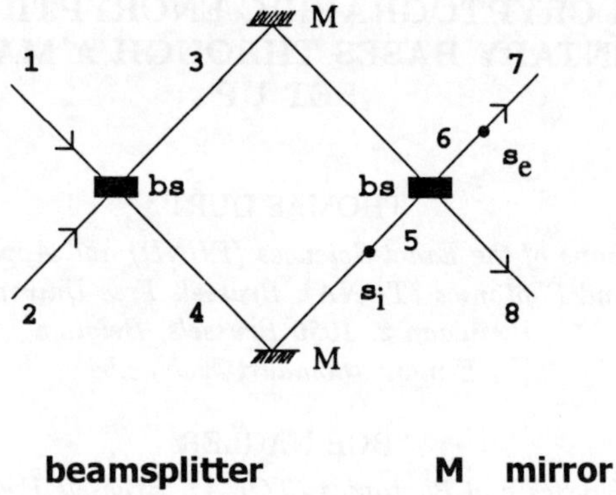

Figure 1. Mach-Zehnder device

beams of $BS1$ into the two input ports of the exit beam splitter $BS2$, plus an internal and an external phase-shifter. Such a device possesses two input and two output ports. Therefore, if there are no losses involved, the conservation of probability implies that its action is described by a unitary transformation. That is, its transformation properties are described by a two-dimensional unitary matrix $U(MZ)_{ij}$, with $i, j = 1, 2$. The indices enumerate the input and output ports. The full transformation[5] performed by such a device (see fig. 1) is a product of successive unitary transformations of the first beam splitter, the internal phase shifter, the second beam splitter and the phase shifter in one of the exit ports :

$$\hat{U}(MZ) = \hat{U}_u(s_e)\hat{U}(BS)\hat{U}_l(s_i)\hat{U}(BS) \tag{1}$$

which gives

$$\hat{U}(MZ) = e^{is_i/2} \begin{pmatrix} e^{is_e} \cos s_i & -ie^{is_e} \sin s_i/2 \\ -i \sin s_i/2 & \cos s_i/2. \end{pmatrix} \tag{2}$$

In this approach, the state of the light pulse at any port of the device is represented by its amplitudes of probability of presence in the up (+) and down (-) channels: $|\psi\rangle = \alpha|+\rangle + \beta|-\rangle, |\alpha|^2 + |\beta|^2 = 1$.

It is useful to represent such a state on the Poincaré sphere (the unit 3-sphere), thanks to the Pauli mapping. This mapping maps bijectively the set of physical states[a] of a two-level system (normalised states of a 2-dimensional

[a] Such states are defined up to an arbitrary global phase that we fix here according to the convention that $\frac{\alpha^*}{\beta}$ is purely real.

Hilbert space) onto the Poincaré sphere, according to the following transformation law:

$$|\psi\rangle \rightarrow \vec{n} = \left(2\mathrm{Re}\,\alpha^*\beta, 2\mathrm{Im}\,\alpha^*\beta, |\alpha|^2 - |\beta|^2\right) \tag{3}$$

(in Cartesian coordinates) and conversely:

$$\alpha = \cos\frac{\theta}{2}e^{-\frac{i\varphi}{2}} \qquad \beta = \sin\frac{\theta}{2}e^{\frac{i\varphi}{2}} \tag{4}$$

where θ, φ are the polar angles of $\vec{n}$.
For instance, the state $|+\rangle$ is sent, through this mapping, on the North pole, and the state $|-\rangle$ on the South Pole. The unitary transformation induced by the Mach-Zehnder between the in-going and outgoing amplitudes can be visualised on the Poincaré sphere as the composition of a rotation around the X-axis of angle s_i, followed by a rotation around the Z-axis, of angle $-s_e$. This can be easily proven if one uses the properties of SU2, considered as a representation of the group of rotations of the three dimensional Euclidean space or checked by direct computation. These properties will now be exploited in order to realise cryptographic encryption and decryption in the three complementary bases associated to the directions X, Y, and Z on the sphere[b].

3 Encoding protocol

Let us now consider a protocol during which Alice (the transmitter) sends encoded information to Bob (the receiver) thanks to the device presented in the figure 2. The source belongs to the upper channel which means that, represented on the Poincaré sphere, the initial state is prepared along the positive Z-axis. By performing convenient choices of s_i and s_e, Alice can "rotate" this state and map it, at will, along an arbitrary direction (θ, ϕ) on the Poincaré sphere. It is sufficient therefore to combine a rotation of angle θ around the X-axis with a rotation of angle $\phi + \frac{\pi}{2}$ around the Z-axis. This implies that $s_i = \theta$ and $-s_e = \phi + \frac{\pi}{2}$, which determines the phase-shifts of the Mach-Zehnder. This simple geometrical reasoning is confirmed by direct computations. Thanks to convenient choices of the phase-shifts, Alice can thus encode the signal among three complementary bases (the X, Y, and Z-axes), instead of two as is usually the case.

[b]This is a special case of a general result [5] according to which a Mach-Zehnder with one internal phase-shifter and two external phase-shifters suffices to generate an arbitrary SU2 transformation, so to say an arbitrary rotation of the Poincaré sphere.

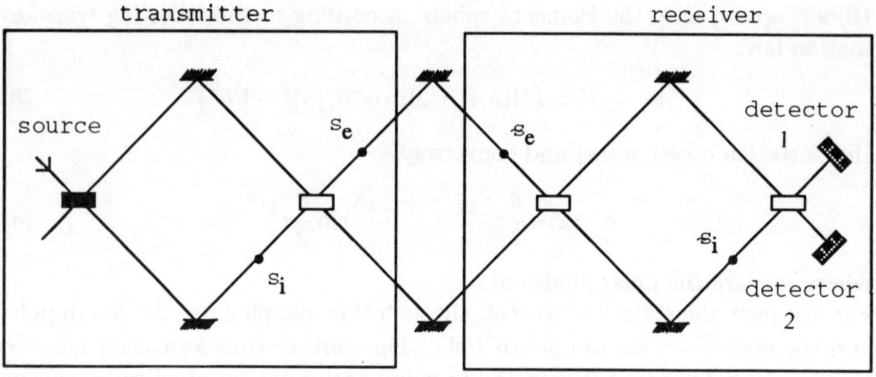

Figure 2. Transmission device

Let us denote $|X_{+/-}\rangle$, $|Y_{+/-}\rangle$, and $|Z_{+/-}\rangle$ respectively the states which correspond, after the Pauli mapping, to the intersection of the positive and negative X, Y, and Z semi-axes with the Poincaré sphere. One gets directly that the values of s_e and s_i necessary in order to encode the states (1) $|X_+\rangle$, (2) $|X_-\rangle$ (3) $|Y_+\rangle$, (4) $|Y_-\rangle$, (5) $|Z_+\rangle$, and (6) $|Z_-\rangle$, are respectively equal to (1) $\frac{-\pi}{2}$ and $\frac{\pi}{2}$, (2) $\frac{\pi}{2}$ and $\frac{\pi}{2}$, (3) π and $\frac{\pi}{2}$, (4) 0 and $\frac{\pi}{2}$, (5) s_e and 0, and (6) s_e and π. In fact, it can be shown that, in the situations 5 and 6, the Mach-Zehnder acts as a perfect mirror and as a transparent device respectively[6].

4 Decoding protocol

We must now determine the values of $\tilde{s}_e$ and $\tilde{s}_i$ necessary for decoding in the X, Y, and Z-bases. This means that when the incoming pulse was prepared in a peculiar state chosen among these bases Bob must send it along a peculiar output port of his interferometer so to say onto a peculiar detector. We obtain by computations similar to those relative to the encoding protocol that, in order to decode the signal in the X, Y and Z bases, Bob must choose the phase shifts $\tilde{s}_e$ and $\tilde{s}_i$ (see figure 2) to be respectively equal to $\frac{\pi}{2}$ and $\frac{-\pi}{2}$, π and $\frac{-\pi}{2}$, $\tilde{s}_e$ and 0. It is easy to show that when Alice's and Bob's bases are different but belong to the X, Y, and Z-bases, the correlation between the value of the encoded bit and the beep in the detector vanishes completely. When they are the same, this correlation is perfect in principle. Therefore if a spy, who ignores (like Bob) Alice's choice of encoding basis, tries to intercept the signal he will necessarily destroy a part of it and leave a trace of his passage, which is the essence of quantum cryptography.

It is interesting to note that when one encodes the signal in the X and in the Y-bases, the outgoing pulse is maximally delocalised in the sense that it is present in each outgoing channel with probability fifty-fifty. The internal phase-shift is equal to $\frac{\pi}{2}$ and does not vary during the encoding procedure. It is the external phase-shifter that induces a controllable phase-difference between the outgoing channels. This situation is then essentially equivalent to the one encountered in the coding in phase protocol (with two encoding bases only [2,3,4]). The passage to the Z-basis, where the state is localised inside one of the channels only, is made possible by the presence of the extra-degree of freedom of the internal phase-shifter. The Z-basis is complementary relatively to the two other bases in the same sense that, in the double slits experiment, the "particle" basis in which the localisation of the particles through one of the slits is unambiguously predetermined is complementary relatively to the "wave" basis in which it is rather the phase shift between the amplitudes related to both slits that is unambiguously predetermined.

5 A realistic variant of the technique

Note that with the transmission set-up presented in the figure 2 the protocol of key distribution is not realisable practically because for realistic distances of transmission (some kilometers), it is impossible to avoid that the fluctuations of the relative length of the transmission arms become larger than the wave-length of the light used to carry the signal, which means that the coherence gets lost underway. This problem can be overcome provided one uses the device presented in the figure 3. In this device, non-symmetric half-interferometers were added, the effect of which is, roughly speaking, to replace two signals in spatially separated arms by two signals on the same arm but now separated in time. When one encodes in the "wave" basis, then, at the end of the line, the signal will consist of a superposition of three successive components: (A) the twice advanced component of the initial pulse, (B) an intermediate component that consists of a superposition of the component that was advanced in Alice's device and delayed in Bob's one with the firstly delayed and then advanced component, and (C) the twice delayed component. Their respective firings can be clearly distinguished experimentally provided the timing of the transmissions and receptions is carefully controlled. Let us represent the renormalised amplitudes of probability of presence of the intermediate (B) component inside the detectors 1 and 2 by a direction on the unit 2-sphere thanks to the Pauli mapping. This oriented direction on the sphere can be obtained by applying two successive rotations to the Z_+-direction (the

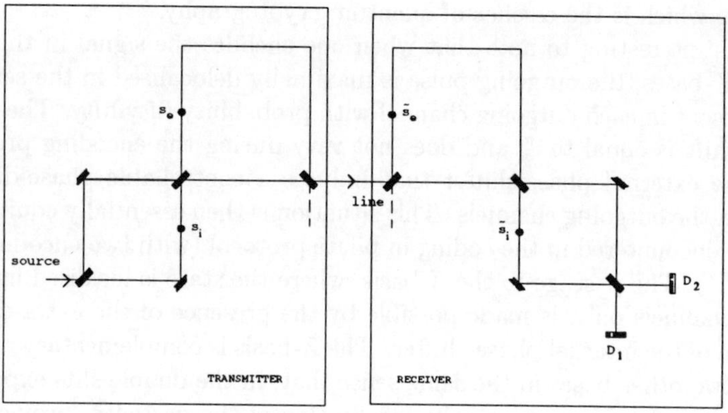

Figure 3. Realistic transmission device

North Pole) which correspond to the successive actions of the Mach-Zehnder devices of Alice and of Bob. As we showed already, the first rotation allows us to send the North Pole on an arbitrary direction (θ, ϕ) on the Poincaré sphere (when $s_i = \theta$ and $-s_e = \phi + \frac{\pi}{2}$), and the second one can be shown by similar reasonings to send an arbitrary direction $(\tilde{\theta}, \tilde{\phi})$ on the North Pole (when $\tilde{s}_i = \tilde{\theta}$ and $-\tilde{s}_e = \tilde{\phi} + \frac{\pi}{2}$ in this case). Thus, when one encodes and decodes the signal in the wave basis, the information is contained in the intermediate component of the pulse. Reciprocally, when one encodes and decodes the signal in the particle basis, the information is carried in the twice advanced or in the twice delayed components of the pulse. The advantage of this scheme is that instead of letting interfere two components that followed different transmission lines, we let interfere components that travelled successively along the same line. Two successive components are likely to encounter exactly the same stresses and thermal inhomogeneities throughout the line provided they are separated by a sufficiently short delay-time. For instance, when the difference of the arms lengths in the non-symmetric half-interferometers is equal to 30 cm which means that each vertical (as on the figure 3) arm of it is 15 cm long, this delay-time is equal to 1 ns, which is quite less than the typical temporal fluctuation scale inside the line. This trick is well-known and was already implemented in quantum cryptography before (see for instance the references [2,7,3,4]). When one encodes the signal in the X and in the Y-bases, it will be robust against decoherence whenever the fluctuations of the arms lengths at the level of the Alice's and Bob's stations remain shorter than the wave-length of the light pulses, a standard condition in interferometry. The essential novelty of our technique is that when the signal gets encoded in the localised basis it is even insensitive to such fluctuations because then the pulses are not split along the transmission line (one of the components (A) or (C) is then empty). Therefore, we must expect that, in comparison to the coding in phase technique, an enhancement of the bit rate will occur because the encryption in localised pulses is quite more robust. As it was shown in [8], one is free to use one basis more often than the other ones without decreasing the level of safety (provided the number of signals sent in each basis separately is sufficiently high in order to be able to make use of the law of large numbers). We could even do more, and encrypt the raw key in the localised basis only, while we should use the two other bases as a control device, in order to reveal whether or not a spy is eavesdropping the communication. The effective bit rate of the raw key would then be the same as the one that can be reached with a classical digital telegraph (in which all pulses are de facto localised), a significant improvement. Besides, our protocol is also advantageous from the point of view of security. It was shown indeed in [9,10] that when one encodes the signal in three complementary bases instead of two, the safety of the cryptographic protocol is enhanced. Note that some of our ideas are already present in the protocol described in [7], but with pairs of entangled photons in their case instead of single photons in ours. Roughly speaking, the decoding is, in their case, alternatively realised in the particle basis and in one of the wave bases, the choice between these bases being made at random and depending on the time of registration of the photons in the detectors. One could seriously question the relevance of introducing entanglement in the way that is described there, because it can be shown [11] that a local hidden variable model exists that makes it possible to simulate the correlations between Alice's and Bob's firings in their set-up. Essentially, this is due to the fact that the correlation disappears when the bases of Alice and Bob differ, which happens in 50 % of the cases. This prevents any possibility of effective violation of Bell's inequalities. Nevertheless, the advantage of their technique is that nature chooses itself in which basis the signal is encoded and decoded (this is an aspect of non-locality already emphasised by Bohr: prior to the measurement, elements of reality do not exist objectively). The advantage of our approach is that we economise the costs (in terms of simplicity, losses and so on) of a PDC source, of a detection station with two detectors (which means a serious increase of the effective bit rate) and that we can overweight the frequency of encoding in the most robust coding basis at will. Note that we prefer to talk about position-phase instead of time-energy complementarity, which is equivalent from a relativistic point of view but is maybe better adapted to human psychology: is it not easier to understand a movie by following it image after image instead of looking at all the images relative to its different locations simultaneously?

The generalisation of such devices to higher dimensional systems is outlined in [5]. The three dimensional systems (qutrits) are particularly interesting from the point of view of quantum cryptography because, in comparison to qubits, they offer improved performances for what concerns non-locality [12] and security [13] as well.

Acknowledgements

One of the authors (T.D.) is affiliated to the Flemish Fund for Scientific Research (FWO), as a post-doctoral fellow, B.N. is a FWO Research Assistant. This paper was written in the framework of the Flemish-Polish Scientific Collaboration Program No. 007. Sincere thanks to Marek Zukowski for comments and discussions. This research was supported by the Belgian Office for Scientific, Technical and Cultural Affairs in the framework of the Interuniversity Attraction Pole Program.

References

1. C.H. Bennett and G. Brassard, "Quantum cryptography: public key distribution and coin tossing", *IEEE International Conference on Computing, Systems and Signal Processing*, Bangalore, India, 175-179 (1984).
2. C. H. Bennett,*Phys. Rev. Lett.*, **68**, 3121-3124 (1992).
3. S. J. D. Phoenix and P. D. Townsend, *Cont. Phys.*, **36**, 165-195 (1995).
4. J. G. Rarity, P. C. M. Owens and P. R. Tapster, *J. Mod. Optics*, **41**, 2435-2444 (1994).
5. M. Reck, A. Zeilinger, H. J. Bernstein and P. Bertani, *Phys. Rev. Lett.*, **73**, 58-61 (1994).
6. M. Zukowski, R. Horodecki, M. Horodecki and P. Horodecki, *Phys. Rev. A*, **58**, 1694-1698 (1998).
7. J. Brendel, N. Gisin, W. Tittel and H. Zbinden, *Phys. Rev. Lett.*, **82**, 2594-2597 (1999).
8. M. Ardehali, H. F. Chau, and H. K. Lo, "Efficient quantum key distribution", *quant-ph/9803007* (1998).
9. H. Bechmann-Pasquinucci and N. Gisin, *Phys. Rev. A*, **59**, 4238-4248 (1999).
10. D. Bruss, *Phys. Rev. Lett.*, **81**, 3018-3021 (1998).
11. S. Aerts, P. Kwiat, J-Å. Larsson and M. Żukowski, "Two-photon Franson-type experiments and local realism, *Phys. Rev. Lett.*, **83**, 2872 (1999).

12. D. Kaszlikowski, P. Gnacinski, M. Zukowski, A. Zeilinger and W. Miklaszewski, *Phys. Rev. Lett.*, **85**, 4418 (2000).
13. N. Cerf, T. Durt and N. Gisin, to appear in *Journal of Modern Optics* for the special issue on *Quantum Information: Theory, Experiment and Perspectives, Proceedings of the ESF Conference*, Gdansk, July 10-18 (2001).

QUANTUM CRYPTOGRAPHY WITHOUT QUANTUM UNCERTAINTIES

THOMAS DURT

Foundations of the Exact Sciences (FUND) and Applied Physics and Photonics (TONA), Brussels Free University, Pleinlaan 2, 1050 Brussels, Belgium

E-mail: thomdurt@vub.ac.be

Quantum cryptography aims at transmitting a random key in such a way that the presence of a spy eavesdropping the communication would be revealed by disturbances in the transmission of the message. In standard quantum cryptography, this unavoidable disturbance is a consequence of the uncertainty principle of Heisenberg. We propose in this paper to replace quantum uncertainties by generalised, technological uncertainties, and discuss the realisability of such an idea. The proposed protocol can be considered as a simplification, but also as a generalisation of the standard quantum cryptographic protocols.

1 Introduction

Roughly speaking, the essential non-classical feature that differentiates quantum cryptography from conventional cryptographic techniques is that, in the quantum world, it is impossible to acquire information about the state in which a system is prepared without disturbing it, when this state is chosen at random among conjugate bases. More technically, this property is a consequence of the no-cloning theorem and of the uncertainty principles, which reflect typical non-classical features of the quantum theory. In the present paper, we generalise these ideas and consider a situation in which the knowledge that we can get about a system is limited by the unefficiency of our measuring device. Furthermore, we consider that, even when we know the state of the system, our ability to reproduce a copy of this state in order to dissimulate our intervention is also limited by the unefficiency of our source. In order to illustrate these ideas, we describe in detail a cryptographic protocol which generalises the quantum cryptographic protocols, and which is essentially equivalent to a classical telegraphic protocol of key distribution (transmission of binary classical information). The security of this protocol is conditioned by the technological limitations of the presently available sources and detectors of low intensity light pulses (not more than one photon in average). We also evaluate the menace for the security of this protocol imposed by the recent progresses in the technology of available detectors and sources in the low intensity regime. Finally, by taking account of the fact that our ideas remain valid in principle whenever it is not possible to detect and to reproduce a signal with perfect accuracy, we propose, in a more speculative approach, semi-classical cryptographic schemes which exploit the stochastic nature of spontaneous nuclear decay processes or the evanescent nature of particles such as the neutrino. The common feature of these various encryption schemes is the appearance of a Poissonian distribution for the signal, during the emission and reception processes as well. We derive a systematic treatment of the properties inherent to such a protocol in connection with cryptographic applications.

2 Quantum Cryptography

Townsend and Thompson [1] define the cryptography techniques as techniques which aim at scrambling a message in such a way that only the authorised users of the channel can easily recover the message. Usually, such techniques work because the users of the channel possess a shared secret random sequence of bits called the key which can be used with a specified algorithm to scramble and unscramble the content of a message.

The aim of the techniques that we shall describe in this paper is to transmit this random key with a guarantee of confidentiality based on the technological limitations of existing sources and detectors. Once this is done, various methods exist to scramble and unscramble the message. For instance, the Vernam's cipher [2], better known as the one-time-pad, is proven to possess the highest degree of safety (in fact absolute safety). It happens when the key is as long as the message to transmit.

The quantum technique is non-conventional in the sense that the carrier of the signal is so light (less than one photon in average) that if a spy intercepts it, he modifies a part of the information in a stochastic and irreversible manner. Because of his presence, errors appear in the transmitted information. After the transmission, by a discussion afforded on a public channel during which they compare a part of their respective signals (this is called the reconciliation protocol [3,4]), the transmitter and the receiver can then suppress the errors (with high probabillity), and by doing so evaluate the error rate and thus the maximal quantity of information possessed by the spy. They can then, on the public channel, transform the remaining, error-free, signal shared by them (this is called the process of amplification of privacy [3,4,5]) in order to be sure (with a probability very close to hundred per cent) that the spy is nearly totally ignorant of the content of the remaining signal, which constitutes the cryptographic key. In fact, each bit of the final key is a function of (on average) half the bits of the "raw" key. We shall not describe these techniques in detail.

They are essentially classical techniques, based on classical probability theory and they require a classical protocol of transmission of the signal (on the public channel). We invite the interested reader to consult the reference [5]. Let us now illustrate quantum key distribution by a particular example, the BB84 protocol based on polarisation [6].

This protocol is defined as follows. The transmitter produces photons which are linearly polarised along one direction chosen among four possible directions (0, 45, 90 and 135 degrees), and sends them to the receiver after having assigned a conventional binary value to each direction (for instance (0°, 45°, 90°, 135°)→(1, 1, 0, 0)), in accordance with the receiver. The series constituted by the polarisations successively measured by the receiver constitutes the signal on the basis of which the cryptographic key will be built. Let us now assume that the transmitter chooses the direction of polarisation of the signal at random between the four possible directions, with the same probability for each of them and that the receiver chooses the direction of his eigen-basis of polarisation at random between the two canonical bases ((0°, 90°) and (45°, 135°)).

If the basis chosen by the transmitter coincides with the basis chosen by the receiver, the information is correctly measured. Otherwise, when these bases are different, the information is totally at random (standard quantum computations show that the probabilities of getting the answer yes and the answer no are then equal to 50 %, whatever the polarisation of the incoming photon could be).

After the transmission of a whole series of pulses, the transmitter and the receiver communicate through a public channel what were their respective choices of basis. They drop then the information associated to pulses emitted and received in different bases (it is in average half of the series). They only consider the remaining series of binary numbers, which serves to build the key. This is made possible because no confidential information is directly contained in the key; otherwise 50 % of this information would be lost at this step.

Let us now assume that a spy intercepts the pulses, measures them and reemits them (this is what is called the intercept-resend strategy in [4]). Then, if we consider the part of the signal that was transmitted and received in the same bases, it is disturbed by the action of the spy. For instance, if the spy intercepts and resends the signal in the canonical (horizontal and diagonal) bases, he will choose the right detection basis in 50 % of the cases, and the wrong basis otherwise. Therefore, even if the spy possesses a perfect detector and a perfect source, he transmits the correct signal in 50 % of the cases and a random signal otherwise. This means that the average error of the signal obtained by the receiver (due to the perturbation caused by the spy) is equal to 25 %. If the spy intercepts a fraction r of the signal, he knows a fraction r/2 of it, and the error-rate is equal to r/4. This illustrates the peculiarity of quantum cryptography: there exists an upper bound between the information gained by the spy and the error that he causes in the transmission of the signal. Note that the information of the spy is deterministic only when the basis of measurement coincides with the basis of the transmitter. For instance, it can be shown that by performing an intercept-resend attack in the (non-canonical) Breidbart basis, which lies "in-between" the canonical bases, the spy intercepts the signal correctly with a probability close to 85 % [4] and causes the appearance of an error rate of 25 % in the transmission between the transmitter and the receiver. Many other attacks (collective, coherent, translucent, and opaque attacks for instance) are possible in principle [7,8,4,9,10]. We shall discuss some of these possibilities later. For instance, in the reference [10], optimal attacks are described during which the spy entangles a probe with the transmitter's signal and waits until the transmitter and the receiver reveal their respective choices of basis publically in order to optimise his attack. It can be shown that during such attacks the spy is able to intercept the signal correctly with a probability close to 85 % (see also ref.[11] about puzzling connections between this attack and non-locality) while the error rate in the transmission between the transmitter and the receiver is only equal to 15 %. An absolute bound equal to 11 % was derived by Mayers (see ref.[11] section VI G and references therein).

After having communicated through a public channel their choices of bases, transmitter and receiver can apply the protocol of reconciliation, during which they will eliminate the errors of transmission and, indirectly, evaluate the error rate [3,4]. This error rate multiplied by 2 constitutes the maximal deterministic information possessed by an eventual spy directly after an intercept-resend attack performed in the canonical bases. In principle, it is possible that the spy gains more information during the reconciliation process. It can be shown that it is advantageous for the spy to perform non-canonical attacks, for which the information, measured by the Shannon [12] or the Rényi [5] entropies, is not deterministic. Roughly speaking, this shows that it is better to know a little bit about much than to know much about a little bit.

During the protocol of amplification of privacy[3,5], the receiver and the transmitter recombine a part of the key in order to make sure that the spy is nearly totally ignorant of the content of the remaining key (with probability close to 100 %). Actually, it can be shown that a part of the raw key gets sacrificed during the protocol of privacy amplification and that after a sacrifice of an amount of information equal to that of the spy plus n+1 bits, the spy knows in average 2 to the power - n bits after the protocol of privacy amplification [3]. At this step, the transmitter and the receiver share thus a key (the remaining and rearranged part of the signal) with an arbitrary high level of confidentiality. This was the goal of the whole technique.

In order to compare the quantum technique and the semi-classical technique that we shall introduce in the next section, it is useful to introduce a quality factor defined as follows.

Definition 1 *The quality factor Q of a protocol of key distribution is the ratio between the error that a spy causes in the transmission of the signal and the deterministic information gained by him, during intercept-resend attacks.*

When this factor increases, the safety of the technique is better, because, for the same gain of information, the spy will leave more trace of his intervention, and for the same trace left by him, he will get less information. The quality factor that we consider here is not the most general one: it is representative of the deterministic information gained by the spy during deterministic attacks only, but it is useful to introduce this parameter in order to compare standard quantum protocols such as the BB84 protocol with our semi-classical protocol, in which the information is classically (deterministically) encoded as we shall show.

In the quantum technique based on encryption in polarisation, it is only when the spy chooses the detection basis between the canonical bases that his information is deterministic (zero or one). Then, his deterministic information is zero in 50 % of the cases and one otherwise, and the error rate is equal to 25 %. The intrinsic quality factor of the BB84 encryption scheme relatively to intercept-resend attacks in the canonical bases is thus equal to $\frac{1}{2}$. Note that, although non-deterministic information plays an important role in quantum cryptography, we prefear to consider only deterministic information here, in order to compare what is comparable and to avoid confusions of logical level. For information, if we estimate thanks to the Shannon function the probabilistic information gained by the spy during representative non-deterministic attacks, we get similar results. For instance, we get that the quality factor is close to 60 %

$$\frac{\frac{1}{4}}{1+(\frac{1}{2}-\frac{1}{2\sqrt{2}})log_2(\frac{1}{2}-\frac{1}{2\sqrt{2}})+(\frac{1}{2}+\frac{1}{2\sqrt{2}})log_2(\frac{1}{2}+\frac{1}{2\sqrt{2}})} \simeq 60\%$$

in the case of an intercept-resend attack in the Breidbart basis [4] and close to 40 %

$$\frac{\frac{1}{2}-\frac{1}{2\sqrt{2}}}{1+(\frac{1}{2}-\frac{1}{2\sqrt{2}})log_2(\frac{1}{2}-\frac{1}{2\sqrt{2}})+(\frac{1}{2}+\frac{1}{2\sqrt{2}})log_2(\frac{1}{2}+\frac{1}{2\sqrt{2}})} \simeq 40\%$$

in the case of an optimal attack described in [10]. We can thus consistently limit ourselves to the possibility of intercept/resend attacks realised in the canonical bases without losing much generality.

Remark that our evaluation of the information possibly possessed by the spy is correct only in average and that it is necessary to compute the standard deviation of this number in order to guarantee a safety margin. For instance, in [4], the authors add five standard deviations to their estimation of the average number of bits of information of the spy in order to make sure that they do not underestimate the information possessed by him. We generalised their computation of the standard deviation (see appendix) in order to deal with our semi-classical protocol.

Let us now briefly mention some technological limitations that restrict the practical implementation of the quantum technique. Some of them will be exploited in our generalised, semi-classical, protocol of key distribution.

3 Some Technological Limitations of the Quantum Technique

3.1 Limitation of the Detector - The Dark Count

The efficiency of the detectors commonly used in quantum cryptographic set-ups is rather low[a]. This means that a non-negligible percentage of the incoming pulses gets "killed" during the detection. This is not a problem for what concerns quantum key distribution, because the key is composed of the surviving part of the signal only and does not contain any sensible information. Nevertheless, the existence of the so-called dark count, an unavoidable intrinsic noise at the level of the detector due to the amplification of thermal photons and of all kinds of noises imposes indirectly limitations on the maximal distance of transmission. Effectively, in every fibre, some loss occurs, so that a part of the signal is irremediably lost. A decrease of the intensity due to fibre-losses implies that the time of opening of the receiver's detector increases (exponentially with the distance), in order to obtain a constant number of effectively transmitted signals. The dark count rate and thus the

[a]Standard photon counting detectors have efficiencies around 10 % at telecom wavelength but much better detectors have already been demonstrated in various labs, that would reach very high efficiencies, around 90 % [13]. This shows that the technology of detection is evolving quickly. Nevertheless it seems well that the majority of laboratories active in the field has no such nearly perfect detector at their disposal. This can be infered from the fact that, if such efficiencies were available, the old polemic about Bell's inequalities would be definitively closed. It can be shown indeed that the loophole due to unefficiencies of available detectors can be avoided provided their efficiency is higher than 82.83 % [14]. Presently, this polemic has been open for more than twenty years.

error rate increase proportionally. Now, it can be shown that if the error rate is of the order of 15 %, the amount of information gained by the spy could be such that in principle the safety of the protocol is not guaranteed anymore[b]. This means that 15 % is the maximal error rate that can be tolerated if one wants to guarantee that the protocol is safe against eavesdroppers attacks. Because of this limitation of the error rate, the rate of transmission along the line is bounded by below. This constitutes the main (and unavoidable) limit on the distance of transmission.

3.2 Limitation of the Source - Translucent Attacks

In the description of the quantum technique, we implicitly considered that it was technically possible to emit pulses which contain exactly one photon. In fact, the sources which are used in concrete quantum key distributions consist principally[c] of laser sources, which produce Poissonian pulses at a very high rate (some Mhz). This means that the probability of finding N photons is equal to $exp(-\mu) \cdot \mu^N/N!$. With such a source, it is in principle possible that, when more than one photon is present in the pulse, the spy splits the incoming beam, stores one photon and transmits the other ones without changing their state of polarisation. Afterwards, when the transmitter and the receiver publicly communicate their choices of bases, the spy could measure, in the good basis, the polarisation of the photon made prisoner. A possible parade to this attack (sometimes called translucent in the litterature) is to reduce the population of the pulses containing more than one photon, so to say, to reduce μ. Then the spy can, in principle, eavesdrop without risk of being discovered a fraction of the signal of the order of $\mu/2$ $((\mu^2/2)/\mu)$. Besides, the limitations due to the dark count rate impose that the value of μ must be higher than a minimal value, so that it is impossible to suppress totally the possibility of eavesdropping by beamsplitting. For practical purposes, μ is taken to be of the order of 10 %.

---

[b]For instance, during an optimal attack described in [10], the information of the spy about the transmitted signal becomes comparable to the information of the receiver when the error rate gets close to 15 %. A limiting error rate equal to 15 % is also obtained in the reference [12] where one investigates the possibility that the spy eavesdrops the communications publically exchanged between the authorised users of the line during the reconciliation process in order to gain more information about the key.

[c]The possibility of other sources is discussed in the sections 4 and 5.

4 A Semi-Classical Technique

We shall now introduce a semi-classical protocol of key distribution in which we replace the quantum uncertainties by technological uncertainties related to the stochastic and uncontrollable behaviour of the detectors and sources available today. Nevertheless, some aspects of the technique are specifically quantum: the signal consists of weak laser pulses, as in the quantum technique. When detectors are not fully efficient, a measurement of these pulses destroys with non-negligible probability the information that they contain, and, as we shall show, it is impossible to copy these pulses without producing a non-negligible percentage of empty signals. These features justify our adoption of the label "semi-classical" for characterizing this protocol. In order to improve the clearness of the text, we shall first present a technique in which the detector of the spy is imperfect but his source is perfect (1), and a technique in which his detector is perfect while his source is imperfect (2). We shall then present (3) the technique in which his detector and his source are both imperfect.

4.1 Perfect Source and Imperfect Detector

Let us assume that the spy (and the transmitter as well) is able to produce pulses containing exactly one photon, but that the efficiency of his detector is not equal to 100 %, but is equal to ϵ ($\epsilon < 1$). The protocol is defined as follows:

-The transmitter sends either nothing, or a single pulse, inside a temporal window $[0, \tau]$.

-Then, he reproduces the operation with a second signal, inside the window $[t, t + \tau]$. The transmitter sends at most one pulse (a monophotonic pulse if he makes use of a perfect source) during the two operations. He and the receiver agree about the convention that, for instance, the first pulse means yes (1), the second one means no (0). This is essentially a classical protocol of transmission of binary information.

All this procedure, which requires a time $2t$ ($t > \tau$), constitutes thus the transmission of one binary symbol, and contains the essence of our new technique. Now, if a spy tries to intercept a fraction r of the signal, he will only detect correctly a fraction $r \cdot \epsilon$ of it. Nevertheless, he must replace the signal destroyed by him otherwise the average number of pulses detected by the receiver will decrease and his presence will be revealed. This number is in fact proportional to the product of three factors: the intensity of the source of the transmitter, the transmission rate of the fibre, and the efficiency of the detector of the receiver. These quantities can be evaluated when no spy is present, and the receiver can check that the intensity measured by him corresponds to this evaluation. Note that the assumption under which the spy may not change the intensity of the pulses is also necessary for what concerns the quantum technique because of the possibility of so-called opaque and translucent attacks [4,9].

In the situation considered here, a fraction $r \cdot (1 - \epsilon)$ of the signal detected by the receiver is totally at random, which means that the presence of the spy causes an average error of

$$\frac{r \cdot (1 - \epsilon)}{2}$$

In virtue of our definition (section 2), the quality factor of this protocol of transmission is then equal to

$$\frac{(1 - \epsilon)}{2\epsilon}$$

For instance, if, for the detectors existing in the world, the maximal efficiency available at the considered wavelength was equal to 50 % (this question is addressed in the footnote 2), this intrinsic quality factor would also be equal to 50 %, the same factor as for the BB84 protocol already described.

Note that we neglected here the unavoidable errors due to the dark count rate, in order not to overload the formulation of the problem. We consider the detector to behave deterministically in the sense that it is a "'all or nothing" detection process. If we took the randomness due to the dark count into account, this would still improve the quality factor of the technique so that we can consistently simplify the presentation of the problem and neglect it.

4.2 Imperfect Source and Perfect Detector

We shall now assume that the spy as well as the transmitter make use of a laser source. As we mentioned already when we spoke about the limitations of the quantum techniques, the distributions obtained by laser sources are Poissonian. For a Poissonian distribution of constant μ, the ratio between the population of non-empty pulses and the whole population of (empty and non-empty) pulses is equal to $1 - exp(-\mu)$ and the ratio R between the population of pairs of photons and the population of single photons is equal to $\mu/2$. By measuring these populations, the receiver imposes a serious constraint on the source used by the spy. Effectively, provided the source used by the spy is a laser source, the application of this control procedure allows the authorised users to evaluate the maximal ratio between the population of non-empty pulses and the whole population produced by the source of this spy. This ratio is equal to 1 - $exp(-2R)$. We shall denote this ratio by μ_{eff} in the following reasoning in order to have a first estimation of the quality factor. We have that $\mu_{eff} = 1 - exp(-\mu)$, and when μ is small compared to unity, $\mu_{eff} = 2R$ in good approximation.

Note that the present analysis is based on the (non-conservative) hypothesis that the spy has not at his disposal a better source than a laser source. We shall estimate in the forthcoming sections the possibilities offered by sources of squeezed, non-Poissonian, light. Let us now assume that the spy is able to detect the incoming pulses with efficiency 100 %. The foregoing discussion implies that, even if he detects correctly a pulse sent by the transmitter, the spy is able to reemit it with a probability only equal to μ_{eff}.

Here again, a decrease of the intensity can be measured by the receiver, so that the spy must reemit a random signal in order to compensate the loss in intensity (1 - μ_{eff}) due to his intervention. For short distances, μ_{eff} is comparable to the corresponding ratio at the level of the source of the transmitter and the spy can only compensate the intensity loss by sending pulses which fill the gaps corresponding to the empty pulses sent by the transmitter. If the spy intercepts a fraction r of the non-empty signal, the error rate is thus equal to

$$\frac{r \cdot (1 - \mu_{eff})}{2}$$

and the information obtained by him is equal to r, of which only a fraction μ_{eff} is effectively transmitted. The quality factor of the technique is then equal to

$$\frac{1 - \mu_{eff}}{2\mu_{eff}}$$

Note that our technique differs from the standard quantum technique in two aspects: the simplification in the method of coding (directly through the photonic localisation in our case) and the fact that the receiver can control the quality of the signal received by him by measuring the populations of incoming photons. This provides a severe constraint on the source used by the spy and is an essential guarantee of safety for our semi-classical protocol. For instance, in [15], the author already introduced the idea of a classical technique where the colour of the photon carries the information. He also showed why this technique was not safe by describing an explicit technique of eavesdropping which does not disturb the transmitted signal. This attack is actually not relevant in our case because the author made implicit assumptions about the quality of the sources available for the spy that we do not make in our approach. The author assumed that the spy could emit bright pulses, containing more than one photon in average, in order to compensate the unefficiency of his detector. This possibility is now forbidden to the spy because we assume that the receiver checks the relative population of pairs. This relative population is higher for bright laser pulses than for the weak laser pulses emitted by the source of the transmitter.

4.3 Imperfect Source and Imperfect Detector

In this case, we obtain by the same reasonings as in the two previous sections a quality factor Q equal to

$$\frac{(1 - \epsilon_{eff} \cdot \mu_{eff})}{2\epsilon_{eff}\mu_{eff}}$$

where

$$\epsilon_{eff} = \frac{1 - exp(-\epsilon \cdot \mu)}{1 - exp(-\mu)}$$

and

$$\mu_{eff} = 1 - exp(-\mu)$$

If, for instance, ϵ_{eff} is equal to 50 % and μ_{eff} is equal to 10%, the quality factor is equal to 9.5, which is nearly twenty times "better" than for the quantum technique. If $\mu_{eff} = 60$ %, the quality factor of the quantum technique (50 %) is reached for an effective efficiency ϵ_{eff} of 5/6.

In order to deduce this expression of the quality factor, one must note that if the spy possesses a detector with efficiency ϵ, he will observe, instead of a Poissonian distribution of constant μ, a Poissonian distribution of constant equal to $\epsilon \cdot \mu$. This means that he will detect a population $1 - exp(-\epsilon \cdot \mu)$ of non-empty pulses. A fraction $1 - exp(-\mu)$ of the intercepted pulses is not empty. Thus, the spy detects a fraction

$$\frac{1 - exp(-\epsilon \cdot \mu)}{1 - exp(-\mu)}$$

of the non-empty pulses. The appearance of these exponential factors has nothing mysterious, it simply expresses that, with our choice of coding through the presence or absence of photons in a pulse, the average number of photons (μ) per pulse is not equal to the average number of bits per pulse $(1 - exp(-\mu))$. Note also that the expression

$$Q = \frac{exp(-\epsilon \cdot \mu)}{2(1 - exp(-\epsilon \cdot \mu))}$$

in the case of weak laser pulses, reduces to a simpler form. When μ is low, we can write the following approached expression for the quality factor:

$$Q = \frac{(1 - \epsilon \cdot \mu)}{2\epsilon\mu}$$

Both expressions emphasise the symmetrical role played by the processes of detection and emission during the intercept-resend attack. Note that this feature enhances the flexibility of the protocol: even if, let us say, the quality of existing sources is low while existing detectors are nearly perfect, the safety of the technique is still guaranteed.

In the previous sections, we did not consider all the possible attacks, for instance it could happen that the spy makes use of non-laser sources. This possibility is analysed in the next section.

4.4 Opaque Attacks Realised with Non-Laser Sources in the Case of Lossy Transmission Lines

As we already mentioned, in [15], the author commented upon a possible attack against classical protocols similar to ours in which he assumed that the spy could emit bright pulses, containing more than one photon in average, in order to compensate the unefficiency of his detector. The control device limits this possibility, essentially because the relative population of pairs is higher for bright laser pulses than for the weak laser pulses emitted by the source of the transmitter.

Besides, it is also possible to prove that the control device is still efficient when the spy eavesdrops the signal at the source, communicates the information on a classical channel (without absorption) to a second spy located at the end of the line, in order to compensate the unefficiency of his detector by a gain in transmission (this kind of attacks where the spy compensates a decrease of intensity by a gain in transmission is similar to so-called opaque attacks [9] so that we shall from now on call them opaque attacks too). We generalised the deduction of the quality factor performed in the previous section in order to cover this situation. We obtained by a lenghty but straightforward computation that, provided the second spy makes use of a laser source when he reemits the signal, the constraints imposed by the presence of the control device imply that the quality factor is given by the following expression:

$$Q = \frac{exp(-\epsilon \cdot \mu T_{tr1})}{2(1 - exp(-\epsilon \cdot \mu T_{tr1}))}$$

where T_{tr1} is the transmission rate between the source of the transmitter and the first spy. Note that the transmission rate between the first and the second spy does not appear in this expression. This is due to the fact that, thanks to the control device, the gain in transmission is lost because the second spy must diminish the intensity of his laser source in the same proportion in order not to reveal his presence. The minimal quality factor, which corresponds to the ideal location for the spies is clearly reached when the first spy is located at the level of the transmitter (T_{tr1} =1). Then, the quality factor is given by the expression

$$Q = \frac{exp(-\epsilon \cdot \mu)}{2(1 - exp(-\epsilon \cdot \mu))}$$

and we recover the expression deduced in the previous subsection in the case of short distances of transmission.

Note that the assumption according to which the source of the second spy is Poissonian is essential in the previous reasonings. As we shall now show, this hypothesis is no longer valid whenever we consider strong absorption rates along the transmission line because of the possibility of subPoissonian sources.

Let us consider for instance a concrete case of application of our semi-classical protocol for which the source is a laser source, with an average number of photons by pulse $\overline{n}$ taken to be equal to 1, while the transmission rate of the whole line of transmission is close to $\frac{1}{1000}$. Let us now consider the following attack. A first spy, located at the beginning of the line, intercepts all the pulses emitted by the transmitter. Then, he knows correctly the value of the encoded bit with a probability $1 - exp(-\epsilon \cdot \overline{n})$, where ϵ is the efficiency of his detector that we assume here to be equal to $\frac{1}{6}$, a typical efficiency for commonly available detectors. Afterwards, he informs, on a public channel, a second spy (or even a robot) located at the end of the line about the bits that he managed to measure correctly. At this level, a non-intercepted signal ought to consist of Poissonly distributed pulses with an average number of photons by pulse equal to $\overline{n} \cdot T_{tr}$, where T_{tr} is the transmission rate along the line. In order to dissimulate the presence of the first spy and to compensate the unefficiency of his detector, the second spy ought to reemit bright copies of the signals measured by him (more intense in a ratio $\frac{\overline{n}}{\epsilon \cdot \overline{n}}$ in average). He would resend nothing when the first spy did not detect any signal (in a fraction $exp(-\epsilon \cdot \overline{n})$ of the cases). Let us denote $P_{res}(N)$ the probability of finding N photons in a pulse resent by the second spy. The global pulse distribution must remain unchanged at the end of the line, otherwise the control device would reveal the presence of the spies. Therefore, the following constraints must be fulfilled:

$$\frac{(\overline{n} \cdot T_{tr})^N}{N!} \cdot exp(-\overline{n} \cdot T_{tr}) = (1 - exp(-\epsilon \cdot \overline{n})) \cdot P_{res}(N)$$

when $N \neq 0$, and

$$exp(-\overline{n} \cdot T_{tr}) = (1 - exp(-\epsilon \cdot \overline{n})) \cdot P_{res}(0) + exp(-\epsilon \cdot \overline{n})$$

In our case,

$$T_{tr} = \frac{1}{1000}$$

and

$$\epsilon = \frac{1}{6}$$

so that

$$P_{res}(N) = \frac{(\frac{\overline{n}}{1000})^N}{(1 - exp(-\frac{\overline{n}}{6})) \cdot N!} \cdot exp(-\frac{\overline{n}}{1000})$$

when $N \neq 0$ and

$$P_{res}(0) = \frac{exp(-\frac{\overline{n}}{1000}) - exp(-\frac{\overline{n}}{6})}{1 - exp(-\frac{\overline{n}}{6})}$$

The resulting distributions and the original one, from the point of view of the authorised users, are then exactly alike and neither errors of transmission nor an intensity decrease do occur, although a spy is present, and knows exactly the whole series of bits transmitted along the line. Such an attack is impossible to realise whenever the second spy makes use of a laser source because the distribution $P_{res}(N)$ is subPoissonian. This confirms our previously mentioned result according to which the semi-classical protocol is still robust against opaque attacks provided the second spy makes use of a laser source. Now, it is not impossible, in principle, to obtain the required subPoissonian distribution provided one exploits recent technological advancements. For instance, devices such as parametric down converters allow us to produce subPoissonian distributions, by squeezing coherent states [16]. In the situation considered here, the proportion of empty pulses at the end of the line P_0 is close to 0.999, the proportion of single photons P_1 is close to 10^{-3}, the proportion of pairs P_2 is close to 5 10^{-7}, and the population $P_{>2}$ of more than two photons is negligible. The protocol is then fragile if the spy possesses a source which produces 6.10^{-3} photons by pulse in average without significantly increasing the relative population of pairs. Squeezed coherent states are good candidates for such a source. The reason is that one can show [16] that for a coherent state characterised by a Poissonian distribution of the photonic population

$$P_N = \frac{\alpha^{2N} \, exp - \alpha^2}{N!}$$

the population of the squeezed coherent state obtained after "amplification" of this coherent state with a "gain" r is defined as follows:

$$P_N = (\frac{c^2-1}{c^2})^{\frac{1}{2}} \frac{c^{-N}}{2^N N!} H_N^2(\frac{c+1}{\sqrt{2c}}\alpha)\, exp - (\frac{c+1}{c}\alpha^2)$$

where H_N is the Nth polynomial of Hermite and $c = coth\, r$. Hence, if one squeezes (with a gain $r = 0.0085$) a coherent state with 6 10^{-3} photons by pulse in average, one obtains the following subPoissonian population: $P_0 = 0.994$, $\frac{P_1}{6} = 10^{-3}$, $\frac{P_2}{6} = 4.4\, 10^{-7}$ $P_{>2} = 5.5\, 10^{-8}$. For sure, this is not exactly what we need but it is very close. The difference between the Poissonian statistics and the statistics of the squeezed state is that 10 % of the pairs of photons approximately are replaced by triplets. This is not easy to detect, taking account of the unavoidable dark countings, and of the relative proportion of pairs (0.05 % of the signal). This example clearly illustrates the fragility of the semi-classical key distribution in the case of lossy media. If one considers realistic situations for which distances are comparable to, say, 30 kilometers, the situation is quite less tragical: the average number of photons by pulse at the end of an optical fibre is then one hundred times higher and the possibility to produce significantly subPoissonian distributions at these intensities is considerably reduced. Effectively, the best gain in the population of single photons that one can obtain by squeezing coherent states, provided one does not enhance the populations of pairs present at these intensities, can be shown to be at most of the order of 25 % (instead of 600 % in the previous example, at low intensity), which does not affect much the quality factor. Furthermore, even these pessimistic estimations, that we obtained by performing theoretical simulations on the computer, are not realistic because they imply parametric gains that are not technologically realisable today. Actually, the domain of validity of the semi-classical protocol is the semi-classical domain for which the coherent sources are sufficiently intense so that squeezing does not affect them too much and at the same time sufficiently weak so that the presence of empty states cannot be avoided (grosso modo, the Poisson constant μ belongs to, let us say, the interval [0,1, 1]). Let us now give some technical precisions about the control device itself.

4.5 The Control Device

Different methods allow us to measure the populations of pairs and of single photons. For instance, we can use detectors-counters for which the ratio between the number of incoming photons and the photo-electrical current that they generate is relatively stable, so that if a pair of photons enters the detector, we measure a double current. In general, in sensitive detectors, strong non-linearities appear during the amplification process of the incoming light so that this ratio is not stable but detectors-counters exist [17], and even if their efficiency is relatively low, it does not matter because the control device is aimed at comparing average values of the photonic populations.

Another possibility is the following. The semi-classical technique requires the use of pulses of which the Poisson constant μ belongs to, let us say, the interval [0,1, 1]. Then, it is easy to show that the average time between two incoming photons is comparable to one halve of the temporal width of the pulse. It is possible to find detectors [18] of which the dead time (of the order of 10^{-9} seconds) is short in comparison to the temporal width of the incoming pulses (of the order of $30 \cdot 10^{-9}$ seconds). If we couple these detectors to precise clocks, we can integrate the number of detections during the temporal window used to code (and to detect) one bit of the signal and obtain in this way information about the photonic populations of pairs and of single photons.

Another safety device against eavesdropping techniques was developed in the framework of quantum cryptography. This device was implemented in free space quantum key distribution [9]. In this experimental realisation, the signal is coded through phase delays according to the B92 protocol [15] and detected by interferometrical techniques. At the end of the line, the signal is beamsplitted and one detector is placed at each arm of the beamsplitter. The basic principle of the safety procedure implemented by Buttler et al. in this set-up is to measure the rate of coincident detections, which can be shown to increase whenever an eavesdropper performs an opaque attack. This idea is very close to our conception, because the rate of coincident detections is in first approximation proportional to the population of pairs, so that we recover the approach developed in our semi-classical technique. The control device in which the pair rate is evaluated thanks to the measurement of the coincident detections at the outcome of a beamsplitter only requires photon detectors and is thus cheaper and easier to perform than a direct counting which requires the use of photon counters.

5 Comparison of the Quantum and Semi-Classical Protocols

In summary, our approach is the following: we assume that, in the field of optical communications, the presently existing technology is limited in the low intensity regime: ϵ is assumed to be the maximal available efficiency of the existing detectors (at the wavelength considered), and μ_{eff} is assumed to be the maximal available ratio between the populations of empty pulses and the populations of non-empty pulses (at the intensities and relative populations of pairs considered).

If we assume that the spy (as well as the transmitter) uses a laser source, and that the Poisson constant of the source of the transmitter is equal to μ, then, as shown by us, the quality factor of the semi-classical protocol is equal to

$$\frac{1 - \epsilon_{eff} \cdot \mu_{eff}}{2\epsilon_{eff}\mu_{eff}}$$

where we must replace ϵ_{eff} by

$$\frac{1 - exp(-\epsilon \cdot \mu)}{1 - exp(-\mu)}$$

and μ by $1 - exp(-\mu)$. Let us now study the advantages and the disadvantages of the semi-classical technique.

5.1 Advantages of the Semi-Classical Technique

We showed in the previous sections that the semi-classical technique is safe against intercept-resend techniques, and against opaque attacks when the absorption along the line is not too strong. Besides, it is simpler to realise (in particular when the control device makes use of a beamsplitter), because it only requires the use of fastly modulable switches and of passive elements such as detectors, beamsplitters and clocks. We must not control quantum degrees of freedom such as polarisation and/or phase, and this implies a gain in simplicity: it is not necessary to introduce polarisers or interferometers. Therefore, our protocol is more robust against quantum decoherence (it is essentially a classical pulsed telegraph). It is not impossible to eliminate decoherence in quantum cryptography, and solutions already exist such as the Plug-and-Play implementation [19], but, in comparison to the semi-classical protocol, the price to pay lies in the complexity of the device. Furthermore, the translucent attacks that we mentioned in the section 3.2 relative to the limitations of the source do not present any particular danger in the framework of the semi-classical technique for which the signal is coded inside the pulse localisation directly, and not in an "extra" degree of freedom such as the polarisation. Therefore, the beamsplitting of the signal does not bring any advantage to the spy, so that a priori nothing imposes an upper bound on μ as is the case with the quantum technique. In fact, the presence of pairs, triplets and so on, was already taken into account in the estimation of the quality factor from the beginning, through the parameter μ that reflects the Poissonian nature of the statistics. Therefore, it is possible to increase the intensity at the level of the source without losing much safety. Nevertheless, we must take account of the fact that the quality factor is a decreasing function of μ. In the classical limit (when μ gets much larger than $\frac{1}{\epsilon}$), it is easy to show that the quality factor of the semi-classical technique converges to zero. This is not surprising, because when the average number of photons by pulse increases, there are nearly never empty pulses, and nearly all the pulses can be detected, even by an unefficient detector. Anyhow, the protocol is still robust when we choose, say, a relatively large Poisson constant, for instance 80 % as show the following estimations. If $\epsilon = 50$ % (the validity of this estimation is discussed in the footnote 2), and $\mu = 80\%$, the quality factor is equal to $exp(-0.5\cdot 0.8))$ / $2(1 - exp(-0.5\cdot 0.8))$ (see previous results), which is approximately equal to one. Note that if we estimate the maximal available efficiency ϵ to be equal to 100 %, we obtain one halve instead of one, which is the same as for the quantum technique. As we already noted in the section 4.3., if $\mu_{eff} = 60$ % ($\mu = 0.92$), the quality factor of the quantum technique (50 %) is reached for an efficiency ϵ_{eff} equal to 5/6 ($\epsilon = 0.76$, not so far from the efficiency treshhold of 82.83 % required for the solution of the efficiency loophole [14] in the violation of Bell's inequalities that we already mentioned in the footnote 2).

All this shows that, presently, the semi-classical protocol remains competitive with quantum protocols, and offers the possibility of sending more intense signals, an advantage from the point of view of the dark count rate and of the bit transfer rate. Nevertheless, things evolve fastly in quantum optics and it could be that nearly perfect detectors will be available within some years. We shall discuss this point in the next paragraph (disadvantages of the semi-classical technique) as well as the implications that progresses in detection could have for what concerns the quality of available sources.

One could object against our protocol that it is not safe in the case of lossy media but it is worth noting that in the case of lossy media, quantum protocols are fragile against aggressive techniques too, a well known fact [20]. For instance, a technique, sometimes called aggressive translucent attack, is described in [4]. The authors mention that this aggressive strategy can be thwarted if μ is kept small compared to $4T_{tr}$, the transmission factor over the whole line: $\mu << 4T_{tr}$. For a a line of 50 km of optical fibre, $4T_{tr} = 0,125$. When $\mu = 0{,}1$, if we impose to the quantum technique to be absolutely safe, among others towards aggressive translucent attacks, the maximal distance is thus less than 50 km when optical fibre is used in order to carry the signal. Note that the situation is not better in the case of atmospheric transmission lines: it is easy to show [21] that, in the experimental conditions described in [9], the B92 protocol [15] is also fragile, in principle, against the opaque attack that we described in the section 4.4, for the same transmission rates and intensities. A transmission rate of 0.999 is precisely the rate that can be expected for earth-satellite key distribution [9]. Besides, if we make the most conservative assumptions and that we consider the possibility of collective or coherent attacks [7,8], the maximal tolerable error rate and thus the maximal distance of safe transmission of quantum cryptography must still be lowered in comparison to the present values. It could seem unfair to consider theoretical possibilities which are irrealisable presently when we estimate the robustness of quantum techniques (this is the case for collective and coherent attacks) and to limit ourselves to practically possible techniques when we estimate the safety of the semi-classical technique but it is consistent in the framework of quantum cryptography to consider all the theoretical possibilities (and thus the possibility of collective attacks for instance) because one aims at providing an absolute safety which relies on the laws of quantum mechanics themselves and not on technological limitations only as in the semi-classical approach. Each protocol defines its own field of application. This brings us to the next paragraph.

5.2 Disadvantages of the Semi-Classical Technique

It could happen that the quality factor of the semi-classical protocol becomes too low because of technological progresses so that this technique loses its interest. The fundamental advantage of the purely quantum techniques is that they are not, unless we discover and control the behaviour of the mythical "hidden variables", conditioned by technological progresses. In this sense, for short distances of transmission (high transmission rates), the quantum techniques provide a level of safety which is incomparably higher than the level offered by our semi-classical technique, and by all conventional techniques (as the ones based on the decomposition of huge numbers into a product of prime numbers for instance). As we mentioned previously, for long distances of transmission, the present technology, thanks to the development of best detectors and of subPoissonian (squeezed) sources represents a potential menace for the applicability of the quantum technique (and also of our semi-classical technique). Now, the translucent and opaque attacks can be thwarted by the users of the quantum protocols provided they also use subPoissonian sources in order to reduce significantly the populations of pairs, triplets and so on [21]. Therefore, technological progresses in the direction of real single-photon sources do not menace the quantum technique because they simultaneously provide the disease and the cure of the disease. This is not true for the semi-classical technique because the development of single-photon sources makes the use of our control device based on the measurement of the populations of pairs totally irrelevant (such a device is useful provided the spy uses a laser, Poissonian source). Besides, the quality of detectors is permanently increasing and one should expect that within some years their efficiency will approach unity. This would be dramatic for the security of our semi-classical technique, even for short distances, because the spy could in principle, as we shall now show, make use of nearly perfect detectors-counters coupled to sources of so-called squeezed vacuum in order to produce single-photon pulses, which would simultaneously enhance the quality of available sources and detectors and reduce proportionnally the quality factor of the semi-classical technique. These sources of squeezed vacuum[d] produce simultaneously two pulses which possess exactly the same number of photons. By measuring the population of one of these pulses with a perfect detector-counter, one could know exactly the population of the other pulse. If one triggers then the transmission of the second pulse only when a single photon is detected in the first pulse, one obtains a real single-photon source. In order to investigate the danger represented by this possibility, let us evaluate the quality factor of our technique provided we assume that the spy uses such sources and that the transmitter uses a laser source. We shall consider that the efficiency of the detector of the spy is equal to ϵ. With a source of squeezed vacuum, the probability of emitting pairs of single photons, pairs of pairs, pairs of triplets and so on is well known to be the geometrical (overPoissonian) distribution ($P(N) = \lambda^N \cdot (1 - \lambda)$). In order to simplify the treatment of the problem, we shall consistently assume that λ and μ are small parameters compared to unity, and consider the populations of empty pulses, of single photons and of pairs only, at the lowest order in these small parameters. If the spy uses a source of squeezed vacuum and triggers the transmission of one of the entangled pulses on the basis of detections realised on the other one, pairs are still likely to be transmitted when the detector "sees" one photon although a pair arrives. This will happen with a probability equal to

$$2 \cdot \epsilon \cdot (1 - \epsilon) \cdot \lambda^2 \cdot (1 - \lambda)$$

Single photons will be emitted with a probability $\epsilon \cdot \lambda \cdot (1 - \lambda)$. The populations of single photons and of pairs obtained with a laser source are, for μ small, equal to μ and $\frac{\mu^2}{2}$, and $\mu_{eff} = \mu = 2R$, while $\epsilon_{eff} = \epsilon$, up to negligible terms of higher order in μ. The corresponding populations obtained by the spy thanks to the use of squeezed vacuum states are close to $\epsilon \cdot \lambda$ and $2.\epsilon \cdot (1 - \epsilon)\lambda^2$,

[d] During the last decade, the performances offered by such sources became comparable to those offered by laser sources [22].

and

$$\mu_{eff} = \epsilon\lambda = \frac{\epsilon \cdot R}{2(1-\epsilon)} = \frac{\epsilon \cdot \mu}{4(1-\epsilon)}$$

up to negligible terms of higher order in ϵ and λ. This implies that

$$\mu_{eff} = \frac{0,7\mu}{1,2}, \ \mu$$

and

$$\frac{0,9\mu}{0,4}$$

when $\epsilon = 70$ %, 80 %, and 90 % respectively. Whenever the critical efficiency (+/-80 %) required for the solution of the efficiency loophole [14] in the violation of Bell's inequalities would be reached, our semi-classical technique would still be safe against the type of attacks described here. Nevertheless, when the efficiency of the detectors approaches unity, the quality factor of our technique rapidly decreases to zero. For instance, when $\mu = 1$, and $\epsilon = 0,8$, the quality factor is close to 0.5, the corresponding factor for the BB84 protocol; when $\mu = 1$, and $\epsilon = 0,85$, the quality factor is close to 0.125, four times less. Note that the majority of presently existing detectors is based on a cascade process [17] during which the energy of the incoming pulse gets macroscopically amplified. It occurs very often that, due to the presence of essential non-linearities at the amplification level, the detector cannot differentiate single photons and pairs of photons, which diminishes the practical efficiency of the detector-counter necessary for the realisation of the single photon source described in the previous section and renders quite hypothetical its concrete implementation. In summary, even if perfect detectors seem weel to exist already (see footnote 2), perfect single photon sources are certainly not available yet, which is sufficient in order to ensure the realisability of the semi-classical protocol described in the previous sections.

6 Generalisations of the Protocol

Even if the protocol presented here became obsolete because of technological progresses, the general ideas on which it is based are still valid. After all, it is sufficient in order to implement our protocol to dispose of a signal which is difficult to measure correctly and to reproduce exactly. In particular, we exploited the intrinsic limitations of Poissonian distributions, and our treatment can be applied systematically to any situation in which such distributions appear. They do not appear only for laser sources, but for a wide class of physical phenomena, in particular for spontaneous decay processes and for rare phenomena. In the former case, the Poissonian behaviour characterizes the temporal dependence of the phenomenon, in the latter case, the rarity of a phenomenon can be due to the weakness of its coupling to the detectors and sources aimed at controlling it. These possibilities suggest interesting generalisations of our scheme, more speculative than the already described protocol based on localised light pulses but nevertheless appealing, that we shall shortly develop now. Let us consider the decay of a radioactive particle. Presently, it is not possible to stimulate the nuclear decay process, which is therefore a purely spontaneous phenomenon, characterised by an exponentially decreasing probability distribution in time. In principle, one could maybe interact with the nucleus in such a way that its decay occurs faster (by the way, this would solve the problem of the treatment of radioactive residues produced in our nuclear centrals) but the complexity of the internal degrees of freedom of atomic nuclei is very high, and presently we are not able at all to control them with a sufficient accuracy. Let us assume that we want to send a cryptographic key by express courrier (the duration of the expedition is then fixed in advance with a high guarantee of accuracy by the companies which realise this kind of jobs). We could encode a bit of the key exactly as in our semi-classical protocol by preparing two identical nuclei in clearly distinguishible locations according to the following protocol: whenever the first nucleus is excited, the bit value is 1, whenever the second nucleus is excited, the bit value is 0. One can choose the half lifetime of the excited state in such a way that the probability that the nucleus decays during the expedition is less than an arbitrary fraction of unity, which means that no spy can learn more than this fraction of the signal underway. This is analogous to the constraint in our semi-classical approach according to which the efficiency of detection is limited by an upper bound. One could object that unstable isotopes differ by the composition of the nucleus and could thus be distinguished by non-invasive measurements, but this objection is no longer valid if one considers isomeric nuclei that are excited states of a same nucleus. Note that even if a spy knows a part of the signal, he cannot replace this part by "brighter" bits because this would change the average number of desexcitations by unit of time at the level of the receiver. This is analogous to the constraint imposed by the control device in our semi-classical approach. It is worthy to note that the decay process is intimately related with time-energy uncertainties (see also [23] for a cryptographic protocol based on these uncertainties). Our protocol, that was largely inspired from the BB84 protocol [6], a protocol that involves discrete (spin or polarisation) complementary variables, brings us back to uncertainties between continuous variables.

Our protocol could also be implemented whenever the physical processes used in order to encode the information are such that the probability to send succesfully a bit as well as the probability to measure it succesfully are very low (rare phenomena). Usually, the statistics which describes such phenomena is Poissonian (provided some reasonable assumptions are made, for instance, that no memory effect occurs). The neutrino furnishes a good example of such rare phenomena. In this case, the reason for this rarity is the very low value of the weak coupling interaction constant, a limitation imposed by the laws of nature. For such phenomena, we are in a situation similar to the one previously described in our semi-classical protocol. The problem with the neutrinos is that the authorised users of the line significantly suffer from their inability to produce neutrinos in large quantities and to guide them. Indirectly, they also suffer from their inability to measure them because of the dark count rate: noise will expectedly dominate the signal itself. In any case, these examples show that our semi-classical treatment is not essentially passeist as one could think after having read the section 5.2 devoted to the disadvantages of the technique. It can be generalised to possibilities of more speculative nature and could be positively conditioned by technological progresses (in the production and control of isomeric nuclei or of neutrinos for instance, as in the previous examples).

Appendix: Corrections Involving the Standard Deviations of the Stochastic Processes Considered.

During the reconciliation protocol, the receiver and the transmitter correct the errors which occurred during the transmission. They can then estimate the amount of information possessed by the spy by dividing the number of errors by the quality factor. This evaluation is correct only in average and it is necessary to compute the standard error associated with it in order to guarantee a safety margin. For instance, in the reference [4], the authors add five standard deviations to the average evaluation in order to make sure that they do not underestimate the information possessed by the spy. The situation is slightly different in quantum cryptography in the sense that the spy, after an intercept-resend attack, does not know exactly where the errors are located and which bits were succesfully intercepted and read by him (he only knows the probability therefore). Nevertheless, the reasoning made in [4] is well adapted to our purpose because the authors derived an upper bound on the standard deviation on the eavesdropper information by treating the case in which he (thanks to the help of an hypothetical omniscient assistant called Big Brother) knew the location of correctly deciphered bits, while the other ones were replaced by a random noise. This is similar to some extent to what occurs in the semi-classical context considered by us. We can thus reproduce and generalise the computation of the standard deviation made in [4] in order to cover also our semi-classical technique.

If k pulses are intercepted by the spy, and that the percentage of bits correctly measured by him is equal to I (I is derived in the section 4.3., it is equal to $\epsilon_{eff}\mu_{eff}$, where

$$\epsilon_{eff} = \frac{1 - exp(-\epsilon \cdot \mu)}{1 - exp(-\mu)}$$

and $\mu_{eff} = 1 - exp(-\mu)$), the probability that he knows correctly n bits is given by the binomial distribution:

$$P(n) = \frac{k!}{n!\,(k - n)!} \cdot I^n\,(1 - I)^{k-n} \tag{1}$$

If $k \cdot I$ is sufficiently large (in practice, larger than 30), this binomial distribution can be approximated by a Gaussian distribution:

$$P(n) = \frac{1}{\sqrt{2\pi\, k\, I\,(1 - I)}} \cdot exp - \left(\frac{(n - kI)^2}{2\,k\,I\,(1 - I)} \right) \tag{2}$$

Similarly, if the average error rate caused by the interception of one bit by the spy is e, the probability that t errors appear when k pulses are intercepted is given by a Gaussian distribution for $k \cdot e$ sufficiently large:

$$P(t) = \frac{1}{\sqrt{2\pi\, k\, e\,(1 - e)}} \cdot exp - \left(\frac{(t - k\,e)^2}{2\,k\,e\,(1 - e)} \right) \tag{3}$$

The transmitter and the receiver measure t during the protocol of reconciliation. They evaluate the information of the spy n by dividing t by the quality factor Q:

$$n_{ev} = t/Q = I\,t/e \tag{4}$$

This quantity admits also a Gaussian distribution:

$$P(n_{ev}) = \frac{1}{\sqrt{2\pi\, \frac{k}{Q^2}\, e\,(1 - e)}} \cdot exp - \left(\frac{(n_{ev}\,Q - k\,e)^2}{2\,k\,e\,(1 - e)} \right)$$

$$= \frac{1}{\sqrt{2\pi\, kI^2\, \frac{(1-e)}{e}}} \cdot exp - \left(\frac{(n_{ev} - k\,I)^2}{2\,k\,I^2\,\frac{(1-e)}{e}} \right) \tag{5}$$

The distributions of the evaluated information n_{ev} and of the correct information n are two Gaussian distributions having the same average value $(k \cdot I)$ but different standard deviations. It can be shown that the distribution of their difference (when they are independently distributed) is a Gaussian distribution, of which the average value is equal to zero and of which the variance is equal to the sum of the variances of the distributions of n_{ev} and n:

$$\sigma^2 = k\,I\,(1 - I) + k\,I^2\,\frac{(1 - e)}{e} \tag{6}$$

k can be evaluated as t/e, so that, if t errors are detected during the reconciliation protocol, the safety margin of five standard deviations is given by the following expression:

$$\begin{aligned} 5\,\sigma &= 5\sqrt{\frac{t}{e} \cdot [I\,(1 - I) + I^2\,\frac{(1 - e)}{e}]} \\ &= 5\sqrt{\frac{t}{Q} \cdot [1 - 2I + \frac{1}{Q}]} \end{aligned} \tag{7}$$

If for instance the error rate is equal to 4 %, and if the signal is a series of ten thousand bits, this safety margin is, for the quantum technique, in the case of an intercept-resend attack performed in the canonical basis ($I = Q = 1/2$), equal to 200 bits, while the average information of the spy is equal to 800 bits. When the attack is performed in the Breidbart basis, we recover the results derived in [4]

$$(I = \frac{1}{\sqrt{2}}, e = \frac{1}{4} \text{ and } \sigma = \sqrt{t \cdot [4 + 2\sqrt{2}]})$$

In the context of the semi-classical technique, this estimation also constitutes an upper bound because the appearance of errors is not independent of the acquisition of information by the spy in our case as it was the case for Big Brother in the reasoning made in the reference [4] so that we overestimate the standard deviation of the information gained by the spy. In the classical limit (when μ gets much larger than $\frac{1}{\epsilon}$), I approaches unity while e and Q approach zero. Then,

$$\sigma \leq \sqrt{\frac{t}{e} \cdot [I\,(1 - I) + I^2\,\frac{(1 - e)}{e}]} \simeq \sqrt{\frac{t}{Q^2}}$$

In the domain where our technique is interesting (μ gets much smaller than $\frac{1}{\epsilon}$, for instance when ϵ approaches zero), I approaches zero, Q goes to infinity and

$$\sigma \leq \sqrt{\frac{t}{e} \cdot [I\,(1 - I) + I^2\,\frac{(1 - e)}{e}]} \simeq \sqrt{\frac{t}{Q}}$$

One can get a better estimation of the standard deviation σ by taking account of the specificity of the semi-classical technique. In the scenario considered in the reference [4], the error rate was assumed to be independent of the information gained by the spy, which is a feature of the intercept/resend attacks in the quantum domain. In our case, transmission errors are assumed to occur when the spy was not able to intercept/resend correctly a bit, and that he replaces it by a random noise. Thus, if we can estimate the standard deviation on the noise, we can also estimate the deviation on the amount of correctly intercepted bits, so to say, on the deterministic information of the spy. In the previous reasoning we had to add both contributions because the processes of error generation and information acquisition were independent. In our case, let us assume that the number of bits that were not correctly intercepted/resent by the spy is l. They were necessarily replaced by random bits so that the standard deviation on the number of errors is obviously equal to

$$\sqrt{l \cdot \frac{1}{2}(1 - \frac{1}{2})}$$

In average, t, the number of errors is at least equal to $\frac{l}{2}$ so that we can estimate the standard deviation on the number of errors by replacing l by $2t$ in the previous expression. By dividing this standard deviation by the quality factor, we get a better upper bound of the standard deviation on the information of the spy

$$\sigma : \sigma \simeq \sqrt{\frac{t}{2Q^2}}$$

For sure these results make sense only if the error rate is sufficiently low so that our evaluation of the information of the spy remains smaller than unity. This fixes a bound on the error rate that can be tolerated during the transmission: $< I > = \frac{e}{Q} < 1$. This means that the critical error rate above which confidentiality is not guaranteed anymore goes to zero in the classical limit (where Q goes to zero), a quite natural result. By considering also the contribution of the standard deviation, we should impose furthermore that

$$\sqrt{\frac{t}{2Q^2}} < N$$

where N is the lenght of the full series of bits shared by the transmitter and the receiver ($\frac{t}{N} = e$). This second constraint imposes that in the classical limit, even when the error rate is low enough so that $\frac{e}{Q} < 1$, it is necessary to send sufficiently long messages (many bits) in order to guarantee some confidentiality.

Acknowledgements

Thanks to Prof. John Corbett (Macquarie University, Sydney) for his reading and comments of the first version of this work. Also thanks to Prof. E. Goovaerts (University of Antwerp, Department of Solid State Physics), for his useful informations about the present state of photonics (detection devices and typical magnitudes). The author is a postdoctoral fellow of the Flemish Fund for Scientific Research (FWO). During the realisation of a part of this work, he enjoyed the support of the Flemish-Polish Scientific Collaboration Program No. 007 entitled "Probing the structure of Quantum Mechanics: New probability models for new experiments on quantum particles".

References

1. P. D. Townsend, and I. Thompson, "A quantum key distribution channel based on optical fibre", *Journal of Modern Optics*, **41**, 2425-2433 (1994).
2. G. S. Vernam, *Journal of the American Institute of Electronicians and Engineers*, **45**, 109-115 (1926).
3. C. H. Bennett, G. Brassard, and J.M. Robert, "Privacy amplification by public discussion", *SIAM Journal of Computing*, **17**, 210-229 (1988).
4. C.H. Bennett, F. Bessette, G. Brassard, L. Salvail,and J. Smolin, "Experimental quantum cryptography", *Journal of Cryptology*, **5**, 3-28 (1992).
5. C.H. Bennett, G. Brassard, C. Crépeau, and U. M. Maurer, "Generalized privacy amplification", *IEEE Transactions on Information Theory*, IT-**41**, 1915-1923 (1995).
6. C.H. Bennett and G. Brassard, "Quantum cryptography: public key distribution and coin tossing", *IEEE International Conference on Computing, Systems and Signal Processing*, Bangalore, India, 175-179 (1984).
7. See for instance E. Biham, M. Boyer, G. Brassard, J. van de Graaf, and T. Mor, "Security of quantum key distribution against all collective attacks", *quant-ph/9801022*, 1-5 (1998), and references therein.
8. H. Bechmann-Pasquinucci and N. Gisin, "Incoherent and coherent eavesdropping in the six-state protocol of quantum cryptography", *Phys. Rev. A*, **59**, 4238-4248 (1999).

9. W.J. Buttler, R.J. Hughes, P.G. Kwiat, S.K. Lamoreaux, G.G. Luther, G.L. Morgan, J.E. Nordholt, C.G. Peterson, and C.M. Simmons, "Practical free-space quantum key distribution over 1km", *Phys. Rev. Lett.*, **81**, 3283-3286 (1998).
10. C.A. Fuchs, N. Gisin, R.B. Griffiths, C-S. Niu, and A. Peres, "Optimal eavesdropping in quantum cryptography I", *Phys. Rev. A.*, **56**, 1163-1172 (1997).
11. N. Gisin, G. Ribordy, W. Tittel and H. Zbinden, "Quantum Cryptography", *quant-ph/0101098*, submitted to *Reviews of Modern Physics* (2001).
12. B. Huttner and A. Ekert, "Information gain in quantum eavesdropping", *Journal of Modern Optics*, **41**, 586-588 (1997).
13. Y. Yamamoto, J. Kim, and H. Kan, "Single photonics: turnstile device and solid-state photomultiplier", *Proceedings of the International Quantum Electronics Conference (IQEC)*, invited paper QThE5 (1998).
14. A. Garg, and N. D. Mermin, "Detector inefficiencies in the Einstein-Podolsky-Rosen experiment", *Phys. Rev. D.*, **10**, 3831-3835 (1987).
15. C.H. Bennett, "Quantum cryptography using any two nonorthogonal states", *Phys. Rev. Lett.*, **68**, 3121-3124 (1992).
16. W. Schleich, and J. A. Wheeler, *Journal of the Optical Society of America B.*, **4**, 1715-1722 (1987).
17. K.J. Resch, J. S. Lundeen, and A. M. Steinberg, "Experimental observation of nonclassical effects on single-photon detection rates", *quant-ph0006056*, 1-8 (2000).
18. S. J. D. Phoenix and P. D. Townsend, "Quantum cryptography: how to beat the code breakers using quantum mechanics", *Contemporary Physics*, **36**, 165-195 (1995).
19. H. Zbinden, J.D. Gautier, N. Gisin, B. Huttner, A. Muller and W. Tittel, "Interferometry with Faraday mirrors for quantum cryptography", *Electronics Letters*, **33**, 2405-2412 (1994).
20. B. Huttner, N. Imoto, N. Gisin, and T. Mor, "Quantum cryptography with coherent states", *Phys. Rev. A*, **51**, 1863-1869 (1995).
21. T. Durt, "Comment on 'Practical free-space quantum key distribution over 1 km'", *Phys. Rev. Lett.*, **83**, 2476 (1999).
22. See for instance T. Jennewein, C. Simon, H. Weinfurter, G. Weihs, and A. Zeilinger, "Violation of Bell's inequalities under strict einstein locality conditions", *Phys. Rev. Lett.*, **81**, 5039-5043 (1998), and references therein.
23. S.N. Molotov and S. S. Nazin, "Quantum cryptography based on the time-energy uncertainty relation", *quant-ph/9612013*, 1-5 (1996).

HOW TO CONSTRUCT DARBOUX-INVARIANT EQUATIONS OF VON NEUMANN TYPE

JAN L. CIEŚLIŃSKI
Instytut Fizyki Teoretycznej, Uniwersytet w Białymstoku,
ul. Lipowa 41, 15-424 Białystok, Poland
E-mail: janek @ alpha.uwb.edu.pl, janek @ fuw.edu.pl

We present an alternative construction of Abelian and non-Abelian integrable equations of von Neumann type recently introduced by Ustinov and Czachor. Our approach generates a more general class of integrable systems. We show that these equations are invariant with respect to the standard Darboux transformation.

1 Introduction

In a series of recent papers M. Czachor and his collaborators [1,2,3,4,5] developed nonlinear quantum mechanics based on integrable generalizations of the von Neumann equation. The integrability is understood as the existence of the Lax pair and the Darboux transformation (a standard method generating particular exact solutions) [6]. In this paper we present a straightforward method to construct nonlinear systems of that type.

Following Ref. [1] we consider a class of Lax pairs of of the form

$$\begin{aligned} z_\lambda \psi &= \psi X(\lambda) \\ -i\dot{\psi} &= \psi A(\lambda) \ , \end{aligned} \tag{1}$$

where ψ is an element of a linear space V, $X(\lambda)$ and $A(\lambda)$ are λ-families of linear operators on this space, and λ, z_λ are complex parameters. The compatibility conditions are given by

$$i\dot{X}(\lambda) = [A(\lambda), X(\lambda)] \ . \tag{2}$$

The equation (2) is understood as identity with respect to λ which yields a system of nonlinear equations.

Usual approach consists in some assumptions on the form of $A(\lambda)$ and $X(\lambda)$ which converts (2) into a system of differential equations for coefficients of $A(\lambda)$ and $X(\lambda)$. Ustinov and Czachor simplified this procedure expressing A by X in a Darboux-invariant way (see [2,3]). Thus the number of dependent variables was reduced to the coefficients of $X(\lambda)$. The simplest, but important, case is given by

$$X(\lambda) = \rho - H\lambda \ , \tag{3}$$

($\rho = \rho(t)$, H is a constant operator) and some more complicated expression for A (see (34) below, compare also the formulas (18),(35)). The operator $A(\lambda)$ is expressed by H and ρ, without need to solve any system of nonlinear equations. The compatibility conditions imply that H is a constant operator justifying our assumption. We will show that in this case the system (2) reduces to a single equation of the form

$$i\dot{\rho} = [H, f(\rho, H)] \ . \tag{4}$$

where $f(\rho, H)$ is (in general) a non-Abelian function of ρ (i.e., $[f(\rho, H), \rho] \neq 0$).

The equation (4) can be considered as a generalization of the von Neumann equation $i\dot{\rho} = [H, \rho]$. If X is parameterized by more variables, then one gets a system of equations for these variables (some variables can turn out to be constant parameters), e.g. if

$$\begin{aligned} X &= \rho + \lambda H + \lambda^2 G \ , \\ A &= \frac{f(\rho)}{\lambda} \ , \end{aligned} \tag{5}$$

then the compatibility conditions yield

$$\begin{aligned} G &= \text{const} \ , \\ i\dot{H} &= [f(\rho), G] \ , \\ i\dot{\rho} &= [f(\rho), H] \ . \end{aligned} \tag{6}$$

The main goal of this paper is to reconstruct the results of Ref. [3] within a more general framework.

To complete the introduction we will briefly recall the notion of the Darboux invariance (or covariance). Consider the transformation

$$\psi[1] = \psi D \ , \tag{7}$$

where [1] denotes the image under the transformation (compare [6]). Substituting $\psi = \psi[1]D^{-1}$ into (1) we easily compute

$$X[1](\lambda) = D^{-1} X(\lambda) D \ , \tag{8}$$

$$A[1](\lambda) = -iD^{-1}\dot{D} + D^{-1}A(\lambda)D \ . \tag{9}$$

We say that the Lax pair (1) is Darboux-invariant (or Darboux-covariant), if the structure of $A[1]$ and $X[1]$ is the same as the structure of A and X, respectively. The notion of "the structure" is explained, for instance, in Ref. [7].

The most important point is that the poles of $A(\lambda)$ and $A[1](\lambda)$ should coincide (to be more precise: the divisors of poles, i.e., poles with their multiplicities, for A and $A[1]$ should be the same) [8,9,10]. The same concerns $X(\lambda)$. In this paper we will confine our considerations only to this case: the Darboux-invariant Lax pair is characterized by the divisors of poles of $A(\lambda)$ and $X(\lambda)$ which do not change under the Darboux transformation.

2 A Systematic Method to Generate Darboux-Invariant Lax Pairs

First, we introduce a convenient notation. Let $R(\lambda)$ is a λ-family of operators rational in λ which have exactly M pairwise different poles (at $\sigma_1, \ldots, \sigma_M$ where $\sigma_k \in \mathbf{C} \cup \infty$). Consider the decomposition of $R(\lambda)$ into partial fractions

$$R(\lambda) = (R)_0 + \sum_{k=1}^{M} [R(\lambda)]_{\sigma_k} \tag{10}$$

where $(R)_0$ does not depend on λ and $[R(\lambda)]_{\sigma_k}$ is a component of $R(\lambda)$ divergent at $\lambda = \sigma_k$ (the principal part of $R(\lambda)$ at $\lambda = \sigma_k$). In particular, $(R)_0 + [R(\lambda)]_\infty$ is the polynomial part of R. For example, if $Q(\lambda)$ has a pole (of K_0-th order) at $\lambda = \mu$ and the corresponding Laurent expansion is given by

$$Q(\lambda) = \sum_{k=-K_0}^{\infty} s_k (\lambda - \mu)^k ,$$

then

$$[Q(\lambda)]_\mu := \sum_{k=-K_0}^{-1} s_k (\lambda - \mu)^k .$$

If Q is analytic at $\lambda = \mu$, then $[Q(\lambda)]_\mu = 0$. Note that in general $s_0 \neq (Q(\lambda))_0$. For example, if

$$Q(\lambda) = \frac{2}{\lambda - 1} + 3 + \lambda + \lambda^2 \equiv \frac{2}{\lambda - 1} + 5 + 3(\lambda - 1) + (\lambda - 1)^2 ,$$

then

$$[Q(\lambda)]_1 = 2(\lambda - 1)^{-1} , \quad [Q(\lambda)]_\infty = \lambda + \lambda^2 , \quad (Q(\lambda))_0 = 3 , \quad s_0 = 5 .$$

Assume that D and D^{-1} have poles at $\mu_1, \ldots, \mu_S$ and $\mu_{S+1}, \ldots, \mu_L$, respectively. Therefore conjugation by D cannot add new poles to R except

$\mu_1, \ldots, \mu_L$. In other words:

$$D^{-1}RD = (D^{-1}RD)_0 + \sum_{j=1}^{L}[D^{-1}RD]_{\mu_j} + \sum_{k=1}^{M}[D^{-1}RD]_{\sigma_k} , \tag{11}$$

where $(D^{-1}RD)_0$ is the λ-independent component in the decomposition of $D^{-1}RD$ into partial fraction.

This implies the following (rather obvious) properties

$$[[R]_{\sigma_k}]_{\sigma_k} = [R]_{\sigma_k} , \tag{12}$$

$$[[R]_{\sigma_k}]_{\sigma_j} = 0 \qquad (k \neq j) , \tag{13}$$

$$[D^{-1}[R]_{\sigma_k} D]_{\sigma_j} = 0 \qquad (k \neq j) , \tag{14}$$

$$[D^{-1}RD]_{\sigma_k} = [D^{-1}[R]_{\sigma_k} D]_{\sigma_k} , \tag{15}$$

where j, k run from 1 to M.

Now, we are prepared to define $A(\lambda)$ in a Darboux covariant way. Let $F(X(\lambda), \lambda)$, $G(X(\lambda), \lambda)$ be given functions, rational in λ with operator coefficients, satisfying

$$TF(X, \lambda)T^{-1} = F(TXT^{-1}, \lambda) , \qquad TG(X, \lambda)T^{-1} = G(TXT^{-1}, \lambda) \tag{16}$$

for any invertible operator T, and

$$[X, F(X, \lambda)] = 0 , \qquad [X, G(X, \lambda)] = 0 . \tag{17}$$

Typical examples of functions satisfying both conditions are polynoms and rational functions in X which do not contain operator coefficients, e.g. $\lambda X^2 + \lambda^3 X^{-1}$ or $X + \lambda(I + X)^{-1}$.

¿From among all poles of $g(X(\lambda), \lambda)$ we choose several poles denoting them by $\sigma_1, \ldots, \sigma_K$. Then we define:

$$A(\lambda) = (F(X(\lambda), \lambda))_0 + \sum_{j=1}^{K}[G(X(\lambda), \lambda)]_{\sigma_j} + [F(X(\lambda), \lambda)]_\infty . \tag{18}$$

Theorem 1 *If $(D)_0$ is time-independent, then the Lax pair (1) with A given by (18) is Darboux covariant.*

Proof: Substituting (18) into (9), we get

$$A[1] = -iD^{-1}\dot{D} + D^{-1}(A)_0 D + D^{-1}\sum_{j=1}^{K}[G(X(\lambda), \lambda)]_{\sigma_j} D + D^{-1}[F(X(\lambda), \lambda)]_\infty D . \tag{19}$$

It is enough to show that

$$A[1] = (F(X[1](\lambda),\lambda))_0 + \sum_{j=1}^{K}[G(X[1](\lambda),\lambda)]_{\sigma_j} + [F(X[1](\lambda),\lambda)]_\infty \,. \quad (20)$$

The right hand side of (19) is a rational function with poles at most at $\mu_1, \ldots, \mu_L$ and $\sigma_1, \ldots, \sigma_K, \infty$ (we assume that all these points are pairwise different which is not very essential but simplifies the proof). Therefore

$$A[1] = (A[1])_0 + \sum_{i=1}^{L}[A[1]]_{\mu_i} + \sum_{k=1}^{K}[A[1]]_{\sigma_k} + [A[1]]_\infty \,. \quad (21)$$

The first two terms in the formula (19) can have poles at most at $\mu_1, \ldots, \mu_L$. In other words,

$$\begin{aligned} [A[1]]_{\sigma_k} &= [D^{-1}\textstyle\sum_{j=1}^{K}[G(X(\lambda),\lambda)]_{\sigma_j}D]_{\sigma_k} + [D^{-1}[F(X(\lambda),\lambda)]_\infty D]_{\sigma_k} \,, \\ [A[1]]_\infty &= [D^{-1}\textstyle\sum_{j=1}^{K}[G(X(\lambda),\lambda)]_{\sigma_j}D]_\infty + [D^{-1}[F(X(\lambda),\lambda)]_\infty D]_\infty \,, \end{aligned} \quad (22)$$

and, taking into account (14) and (15) we have

$$\begin{aligned} [A[1]]_{\sigma_k} &= [D^{-1}G(X(\lambda),\lambda)D]_{\sigma_k} \,, \\ [A[1]]_\infty &= [D^{-1}F(X(\lambda),\lambda)]_\infty \,. \end{aligned} \quad (23)$$

The Darboux transformation preserves, by definition, the divisor of poles of the operator $A(\lambda)$. Therefore the assumption that D is the Darboux matrix means that $[A[1]]_{\mu_i}$ have to vanish for $i = 1, \ldots, L$. Computing $(A[1])_0$ it is crucial to notice that $(D)_0 = \lim_{\lambda\to\infty} D(\lambda)$ (because, by assumption, D does not have a pole at infinity). Thus, taking also into account that $(D)_0 = \text{const}$, we have

$$(A[1])_0 = (-iD^{-1}\dot{D} + D^{-1}(F(X(\lambda),\lambda))_0 D)_0 = (D^{-1}F(X(\lambda),\lambda)D)_0 \,. \quad (24)$$

Then

$$A[1] = (D^{-1}F(X(\lambda),\lambda)D)_0 + \sum_{k=1}^{K}[D^{-1}G(X(\lambda),\lambda)D]_{\sigma_k} + [D^{-1}F(X(\lambda),\lambda)]_\infty \,, \quad (25)$$

and, using (16) and (8), we get (20) which ends the proof. □

The assumption $(D)_0 = \text{const}$ means that the normalization matrix (see Ref. [7]) is constant and such condition is not very restrictive (compare Section 5).

In the proof we assumed that D is the Darboux matrix. The construction of the Darboux matrix is a separate problem (see Section 5).

3 Examples of Darboux-Invariant Equations

The procedure described in the previous section contains a lot of parameters (e.g. rational functions F and G with any prescribed poles) and fields. Therefore one can easily produce any number of Darboux-covariant Lax pairs.

The compatibility conditions apparently form a system of differential and algebraic equations. We are going to show that for X linear in λ (see (3)) all equations except one are identically satisfied.

Theorem 2 *The compatibility conditions for the Lax pair defined by (1), (3) and (18) reduce to a single equation:*

$$i\dot{\rho} = [H, f(\rho, H)] + [\varphi(\rho, H), \rho] , \tag{26}$$

where $\dot{H} = 0$ *and* $f(\rho, H)$, $\varphi(\rho, H)$ *are given by*

$$f(\rho, H) := \sum_{k=1}^{K} G_{\sigma_k} ,$$

$$\varphi(\rho, H) := (F(\rho - \lambda H, \lambda))_0$$

where G_{σ_k} *is the residuum of* $G(\rho - \lambda H, \lambda)$ *at* $\lambda = \sigma_k$ *and* F, G *are arbitrary functions satisfying (16) and (17) (where* $X = \rho - \lambda H$*).*

Proof: The compatibility conditions read

$$i\dot{X} = [A, X] = [(F)_0, X] + [\sum_{j=1}^{K} [[G]_{\sigma_j}, X] + [[F]_\infty, X] . \tag{27}$$

We are going to simplify the right hand side of (27) using (17). First, we decompose F and G into partial fractions:

$$G(X, \lambda) = \sum_{k=1}^{K} [G(X, \lambda)]_{\sigma_k} + [G(X, \lambda)]_\infty + (G(X, \lambda))_0 + \sum_m [G(X, \lambda)]_{\beta_m} ,$$
$$F(X, \lambda) = [F(X, \lambda)]_\infty + (F(X, \lambda))_0 + \sum_m [F(X, \lambda)]_{\alpha_m} , \tag{28}$$

where α_m are poles of F and σ_k, β_m are poles of G. Then we use (17):

$$0 = [G, X] = [\sum_{k=1}^{K} [G]_{\sigma_k}, X] + [[G]_\infty, X] + [\sum_m [G]_{\beta_m}, X] + [(G)_0, X] ,$$
$$0 = [F, X] = [[F]_\infty, X] + [\sum_m [F]_{\alpha_m}, X] + [(F)_0, X] . \tag{29}$$

From now on it will be essential that X is linear in λ. It is also convenient to write $X = \rho - \lambda H = \rho - \sigma_i H - (\lambda - \sigma_i) H$, e.g.:

$$[\sum_{k=1}^{K} [G]_{\sigma_k}, X] = [\sum_{k=1}^{K} [G]_{\sigma_k}, \rho - \sigma_k H - (\lambda - \sigma_k) H] = \sum_{k=1}^{K} [[G]_{\sigma_k}, X]]_{\sigma_k} - \sum_{k=1}^{K} [G_{\sigma_k}, H] \tag{30}$$

where G_{σ_k} is the residuum of G at $\lambda = \sigma_k$. Equating to zero coefficients by partial fractions in (29)) we obtain (among others):

$$\begin{aligned}[][G]_{\sigma_k}, X]_{\sigma_k} &= 0 \, , \\ [[F]_\infty, X] - \lambda[(F)_0, H] &= 0 \, . \end{aligned} \tag{31}$$

Using (30) and (31) we rewrite (27) as follows

$$i\dot{X} = [(F)_0, X] - \sum_{k=1}^{K} [G_{\sigma_k}, H] + \lambda[(F)_0, H] \, , \tag{32}$$

and, because of (3), the compatibility conditions reduce to the equation (26) and $\dot{H} = 0$ which ends the proof. □

Ustinov and Czachor considered a class of nonlinear systems of a relatively compact form and with several interesting applications [3]. The corresponding Lax pairs are given by (1) where

$$X(\lambda) = \sum_{k=0}^{N} \lambda^k H_k, \tag{33}$$

$$\begin{aligned} A(\lambda) = & \sum_{k=0}^{L} \frac{\lambda^{L-k}}{k!} \left(\frac{d^k}{d\zeta^k} f(\zeta^N X(\frac{1}{\zeta}), \frac{1}{\zeta}) \right) \Big|_{\zeta=0} \\ & + \sum_{k=0}^{M-1} \frac{\lambda^{k-M}}{k!} \left(\frac{d^k}{d\varepsilon^k} g(X(\varepsilon), \varepsilon) \right) \Big|_{\varepsilon=0} \end{aligned} \tag{34}$$

We will show that these Lax pairs can be generated by the formula (18).

Theorem 3 *The Lax pairs defined by (1), (33), (34) are particular examples of the Lax pairs described in Theorem 1.*

Proof: The formula (34) can be rewritten as follows:

$$A(\lambda) = \left[\lambda^L f(\lambda^{-N} X(\lambda), \lambda)\right]_\infty + \left(\lambda^L f(\lambda^{-N} X(\lambda), \lambda)\right)_0 + \left[\lambda^{-M} g(X(\lambda), \lambda)\right]_0 \, . \tag{35}$$

To complete the proof we just identify (35) as a particular case of the formula (18) where

$$\begin{aligned} F(X(\lambda), \lambda) &:= \lambda^L f(\lambda^{-N} X(\lambda), \lambda) \, , \\ G(X(\lambda), \lambda) &:= \lambda^{-M} g(X(\lambda), \lambda) \, , \\ K = 1 \, , \quad & \sigma_1 = 0 \, . \end{aligned}$$

□

The proof that A of the form (34) is Darboux-invariant given in Ref. [3] seems to be much more complicated and is restricted to the simplest (binary) Darboux transformations (given by the formula (41) in Section 5). The main advantage of our approach consists in its generality. It can be applied to any D such that both D and D^{-1} are regular outside a finite number of fixed poles.

4 A Nonisospectral Case

In this section we present another possibility to generate integrable systems of the von Neumann type. Consider the spectral problem of the form (1) where

$$\begin{aligned} X(\lambda) &= \rho - \lambda H \ , \\ A(\lambda) &= \frac{\rho}{\lambda} \ , \end{aligned} \tag{36}$$

and the spectral parameter is allowed to be time-dependent, $\lambda = \lambda(t)$ [11]. Compatibility conditions read

$$\dot{\rho} - \lambda \dot{H} - \dot{\lambda} H = i[f(\rho), H] \tag{37}$$

We see that $\dot{\lambda}$ has to be linear function of λ:

$$\dot{\lambda} = a(t)\lambda + b(t) \ , \tag{38}$$

Hence (37) splits into two equations:

$$\dot{\rho} - bH = i[f(\rho), H] \ , \qquad \dot{H} + aH = 0 \ . \tag{39}$$

Finally, we obtain the following nonlinear system:

$$\begin{aligned} \dot{\rho} &= i[f(\rho), H] + bH \ , \\ H &= H_0 \exp(\textstyle\int_0^t a(\tau) d\tau) \ , \end{aligned} \tag{40}$$

where $a = a(t)$, $b = b(t)$ are arbitrary given functions.

Choosing $X(\lambda)$ in the form of higher order polynoms (like (5)) one can obtain a hierarchy of nonisospectral linear problems. The construction of the Darboux matrix for nonisospectral problems is presented and discussed in Ref. [7].

5 The Darboux-Bäcklund Transformation

In this section we recall a standard approach to the construction of the Darboux matrix [7,8,9]. Let assume the following form of the Darboux matrix D:

$$D=\left(I+\frac{\mu-\nu}{\lambda-\mu}P\right)N\,, \tag{41}$$

where $P^2=P$ and N (the normalization matrix) is invertible, which implies

$$D^{-1}=N^{-1}\left(I+\frac{\nu-\mu}{\lambda-\nu}P\right)\,. \tag{42}$$

D given by (41) is not invertible for $\lambda=\nu$ and has a singularity at $\lambda=\mu$. At other points of complex plane $D(\lambda)$ and $D^{-1}(\lambda)$ are regular. Therefore $D^{-1}AD$ is a rational function with poles at most at $\sigma_1,\ldots\sigma_K,\infty$, μ and ν. In this case

$$\begin{aligned}A[1](\lambda)&=-iN^{-1}\left(I+\frac{\nu-\mu}{\lambda-\nu}P\right)\frac{\mu-\nu}{\lambda-\mu}\dot{P}N\\&+N^{-1}\left(I+\frac{\nu-\mu}{\lambda-\nu}P\right)A(\lambda)\left(I+\frac{\mu-\nu}{\lambda-\mu}P\right)N-iN^{-1}\dot{N}\,.\end{aligned} \tag{43}$$

Now, let us note that

$$\frac{(\mu-\nu)}{(\lambda-\mu)(\lambda-\nu)}\equiv\frac{1}{\lambda-\mu}-\frac{1}{\lambda-\nu} \tag{44}$$

and denote $P_\perp:=I-P$, which yields

$$\begin{aligned}A[1](\lambda)&=-iN^{-1}\dot{N}+N^{-1}A(\lambda)N\\&+(\mu-\nu)N^{-1}P_\perp\left(\frac{A(\lambda)-A(\mu)}{\lambda-\mu}\right)PN+\frac{\mu-\nu}{\lambda-\mu}N^{-1}\left(-i\dot{P}P+P_\perp A(\mu)P\right)N\\&+(\nu-\mu)N^{-1}P\left(\frac{(A(\lambda)-A(\nu)}{\lambda-\nu}\right)P_\perp N+\frac{\nu-\mu}{\lambda-\nu}N^{-1}\left(iP\dot{P}+PA(\nu)P_\perp\right)N\,.\end{aligned} \tag{45}$$

Darboux covariance of $A(\lambda)$ means that the poles at $\lambda=\mu$ and $\lambda=\nu$ vanish, i.e.

$$\begin{aligned}-i\dot{P}P+P_\perp A(\mu)P&=0\,,\\iP\dot{P}+PA(\nu)P_\perp&=0\,.\end{aligned} \tag{46}$$

Subtracting these two equations we get

$$i\dot{P}=P_\perp A(\mu)P-PA(\nu)P_\perp\,. \tag{47}$$

Similarly, considering $X[1]$, we get the second condition for P:

$$P_\perp X(\mu)P - PX(\nu)P_\perp = 0 \ . \tag{48}$$

Note, that multiplying (47) by P from the left and from the right, respectively, we obtain the equations (46). Therefore the system (46) is equivalent to the equation (47). Any solution of the equation (47) defines a Darboux matrix.

Note that there is no constraint on the normalization matrix N and we can put, for instance, $N = I$ (the canonical normalization). Therefore Darboux matrices of the form (41) can satisfy assumptions of Theorem 1.

Suppose that we know the fundamental (invertible) solution Ψ of the linear system (1) for any λ (in the finite dimensional case the fundamental solution is an non-degenerate matrix and its columns are linearly independent vector solutions). Then the solution of (47) can be constructed in a standard way [8,10]:

$$\begin{aligned} \ker P &= \Psi(\mu)V_{ker} \ , \\ \mathrm{im} P &= \Psi(\nu)V_{im} \ , \end{aligned} \tag{49}$$

where V_{ker} and V_{im} are any constant linear spaces such that $V_{ker} \oplus V_{im} = V$. Knowing the image and the kernel we can construct explicitly the corresponding projector (see, for instance, Ref. [7]). However the iteration of two or more Darboux transformations is usually quite cumbersome. In this context we suggest a new method which seems to be more convenient [12,13].

6 Conclusions

The approach presented in this paper is quite general and gives a lot of possibilities to construct integrable equations. The paper presents just a starting point of the research (e.g., further generalizations can be easily constructed). We reconstructed in a more elegant way families of integrable equations introduced by Ustinov and Czachor. The interesting question is to study in our framework other examples of physically interesting integrable systems.

Acknowledgements

I am grateful to Marek Czachor for turning my attention to the interesting class of Darboux-invariant equations of von Neumann type and for helpful discussions.

References

1. S.B.Leble, M.Czachor, "Darboux-integrable nonlinear Liouville-von Neumann equation", *Phys. Rev.* E **58**, 7091-7100 (1998).
2. N.V.Ustinov, M.Czachor, M.Kuna, S.B.Leble, "Darboux integration of $i\dot{\rho} = [H, f(\rho)]$", *Phys. Lett.* A **279**, 333-340 (2001).
3. N.V.Ustinov, M.Czachor, "New class of integrable nonlinear von Neumann-type equations", nlin.SI/0011013; see contribution in this volume.
4. M.Kuna, M.Czachor, S.B.Leble, "Nonlinear von Neumann-type equations, Darboux invariance and spectra", *Phys. Lett.* A **255**, 42-48 (1999).
5. M.Czachor, J.Naudts, "Microscopic foundation of nonextensive statistics", *Phys. Rev.* **E 59**, R2497-R2500 (1999).
6. V.B.Matveev, M.A.Salle, *Darboux Transformations and Solitons*, Springer-Verlag, Berlin-Heidelberg (1991).
7. J.Cieśliński, "An algebraic method to construct the Darboux matrix", *J.Math.Phys.* **36**, 5670-5706 (1995).
8. V.E.Zakharov, A.B.Shabat, "Integration of nonlinear equations of mathematical physics by the inverse scattering method. II", *Funk. Anal. Pril.* **13**, 13-22 (1979) [in Russian].
9. V.E.Zakharov, S.V.Manakov, S.P.Novikov, L.P.Pitaievsky, *Theory of solitons*, Nauka, Moscow 1980 [in Russian], Consultants Bureau, New York 1984.
10. D.Levi, O.Ragnisco, M.Bruschi, "Extension of the Zakharov-Shabat Generalized Inverse Method to Solve Differential-Difference and Difference-Difference Equations", *Nuovo Cim.* A **58**, 56-66 (1980).
11. S.P.Burtsev, V.E.Zakharov, A.V.Mikhailov, "Inverse scattering method with variable spectral parameter", *Teor. Mat. Fiz.* **70**, 323-341 (1987) [in Russian].
12. J.L.Cieśliński, "The construction of the Darboux-Bäcklund transformation without using a matrix representation", *J. Phys.* A **33**, L363-L368 (2000).
13. W.Biernacki, J.L.Cieśliński, "A compact form of the Darboux-Bäcklund transformation for some spectral problems in Clifford algebras", *Phys. Lett.* A **288**, 167-172 (2001).

DARBOUX-INTEGRABLE EQUATIONS WITH NON-ABELIAN NONLINEARITIES

NIKOLAI V. USTINOV

Theoretical Physics Department,
Kaliningrad State University, Al. Nevsky street 14, 236041,
Kaliningrad, Russia
E-mail: n_ustinov@mail.ru

Katedra Fizyki Teoretycznej i Metod Matematycznych,
Politechnika Gdańska, ul. Narutowicza 11/12,
80-952 Gdańsk, Poland

MAREK CZACHOR

Katedra Fizyki Teoretycznej i Metod Matematycznych,
Politechnika Gdańska, ul. Narutowicza 11/12,
80-952 Gdańsk, Poland

Department of Physics, Technische Universität Clausthal
38678 Clausthal-Zellerfeld, Germany
E-mail: mczachor@pg.gda.pl

We introduce a new class of nonlinear equations admitting a representation in terms of Darboux-covariant compatibility conditions. Their special cases are, in particular, (i) the "general" von Neumann equation $i\dot{\rho} = [H, f(\rho)]$, with $[f(\rho), \rho] = 0$, (ii) its generalization involving certain functions $f(\rho)$ which are non-Abelian in the sense that $[f(\rho), \rho] \neq 0$, and (iii) the Nahm equations.

1 Introduction

An investigation of collective phenomena in quantum mechanics leads to various nonlinear evolution equations. Nonlinear equation of a Schrödinger type was derived as a phenomenological equation for the order parameter in superfluid He_4 [1,2]. Recent experiments on Bose-Einstein condensation [3] significantly raise an interest in nonlinear generalizations of the Schrödinger equation (for a review see [4]). Another kind of nonlinear Schrödinger equations, a by-product of work on classification of groups of diffeomorphisms [5], was recently related to certain aspects of D-brane dynamics [6].

In more realistic situations, where entanglement between interacting particles is properly taken into account, one does not arrive at nonlinear Schrödinger wave equations but rather at their density matrix (von Neumann-type) nonlinear versions [7]

$$-i\dot{X} = [X, h(X)]. \tag{1}$$

In this case the Hamiltonian $h(X)$ is considered as a "non-Abelian function" of the density operator X. As is well known, equations analogous to (1) are often encountered in quantum optics and field theory if one deals with the Heisenberg-picture evolution of observables.

Still another class of nonlinear von Neumann type equations may be derived in dissipative contexts [8,9] or on the basis of various entropic variational principles [10,11].

Of some interest is the fact that for a special class of $h(X)$ Eq. (1) can be rewritten as

$$i\dot{X} = [H, f(X)]. \tag{2}$$

Nonlinear equations of this type appeared in the frameworks of nonlinear Nambu-type theories [12] and nonextensive statistics [13]. It should be stressed that $f(X)$ does not always take the usual form known from spectral theory [14]. We will refer to the equations of the general form (1), (2) as nonlinear equations of the von Neumann type. Eq. (2) acquires an additional fundamental flavor if one recalls that for $X = |\psi\rangle\langle\psi|$ and for *all* functions constructed via the spectral theorem, which satisfy $f(0) = 0$ and $f(1) = 1$, one finds $f(X) = X$ and therefore the dynamics of X is equivalent to the *linear* Schrödinger equation.

Similar nonlinear equations can be found also in classical theories. The best known physical example is the Euler equation for a freely rotating rigid body

$$\dot{X} = [H, X^2]. \tag{3}$$

The more abstract versions are related to the Euler-Arnold equation for an "N-dimensional rigid body" [15], the Lie-Poisson equations occurring in fluid dynamics [15,16], and the N-wave equations for electromagnetic waves in nonlinear media [17].

Particularly interesting and in recent years very intensively investigated class of nonlinear equations are the Nahm equations [18]. Their solutions are used as an intermediate step in construction of non-Abelian monopoles. One family of solutions is in a one-to-one relationship to the Euler rigid-body equations. In this sense the Nahm equations may be regarded as a kind of generalized von Neumann equations.

Finally, quite recently nonlinear equations on free associative algebras, including the ones of the form (1), (2) with $h(X)$ and $f(X)$ being (noncommutative) polynomials, were considered in the framework of the symmetry approach to classification of integrable ordinary differential equations [19]. It was found in particular that Eq. (2), where $f(X) = iX^3$, is symmetry for Eq. (3). A class of equations discussed in [19] was termed "non-C-integrable". Below we show that some of them belong to our class of integrable equations with Darboux covariant Lax representation. It is worth mentioning that we do not assume a polynomial structure of the RHS of Eqs. (1), (2).

As we can see, the reasons for generalizations of the linear von Neumann equation may be different, but they all finally lead to the same fundamental difficulty: The resulting equations involve a large (often infinite) number of degrees of freedom and effective integration procedures are difficult to find. The situation is additionally complicated by constraints typical of density matrices or Hamiltonians.

It was only recently that soliton methods were applied to the density matrix version of (2) [20,21,22]. The progress was made possible by the observation that there exist Darboux-covariant Lax representations of certain von Neumann-type equations. The technique used in [20,21,22] is an appropriate modification of the dressing method [23,24,25,26,27] or rather of its analogue constructed via a binary Darboux transformation [28,29,30]. The technique is called the Darboux transformation since the construction of generalized gauge transformation is performed with the help of additional solutions of the Lax pairs.

The Darboux-type method of integrating the density-matrix analogue of Eq. (3), introduced in [20] and further generalized in [22], led to discovery of the so-called self-scattering solutions. The process of self-scattering continuously interpolates between two asymptotically linear evolutions. This very characteristic property was found in all the nontrivial solutions of nonlinear von Neumann equation obtained by the above technique.

The paper presents further development of our results. All the equations we derive can be regarded as compatibility conditions for Darboux-covariant Lax pairs. Previously discussed nonlinear von Neumann equations as well as the Nahm equations are particular examples of the class under consideration, but they form just a tip of an iceberg.

Darboux covariance of Lax pairs is proved in detail along the lines of [31]. An alternative proof, involving a more general class of Darboux transformations, is given in this volume in [32]. The two constructions are based on different mathematical techniques, show different aspects of the same problem, and it is not completely clear whether they are entirely equivalent.

The layout of the paper is as follows. The compatibility-condition representation of a family of nonlinear equations of the von Neumann type is given in Sec. II. The compatibility conditions are brought into a closed form by a special choice of operator coefficients of the Lax pair. The coefficients are defined in terms of additional functions satisfying restrictions following from the compatibility conditions. A wide class of such functions is proposed. Examples of nonlinear equations of the von Neumann type that are generated by some of these functions are given in Sec. III. Darboux covariance of the Lax pairs with the operator-valued coefficients is proved in Sec. IV. In the next section we show that the restrictions carrying the compatibility condition into integrable nonlinear von Neumann equations are Darboux-covariant if the functions introduced in Sec. II are used.

2 Darboux-Integrable Equations

We begin with the overdetermined system of linear equations (the Lax pair)

$$\begin{cases} -i\dot{\psi} = \psi A(\lambda) \\ z_\lambda \psi = \psi H(\lambda) \end{cases}, \tag{4}$$

where λ and z_λ are complex numbers, ψ is an element of a linear space $\mathbf{L}$, $A(\lambda)$ and $H(\lambda)$ are linear operators $\mathbf{L} \mapsto \mathbf{L}$ belonging to an associative ring, the dot denotes a derivative (i.e. an operator satisfying the Leibnitz rule). The compatibility condition for the Lax pair is

$$i\dot{H}(\lambda) = [A(\lambda), H(\lambda)]. \tag{5}$$

Assume the operators entering the Lax pair are rational functions of λ with operator coefficients

$$A(\lambda) = \sum_{k=0}^{L} \lambda^k B_k + \sum_{k=1}^{M} \frac{1}{\lambda^k} C_k, \tag{6}$$

$$H(\lambda) = \sum_{k=0}^{N} \lambda^k H_k. \tag{7}$$

The compatibility condition implies two sets of relations between operators B_k, C_k and H_k

$$\sum_{k=\max\{0,m-L\}}^{N} [H_k, B_{m-k}] = 0 \qquad (N < m \le L + N), \tag{8}$$

$$\sum_{k=0}^{\min\{N,m+M\}} [H_k, C_{k-m}] = 0 \qquad (-M \le m < 0) \tag{9}$$

and the system of differential equations

$$-i\dot{H}_m = \sum_{k=\max\{0,m-L\}}^{m} [H_k, B_{m-k}] + \sum_{k=m+1}^{\min\{N,m+M\}} [H_k, C_{k-m}] \tag{10}$$

for $(0 \leq m \leq N)$. In order to reduce Eqs. (10) to equations of the von Neumann type one needs to write them in a closed form. In general, Eqs. (8), (9) are inconvenient for defining operators B_k and C_k in terms of H_k. Nevertheless, one can express B_k and C_k explicitly through operator H_k by imposing on them some additional relations which obey Eqs. (8), (9). It is clear that not all such additional relations have to be consistent with the requirement of Darboux-covariance of the Lax pair.

Consider

$$B_k = \frac{1}{(L-k)!}\left(\frac{d^{L-k}}{d\varsigma^{L-k}} f(\varsigma^N H(\varsigma^{-1}), \varsigma^{-1})\right)\Big|_{\varsigma=0}, \tag{11}$$

$$C_k = \frac{1}{(M-k)!}\left(\frac{d^{M-k}}{d\varepsilon^{M-k}} g(H(\varepsilon), \varepsilon)\right)\Big|_{\varepsilon=0}, \tag{12}$$

where $f(X, \lambda)$ and $g(X, \lambda)$ are properly defined functions of operator X and parameter λ. The operator at the RHS of the first equation of the Lax pair now reads

$$\begin{aligned} A(\lambda) = &\sum_{k=0}^{L} \frac{\lambda^{L-k}}{k!}\left(\frac{d^k}{d\varsigma^k} f(\varsigma^N H(\varsigma^{-1}), \varsigma^{-1})\right)\Big|_{\varsigma=0} \\ &+ \sum_{k=0}^{M-1} \frac{\lambda^{k-M}}{k!}\left(\frac{d^k}{d\varepsilon^k} g(H(\varepsilon), \varepsilon)\right)\Big|_{\varepsilon=0}. \end{aligned} \tag{13}$$

There exists a large class of functions $f(X, \lambda)$ and $g(X, \lambda)$ that results in operators B_k and C_k *identically* satisfying conditions (8) and (9). The class is defined by

$$[f(X(\lambda), \lambda), X(\lambda)] = [g(X(\lambda), \lambda), X(\lambda)] = 0. \tag{14}$$

To prove the covariance of Eqs. (11), (12) under the binary Darboux transformation we also assume that these functions possess an additional property, namely they are covariant with respect to the similarity transformation:

$$f(TXT^{-1}, \lambda) = Tf(X, \lambda)T^{-1}, \quad g(TXT^{-1}, \lambda) = Tg(X, \lambda)T^{-1}, \tag{15}$$

where T is a transformation. The above conditions are satisfied, for example, by polynomials in X and sums of negative powers of polynomials in X. If X is selfadjoint, the same is valid for all $f(X)$ defined via the spectral theorem.

In such a case Eqs. (8) and (9) turn out to be identically fulfilled as a consequence of the trivial identities

$$\frac{d^n}{d\lambda^n}[f(H(\lambda),\lambda),H(\lambda)]\Big|_{\lambda=0}\equiv 0,\quad \frac{d^n}{d\lambda^n}[g(H(\lambda),\lambda),H(\lambda)]\Big|_{\lambda=0}\equiv 0$$

and Eq. (10) can be written in equivalent form

$$i\dot{H}_m=\sum_{k=m+1}^{N}[H_k,B_{m-k}]+\sum_{k=0}^{m}[H_k,C_{k-m}]\quad(0\le m\le N).\tag{16}$$

In the next section we will see that there are two representations of the compatibility condition and they correspond to Eqs. (1) and (2).

3 Examples

Below we present a few examples of integrable nonlinear von Neumann-type equations that correspond to different choices of positive integers N, L, M and functions $f(X,\lambda)$, $g(X,\lambda)$. In what follows we will use the notation

$$H_1=H,\quad H_0=\rho.$$

If $N=1$ Eqs. (10) imply

$$\dot{H}=0.$$

3.1

$N=1$, $L=1$, $f(X,\lambda)=X^n$ $(n\in\mathbf{N})$, $g(X,\lambda)=0$.

The compatibility condition gives the equation

$$i\dot{\rho}=\left[\sum_{k=0}^{n-1}H^{n-k-1}\rho H^k,\rho\right].\tag{17}$$

The Darboux-covariant Lax pair for this equation was found in [21]. For $n=2$ Eq. (17) reads

$$i\dot{\rho}=[\rho H+H\rho,\rho]=[H,\rho^2],\tag{18}$$

which is equivalent to Eq. (3). Mutual replacement of $H(\lambda)=\rho+\lambda H$ and $A(\lambda)=H\rho+\rho H+\lambda H^2$ in the corresponding Lax pair results in the compatibility condition

$$i(H\dot{\rho}+\dot{\rho}H)=[H,\rho^2],$$

which is essentially a form of Euler's top equations given in [15,33].

3.2

$N=1,\ f(X,\lambda)=0,\ M=1,\ g(X,\lambda)=g(X)$.

Here we have

$$i\dot{\rho}=[g(\rho),H]\,.$$

The Lax-pair representation and Darboux covariance properties of this equation have already been established in [22]. It should be stressed that the function $g(X)$ is basically arbitrary. The cases $g(\rho)=i\rho^3$ and $g(\rho)=i\rho^{-1}$ were considered in [19].

3.3

$N=1,\ L=3,\ f(X,\lambda)=\lambda^{-2}(a_0X^2+(b_0+\lambda b_1)X^3+(c_0+\lambda c_1+\lambda^2c_2)X^4)$, $g(X,\lambda)=0$.

The compatibility condition becomes

$$i\dot{\rho}=[h(\rho),\rho]=[H,F(\rho)], \tag{19}$$

where

$$\begin{aligned}h(\rho)&=a_0(\rho H+H\rho)+b_0(\rho H^2+H\rho H+H^2\rho)+b_1(\rho^2H+\rho H\rho+H\rho^2)\\&\quad+c_0(\rho H^3+H\rho H^2+H^2\rho H+H^3\rho)\\&\quad+c_1(\rho^2H^2+\rho H\rho H+\rho H^2\rho+H\rho^2H+H\rho H\rho+H^2\rho^2)\\&\quad+c_2(\rho^3H+\rho^2H\rho+\rho H\rho^2+H\rho^3),\\F(\rho)&=a_0\rho^2+b_0(\rho^2H+\rho H\rho+H\rho^2)+b_1\rho^3\\&\quad+c_0(\rho^2H^2+\rho H\rho H+\rho H^2\rho+H\rho^2H+H\rho H\rho+H^2\rho^2)\\&\quad+c_1(\rho^3H+\rho^2H\rho+\rho H\rho^2+H\rho^3)+c_2\rho^4.\end{aligned}$$

Here a_0, b_0, b_1, c_0, c_1, c_2 are arbitrary complex parameters independent of λ. If the dot is a derivative with respect to a time variable t, they can depend on t. The same is also valid for the next example. Let us note that the map $\rho\mapsto h(\rho)$ is not a function of ρ in the standard sense of the spectral theory [14] (such as $g(\rho)$ of the previous subsection). In particular, $[h(\rho),\rho]\neq 0$. We refer to such maps as *non-Abelian functions*, or *non-Abelian nonlinearities*.

3.4

$N=1,\ f(X,\lambda)=0,\ M=2$,
$g(X,\lambda)=(a_0+\lambda a_1)\left((b_0+\lambda b_1)\mathbf{1}+X\right)^{-1}+(c_0+\lambda c_1)\left((d_0+\lambda d_1)\mathbf{1}+X\right)^{-1}\left((e_0+\lambda e_1)\mathbf{1}+X\right)^{-1}$.

In this case we obtain

$$i\dot{\rho} = [H, F(\rho)], \tag{20}$$

where

$$\begin{aligned} F(\rho) = a_0(b_0\mathbf{1}+\rho)^{-1}(b_1\mathbf{1}+H)(b_0\mathbf{1}+\rho)^{-1} - a_1(b_0\mathbf{1}+\rho)^{-1} \\ +c_0((d_0\mathbf{1}+\rho)^{-1}(d_1\mathbf{1}+H)(d_0\mathbf{1}+\rho)^{-1}(e_0\mathbf{1}+\rho)^{-1} \\ +(d_0\mathbf{1}+\rho)^{-1}(e_0\mathbf{1}+\rho)^{-1}(e_1\mathbf{1}+H)(e_0\mathbf{1}+\rho)^{-1}) \\ -c_1(d_0\mathbf{1}+\rho)^{-1}(e_0\mathbf{1}+\rho)^{-1}. \end{aligned} \tag{21}$$

As opposed to the previous examples $F(\rho)$ is a non-Abelian nonpolynomial function.

For $N = 1$ the nonlinear equations involve only two types of operators: ρ and H. Increasing N we can introduce non-Abelian nonlinearities involving an arbitrary number of different operators.

3.5

$N = 2$, $L = 2$, $f(X, \lambda) = X^2$, $g(X, \lambda) = 0$.

This is the simplest example of $N = 2$ nonlinearity. The compatibility conditions are

$$\begin{aligned} i\dot{\rho} &= [H^2, \rho] + [H_2, \rho^2] = [H^2 + H_2\rho + \rho H_2, \rho], \\ i\dot{H} &= [H_2, H\rho + \rho H], \\ \dot{H}_2 &= 0. \end{aligned} \tag{22}$$

This system is equivalent to a nonlinear von Neumann equation with two types of nonlinearity: One given in an implicit form and the other of the Euler type.

3.6

$N = 2$, $L = 1$, $f(X, \lambda) = X$, $g(X, \lambda) = 0$.

The Lax pair is

$$z_\lambda \psi_\lambda = \psi_\lambda (H_0 + \lambda H_1 + \lambda^2 H_2), \tag{23}$$

$$-i\dot{\psi}_\lambda = \psi_\lambda (H_1 + \lambda H_2) \tag{24}$$

with the compatibility conditions

$$\dot{H}_2 = 0, \tag{25}$$

$$i\dot{H}_1 = [H_2, H_0], \tag{26}$$

$$i\dot{H}_0 = [H_1, H_0]. \tag{27}$$

Defining

$$F_1 = (H_0 - H_2)/(2i), \tag{28}$$
$$F_2 = (H_0 + H_2)/2, \tag{29}$$
$$F_3 = H_1/(2i) \tag{30}$$

and the connection $\nabla f = \dot{f} + [f, F_3]$ we can write the compatibility conditions as

$$\nabla F_1 = i[F_2, F_3], \tag{31}$$
$$\nabla F_2 = i[F_3, F_1], \tag{32}$$
$$\nabla F_3 = i[F_1, F_2]. \tag{33}$$

The connection can be trivialized if we find an invertible solution ξ of the linear problem

$$\dot{\xi} = -\xi F_3. \tag{34}$$

Then

$$f_k = \xi F_k \xi^{-1} \tag{35}$$

satisfies the standard Nahm equations

$$\dot{f}_1 = i[f_2, f_3], \tag{36}$$
$$\dot{f}_2 = i[f_3, f_1], \tag{37}$$
$$\dot{f}_3 = i[f_1, f_2]. \tag{38}$$

4 Binary Darboux Transformation

The first step towards extending the technique of Darboux transformations to integrable nonlinear von Neumann-type equations on associative rings is to establish the Darboux covariance of the Lax pair (4) without any additional constraints. In this section we show that an appropriate formulation of the binary Darboux transformation makes the Lax pair covariant.

Assume χ is a solution of the Lax pair with parameter ν:

$$\begin{cases} -i\dot{\chi} = \chi A(\nu) \\ z_\nu \chi = \chi H(\nu) \end{cases} \tag{39}$$

and φ is a solution of the dual Lax pair with parameter μ:

$$\begin{cases} i\dot{\varphi} = A(\mu)\varphi \\ z_\mu \varphi = H(\mu)\varphi \end{cases} . \tag{40}$$

We further suppose that these systems can be related with an operator P satisfying $P^2 = P$ and

$$-i\dot{P} = PA(\nu)P_{\perp} - P_{\perp}A(\mu)P, \tag{41}$$

where $P_{\perp} = \mathbf{1} - P$. The above assumptions are fulfilled, for example, if $\chi = \langle\chi|$ and $\varphi = |\varphi\rangle$ are some "bra" and "ket" associated with a Hilbert space. Then

$$P = \frac{|\varphi\rangle\langle\chi|}{\langle\chi|\varphi\rangle}.$$

Another example is provided by $m \times n$ matrix χ and $n \times m$ matrix φ. P is then defined by

$$P = \varphi(\chi\varphi)^{-1}\chi.$$

Some realizations of Darboux transformations in infinite dimensional cases were given in [34]. Very recently a new construction of the Darboux transformation in terms of Clifford numbers was described in [35].

Defining

$$\psi[1] = \psi D_{\lambda}, \tag{42}$$

$$D_{\lambda} = \mathbf{1} + \frac{\nu - \mu}{\mu - \lambda}P \tag{43}$$

we come to the following

Theorem 1. *The Lax pair* (4) *with the coefficients defined by Eqs.* (6), (7) *is covariant with respect to the binary Darboux transformation* $\{\psi, A(\lambda), H(\lambda)\} \to \{\psi[1], A(\lambda)[1], H(\lambda)[1]\}$, *where*

$$A(\lambda)[1] = \sum_{k=0}^{L}\lambda^k B_k[1] + \sum_{k=1}^{M}\frac{1}{\lambda^k}C_k[1], \tag{44}$$

$$B_k[1] = B_k + (\mu - \nu)\sum_{m=k+1}^{L}\left(\mu^{m-k-1}P_{\perp}B_mP - \nu^{m-k-1}PB_mP_{\perp}\right) \tag{45}$$

$$C_k[1] = C_k - (\mu - \nu)\sum_{m=k}^{M}\left(\mu^{k-m-1}P_{\perp}C_mP - \nu^{k-m-1}PC_mP_{\perp}\right) \tag{46}$$

and

$$H(\lambda)[1] = \sum_{k=0}^{N}\lambda^k H_k[1], \tag{47}$$

$$H_k[1] = H_k$$

$$+(\mu-\nu)\sum_{m=k+1}^{N}\left(\mu^{m-k-1}P_{\perp}H_mP-\nu^{m-k-1}PH_mP_{\perp}\right) \quad (48)$$

Proof: The condition of covariance of the second equation of the Lax pair with respect to the transformation yields

$$\begin{aligned}
H(\lambda)[1] &= D_{\lambda}^{-1}H(\lambda)D_{\lambda} \qquad (49)\\
&= \left(\mathbf{1}+\frac{\mu-\nu}{\nu-\lambda}P\right)\sum_{k=0}^{N}\lambda^kH_k\left(\mathbf{1}+\frac{\nu-\mu}{\mu-\lambda}P\right)\\
&= \sum_{k=0}^{N}\lambda^kH_k+\frac{\mu-\nu}{\nu-\lambda}\sum_{k=0}^{N}\lambda^kPH_kP_{\perp}+\frac{\nu-\mu}{\mu-\lambda}\sum_{k=0}^{N}\lambda^kP_{\perp}H_kP.
\end{aligned}$$

Taking into account

$$PH(\nu)P_{\perp}=P_{\perp}H(\mu)P=0$$

we are able to rewrite the previous expression in the following manner

$$\begin{aligned}
H(\lambda)[1] &= \sum_{k=0}^{N}\lambda^kH_k+\frac{\mu-\nu}{\nu-\lambda}\sum_{k=0}^{N}(\lambda^k-\nu^k)PH_kP_{\perp}\\
&\quad+\frac{\nu-\mu}{\mu-\lambda}\sum_{k=0}^{N}(\lambda^k-\mu^k)P_{\perp}H_kP\\
&= \sum_{k=0}^{N}\lambda^kH_k+(\nu-\mu)\sum_{k=1}^{N}\sum_{j=0}^{k-1}\lambda^{k-j-1}\nu^jPH_kP_{\perp}\\
&\quad+(\mu-\nu)\sum_{k=1}^{N}\sum_{j=0}^{k-1}\lambda^{k-j-1}\mu^jP_{\perp}H_kP, \qquad (50)
\end{aligned}$$

which is equivalent to Eq. (47).

From the condition of the Darboux covariance of the first equation of the Lax pair we have

$$A(\lambda)[1]=D_{\lambda}^{-1}A(\lambda)D_{\lambda}-iD_{\lambda}^{-1}\dot{D}_{\lambda}.$$

Substitution of Eqs. (41), (43) gives

$$\begin{aligned}
A(\lambda)[1] &= \left(\mathbf{1}+\frac{\mu-\nu}{\nu-\lambda}P\right)A(\lambda)\left(\mathbf{1}+\frac{\nu-\mu}{\mu-\lambda}P\right)\\
&\quad+\frac{\nu-\mu}{\mu-\lambda}\left(\mathbf{1}+\frac{\mu-\nu}{\nu-\lambda}P\right)\left(PA(\nu)P_{\perp}-P_{\perp}A(\mu)P\right)
\end{aligned}$$

$$= A(\lambda) + \frac{\mu-\nu}{\nu-\lambda} P(A(\lambda) - A(\nu))P_\perp + \frac{\nu-\mu}{\mu-\lambda} P_\perp (A(\lambda) - A(\mu))P$$

$$= \sum_{k=0}^{L} \lambda^k B_k + \sum_{k=1}^{M} \frac{1}{\lambda^k} C_k$$

$$+(\mu-\nu)P_\perp \left(\sum_{k=1}^{L} \sum_{j=0}^{k-1} \lambda^{k-j-1} \mu^j B_k - \sum_{k=1}^{M} \sum_{j=0}^{k-1} \lambda^{-j-1} \mu^{j-k} C_k \right) P$$

$$+(\nu-\mu)P \left(\sum_{k=1}^{L} \sum_{j=0}^{k-1} \lambda^{k-j-1} \nu^j B_k - \sum_{k=1}^{M} \sum_{j=0}^{k-1} \lambda^{-j-1} \nu^{j-k} C_k \right) P_\perp .$$

The final expression is (44). □

5 The Main Theorem

Theorem 1 establishes Darboux covariance of Lax pairs involving operators with positive or negative powers of spectral parameters. The compatibility condition is also covariant: Transformed operators $B_k[1]$, $C_k[1]$ and $H_k[1]$ solve Eqs. (8), (9), and (10). However, if there are additional relations between the operators, they do not have to be Darboux covariant.

Theorem 2. *The relations* (11) *and* (12) *are Darboux covariant if Eqs.* (15) *are fulfilled.*

Proof: Let us check the Darboux covariance of Eq. (11), i.e.

$$B_k[1] = \frac{1}{(L-k)!} \left(\frac{d^{L-k}}{d\varsigma^{L-k}} f(\varsigma^N H(\varsigma^{-1})[1], \varsigma^{-1}) \right) \Big|_{\varsigma=0} .$$

It follows immediately that

$$B_L[1] = B_L$$

and

$$\frac{d^k}{d\varsigma^k} D_{1/\varsigma} \Big|_{\varsigma=0} = k!(\mu-\nu)\mu^{k-1} P,$$

$$\frac{d^k}{d\varsigma^k} D_{1/\varsigma}^{-1} \Big|_{\varsigma=0} = k!(\nu-\mu)\nu^{k-1} P.$$

For $k \neq L$ we have, using Eq. (49),

$$\frac{1}{(L-k)!} \left(\frac{d^{L-k}}{d\varsigma^{L-k}} f(\varsigma^N H(\varsigma^{-1})[1], \varsigma^{-1}) \right) \Big|_{\varsigma=0}$$

$$
\begin{aligned}
&= \frac{1}{(L-k)!}\left(\frac{d^{L-k}}{d\varsigma^{L-k}} f(\varsigma^N D_{1/\varsigma}^{-1} H(\varsigma^{-1}) D_{1/\varsigma}, \varsigma^{-1})\right)\bigg|_{\varsigma=0} \\
&= \frac{1}{(L-k)!}\left(\frac{d^{L-k}}{d\varsigma^{L-k}}\left(D_{1/\varsigma}^{-1} f(\varsigma^N H(\varsigma^{-1}), \varsigma^{-1}) D_{1/\varsigma}\right)\right)\bigg|_{\varsigma=0} \\
&= \frac{1}{(L-k)!}\sum_{a=0}^{L-k}\sum_{b=0}^{a}\frac{(L-k)!}{(L-k-a)!a!}\frac{a!}{(a-b)!b!} \\
&\times \left(\frac{d^{L-k-a}}{d\varsigma^{L-k-a}} D_{1/\varsigma}^{-1}\right)\bigg|_{\varsigma=0}\left(\frac{d^{a-b}}{d\varsigma^{a-b}} f(\varsigma^N H(\varsigma^{-1}), \varsigma^{-1})\right)\bigg|_{\varsigma=0}\left(\frac{d^b}{d\varsigma^b} D_{1/\varsigma}\right)\bigg|_{\varsigma=0} \\
&= \sum_{a=0}^{L-k}\sum_{b=0}^{a}\frac{1}{(L-k-a)!b!}\left(\frac{d^{L-k-a}}{d\varsigma^{L-k-a}} D_{1/\varsigma}^{-1}\right)\bigg|_{\varsigma=0} B_{L+b-a}\left(\frac{d^b}{d\varsigma^b} D_{1/\varsigma}\right)\bigg|_{\varsigma=0} \\
&= \sum_{a=1}^{L-k}\sum_{b=0}^{a}\frac{1}{(L-k-a)!b!}\left(\frac{d^{L-k-a}}{d\varsigma^{L-k-a}} D_{1/\varsigma}^{-1}\right)\bigg|_{\varsigma=0} B_{L+b-a}\left(\frac{d^b}{d\varsigma^b} D_{1/\varsigma}\right)\bigg|_{\varsigma=0} \\
&\quad + (\nu-\mu)\nu^{L-k-1} P B_L \\
&= \sum_{a=1}^{L-k}\sum_{b=1}^{a}\frac{(\mu-\nu)\mu^{b-1}}{(L-k-a)!}\left(\frac{d^{L-k-a}}{d\varsigma^{L-k-a}} D_{1/\varsigma}^{-1}\right)\bigg|_{\varsigma=0} B_{L+b-a} P \\
&\quad + \sum_{a=1}^{L-k}\frac{1}{(L-k-a)!}\left(\frac{d^{L-k-a}}{d\varsigma^{L-k-a}} D_{1/\varsigma}^{-1}\right)\bigg|_{\varsigma=0} B_{L-a} + (\nu-\mu)\nu^{L-k-1} P B_L \\
&= \sum_{a=1}^{L-k-1}\sum_{b=1}^{a}\frac{(\mu-\nu)\mu^{b-1}}{(L-k-a)!}\left(\frac{d^{L-k-a}}{d\varsigma^{L-k-a}} D_{1/\varsigma}^{-1}\right)\bigg|_{\varsigma=0} B_{L+b-a} P \\
&\quad + \sum_{b=1}^{L-k}(\mu-\nu)\mu^{b-1} B_{k+b} P \\
&\quad + \sum_{a=1}^{L-k-1}\frac{1}{(L-k-a)!}\left(\frac{d^{L-k-a}}{d\varsigma^{L-k-a}} D_{1/\varsigma}^{-1}\right)\bigg|_{\varsigma=0} B_{L-a} \\
&\quad + B_k + (\nu-\mu)\nu^{L-k-1} P B_L \\
&= \sum_{a=1}^{L-k-1}\sum_{b=1}^{a}(\mu-\nu)\mu^{b-1}(\nu-\mu)\nu^{L-k-a-1} P B_{L+b-a} P \\
&\quad + \sum_{b=1}^{L-k}(\mu-\nu)\mu^{b-1} B_{k+b} P
\end{aligned}
$$

$$+\sum_{a=1}^{L-k-1}(\nu-\mu)\nu^{L-k-a-1}PB_{L-a}+B_k+(\nu-\mu)\nu^{L-k-1}PB_L$$

$$=B_k+(\mu-\nu)\left(\sum_{b=1}^{L-k}\mu^{b-1}B_{k+b}P-\sum_{a=0}^{L-k-1}\nu^{L-k-a-1}PB_{L-a}\right.$$

$$\left.+(\nu-\mu)\sum_{a=1}^{L-k-1}\sum_{b=1}^{a}\mu^{b-1}\nu^{L-k-a-1}PB_{L+b-a}P\right)$$

$$=B_k+(\mu-\nu)\left(\sum_{b=1}^{L-k}\mu^{b-1}P_\perp B_{k+b}P-\sum_{a=0}^{L-k-1}\nu^{L-k-a-1}PB_{L-a}P_\perp+\delta_k\right),$$

where

$$\delta_k=\sum_{b=1}^{L-k}\mu^{b-1}PB_{k+b}P-\sum_{a=0}^{L-k-1}\nu^{L-k-a-1}PB_{L-a}P$$

$$+(\nu-\mu)\sum_{a=1}^{L-k-1}\sum_{b=1}^{a}\mu^{b-1}\nu^{L-k-a-1}PB_{L+b-a}P$$

$$=\sum_{b=1}^{L-k}\mu^{b-1}PB_{k+b}P-\sum_{a=0}^{L-k-1}\nu^{L-k-a-1}PB_{L-a}P$$

$$+\sum_{a=1}^{L-k-1}\sum_{b=1}^{a}\mu^{b-1}\nu^{L-k-a}PB_{L+b-a}P$$

$$-\sum_{a=1}^{L-k-1}\sum_{b=1}^{a}\mu^{b}\nu^{L-k-a-1}PB_{L+b-a}P.$$

Combining, respectively, the first and the third, the second and the fourth terms gives

$$\delta_k=\sum_{a=1}^{L-k}\sum_{b=1}^{a}\mu^{b-1}\nu^{L-k-a}PB_{L+b-a}P-\sum_{a=0}^{L-k-1}\sum_{b=0}^{a}\mu^{b}\nu^{L-k-a-1}PB_{L+b-a}P\equiv 0.$$

Finally, we obtain

$$\frac{1}{(L-k)!}\left(\frac{d^{L-k}}{d\varsigma^{L-k}}f(\varsigma^N H(\varsigma^{-1})[1],\varsigma^{-1})\right)\bigg|_{\varsigma=0}$$

$$=B_k+(\mu-\nu)\left(\sum_{m=1}^{L-k}\mu^{m-1}P_\perp B_{k+m}P-\sum_{m=0}^{L-k-1}\nu^{L-k-m-1}PB_{L-m}P_\perp\right).\quad(51)$$

The last expression coincides with Eq. (45).

Let us prove the Darboux covariance of Eq. (12):

$$C_k[1] = \frac{1}{(M-k)!}\left(\frac{d^{M-k}}{d\varepsilon^{M-k}} g(H(\varepsilon)[1],\varepsilon)\right)\bigg|_{\varepsilon=0}.$$

One can show that

$$\begin{aligned}
\frac{d^k}{d\varepsilon^k} D_\varepsilon\bigg|_{\varepsilon=0} &= \delta_{k0}\mathbf{1} + k!(\nu-\mu)\mu^{-k-1}P, \\
\frac{d^k}{d\varepsilon^k} D_\varepsilon^{-1}\bigg|_{\varepsilon=0} &= \delta_{k0}\mathbf{1} + k!(\mu-\nu)\nu^{-k-1}P.
\end{aligned} \tag{52}$$

Then

$$\begin{aligned}
&\frac{1}{(M-k)!}\left(\frac{d^{M-k}}{d\varepsilon^{M-k}} g(H(\varepsilon)[1],\varepsilon)\right)\bigg|_{\varepsilon=0} \\
&= \frac{1}{(M-k)!}\left(\frac{d^{M-k}}{d\varepsilon^{M-k}} g(D_\varepsilon^{-1}H(\varepsilon)D_\varepsilon,\varepsilon)\right)\bigg|_{\varepsilon=0} \\
&= \frac{1}{(M-k)!}\left(\frac{d^{M-k}}{d\varepsilon^{M-k}} \left(D_\varepsilon^{-1} g(H(\varepsilon),\varepsilon) D_\varepsilon\right)\right)\bigg|_{\varepsilon=0} \\
&= \frac{1}{(M-k)!}\sum_{a=0}^{M-k}\sum_{b=0}^{a} \frac{(M-k)!}{(M-k-a)!a!}\frac{a!}{(a-b)!b!} \\
&\times \left(\frac{d^{M-k-a}}{d\varepsilon^{M-k-a}} D_\varepsilon^{-1}\right)\bigg|_{\varepsilon=0} \left(\frac{d^{a-b}}{d\varepsilon^{a-b}} g(H(\varepsilon),\varepsilon)\right)\bigg|_{\varepsilon=0} \left(\frac{d^b}{d\varepsilon^b} D_\varepsilon\right)\bigg|_{\varepsilon=0} \\
&= \sum_{a=0}^{M-k}\sum_{b=0}^{a} \frac{1}{(M-k-a)!b!} \\
&\quad \times \left(\delta_{(M-k-a)0}\mathbf{1} + (M-k-a)!(\mu-\nu)\nu^{k+a-M-1a}P\right) \\
&\quad \times C_{M+b-a}\left(\delta_{b0}\mathbf{1} + b!(\nu-\mu)\mu^{-b-1}P\right) \\
&= \left(\mathbf{1} + \frac{\mu-\nu}{\nu}P\right) C_k \left(\mathbf{1} + \frac{\nu-\mu}{\mu}P\right) \\
&\quad + (\nu-\mu)\sum_{b=1}^{M-k} \mu^{-b-1}\left(\mathbf{1} + \frac{\mu-\nu}{\nu}P\right) C_{b+k}P \\
&\quad + (\mu-\nu)\sum_{a=0}^{M-k-1} \nu^{k+a-M-1} P C_{M-a}\left(\mathbf{1} + \frac{\nu-\mu}{\mu}P\right)
\end{aligned}$$

$$
- (\mu - \nu)^2 \sum_{a=1}^{M-k-1} \sum_{b=1}^{a} \nu^{k+a-M-1} \mu^{-b-1} P C_{M+b-a} P
$$

$$
= \left(\mathbf{1} + \frac{\mu - \nu}{\nu} P\right) C_k \left(\mathbf{1} + \frac{\nu - \mu}{\mu} P\right)
$$

$$
+(\mu - \nu) \left(\sum_{a=0}^{M-k-1} \nu^{k+a-M-1} P C_{M-a} P_\perp - \sum_{b=1}^{M-k} \mu^{-b-1} P_\perp C_{b+k} P + \Delta_k \right),
$$

where

$$
\Delta_k = \sum_{a=0}^{M-k-1} \nu^{k+a-M} \mu^{-1} P C_{M-a} P - \sum_{b=1}^{M-k} \mu^{-b} \nu^{-1} P C_{b+k} P
$$

$$
+ (\nu - \mu) \sum_{a=1}^{M-k-1} \sum_{b=1}^{a} \mu^{-b-1} \nu^{k+a-M-1} P C_{M+b-a} P
$$

$$
= \sum_{a=0}^{M-k-1} \nu^{k+a-M} \mu^{-1} P C_{M-a} P - \sum_{b=1}^{M-k} \mu^{-b} \nu^{-1} P C_{b+k} P
$$

$$
+ \sum_{a=1}^{M-k-1} \sum_{b=1}^{a} \mu^{-b-1} \nu^{k+a-M} P C_{M+b-a} P
$$

$$
- \sum_{a=1}^{M-k-1} \sum_{b=1}^{a} \mu^{-b} \nu^{k+a-M-1} P C_{M+b-a} P.
$$

Combining, respectively, the first and the third, the second and the fourth terms we obtain

$$
\Delta_k = \sum_{a=0}^{M-k-1} \sum_{b=0}^{a} \mu^{-b-1} \nu^{k+a-M} P C_{M+b-a} P
$$

$$
- \sum_{a=1}^{M-k} \sum_{b=1}^{a} \mu^{-b} \nu^{k+a-M-1} P C_{M+b-a} P \equiv 0.
$$

Finally, we have

$$
\frac{1}{(M-k)!} \left(\frac{d^{M-k}}{d\varepsilon^{M-k}} g(H(\varepsilon)[1], \varepsilon) \right) \bigg|_{\varepsilon=0}
$$

$$
= \left(\mathbf{1} + \frac{\mu - \nu}{\nu} P\right) C_k \left(\mathbf{1} + \frac{\nu - \mu}{\mu} P\right)
$$

$$+ (\mu - \nu) \left(\sum_{a=0}^{M-k-1} \nu^{k+a-M-1} P C_{M-a} P_\perp - \sum_{b=1}^{M-k} \mu^{-b-1} P_\perp C_{b+k} P \right)$$

$$= C_k + \frac{\mu - \nu}{\nu} P C_k P_\perp + \frac{\nu - \mu}{\mu} P_\perp C_k P$$

$$+ (\mu - \nu) \left(\sum_{a=0}^{M-k-1} \nu^{k+a-M-1} P C_{M-a} P_\perp - \sum_{b=1}^{M-k} \mu^{-b-1} P_\perp C_{b+k} P \right)$$

$$= C_k + (\mu - \nu) \left(\sum_{a=0}^{M-k} \nu^{k+a-M-1} P C_{M-a} P_\perp - \sum_{b=0}^{M-k} \mu^{-b-1} P_\perp C_{b+k} P \right).$$

The last expression coincides with Eq. (46). □

6 Conclusions

We have established the Darboux-covariance of a large class of nonlinear von Neumann-type equations. The next step is to employ this fact in construction of explicit solutions of such equations. Some classes of solutions have already been found in [20,22,36] for nonlinearities

$$i\dot{\rho} = [H, f(\rho)] \tag{53}$$

with $f(\rho) = \rho^2$ and $f(\rho) = \rho^q - 2\rho^{q-1}$ (q is an arbitrary real number). Both finite- and infinite-dimensional cases were treated by this technique in [22].

In a forthcoming paper we will describe other classes of solutions of the integrable equations we have introduced. It also seems that Lax pairs that allow us to reduce the compatibility conditions to nonlinear equations in a closed form can be still generalized. We hope in the future work to develop a description of integrable equations of the von Neumann type by taking into consideration the Mikhailov method of automorphisms [25]. This type of generalization is particularly important if reductions characteristic of Nahm-type equations are involved.

Acknowledgments

M.C. is indebted to Jan L. Cieśliński for his comments and, in particular, for the suggestion of using the similarity-transformation form of the Darboux transformation. The work of M.C. was supported by the Alexander von Humboldt Foundation and the KBN Grant 5 P03B 040 20. The work of N.V.U. was supported by Nokia-Poland.

References

1. E.P. Gross, *Nuovo Cim.* **20**, 454 (1961).
2. L.P. Pitaevskii, *Sov. Phys. JETP* **13**, 451 (1961).
3. M.H. Anderson *et al.*, *Science* **269**, 198 (1995); K.B. Davies *et al.*, *Phys. Rev. Lett.* **75**, 3969 (1995); D.S. Jin *et al.*, *Phys. Rev. Lett.* **77**, 420 (1996); **78**, 764 (1997); M.O. Mewes *et al.*, *Phys. Rev. Lett.* **77**, 988 (1996); D.M. Stamper-Kurn *et al.*, *Phys. Rev. Lett.* **81**, 500 (1998); M.R. Matthews *et al.*, *Phys. Rev. Lett.* **83**, 2498 (1999).
4. F. Dalforo, S. Georgini, L.P. Pitaevskii and S. Stringari, *Rev. Mod. Phys.* **71**, 463 (1999).
5. H.-D. Doebner and G.A. Goldin, *Phys. Lett.* A **162**, 397 (1992); G.A. Goldin, *Int. J. Mod. Phys.* **6B**, 1905 (1992); H.D. Doebner and G.A. Goldin, *J. Phys.* A **27**, 1771 (1992).
6. N.E. Mavromatos and R.J. Szabo, *Int. J. Mod. Phys.* A **16**, 209 (2001).
7. P. Ring and P. Schuck, *The Nuclear Many-Body Problem*, Springer, New York (1980).
8. J. Messer and B. Baumgartner, *Z. Phys.* B **32**, 103 (1978).
9. J. Korsch and H. Steffen, *J. Phys.* A **20**, 3787 (1987); M. Hensel and H.J. Korsch, *J. Phys.* A **25**, 2043 (1992)
10. P. Beretta, E. P. Gyftopoulos, J. L. Park and G. N. Hatsopoulos, *Nuovo Cim.* B **82**, 169 (1984).
11. S. Gheorghiu-Svirschevski, *Phys.Rev.* A **63**, 022105 (2001).
12. M. Czachor, *Phys. Lett.* A **225**, 1 (1997).
13. M. Czachor and J. Naudts, *Phys. Rev.* E **59**, R2497 (1999).
14. W. Thirring, *Lehrbuch der Mathematischen Physik*, vol. 3, Springer, Wien (1979). **37**, 117
15. V.I. Arnold, *Mathematical Methods of Classical Mechanics*, Springer, Berlin (1989).
16. J.E. Marsden and T. Ratiu, *Introduction to Mechanics and Symmetry*, Springer, New York (1994).
17. V.E. Zakharov and S.V. Manakov, *JETP* **69**, 1654 (1975).
18. W. Nahm, in *Monopoles in Quantum Field Theory*, World Scientific, Singapore (1982); N.J. Hitchin, *Commun. Math. Phys.* **89**, 145 (1983); N. A. Nekrasov, *Trieste Lectures on Solitons in Noncommutative Gauge Theories*, hep-th/0011095.
19. A.V. Mikhailov and V.V. Sokolov, *Theor. Math. Phys.* **122**, 72 (2000).
20. S.B. Leble and M. Czachor, *Phys. Rev.* E **58**, 7091 (1998).
21. M. Kuna, M. Czachor and S.B. Leble, *Phys. Lett.* A **255**, 42 (1999).
22. N.V. Ustinov, M. Czachor, M. Kuna, and S.B. Leble, *Phys. Lett.* A **279**,

333 (2001).

23. S.P. Novikov, S.V. Manakov, L.P. Pitaevski and V.E. Zakharov, *Theory of Solitons, the Inverse Scattering Method*, Consultants Bureau, New York (1984).
24. D. Levi, O. Ragnisco, and M. Bruschi, *Nuovo Cim.* A **58**, 56 (1980).
25. A.V. Mikhailov, *Physica* D **3**, 73 (1981).
26. G. Neugebauer and D. Kramer, *J. Phys.* A **16**, 1927 (1983).
27. J.L. Cieśliński, *J. Math. Phys.* **32**, 2395 (1991); ibid. **36**, 5670 (1995).
28. S.B. Leble and N.V. Ustinov, "Solitons of nonlinear equations associated with degenerate spectral problem of the third order", in: *Nonlinear Theory and its Applications*, eds. M. Tanaka and T. Saito, World Scientific, Singapore (1993) v.2, pp.547-550.
29. N.V. Ustinov, *J. Math. Phys.* **39**, 976 (1998).
30. S.B. Leble, *Computers Math. Applic.* **35**, 73 (1998).
31. N.V. Ustinov, M. Czachor, "New class of integrable nonlinear von Neumann-type equations", nlin.SI/0011013.
32. J. L. Cieśliński, "How to construct Darboux-invariant equations of von Neumann type", contribution in this volume.
33. A.S. Mishchenko, *Funct. Anal. Appl.* **4**, 232 (1970).
34. V.B. Matveev and M.A. Salle, *Darboux Transformations and Solitons*, Springer, Berlin (1991).
35. J.L. Cieśliński, *J. Phys.* A **33**, L363 (2000).
36. M. Czachor, M. Kuna, S.B. Leble and J. Naudts, "Nonlinear von Neumann-type equations", in *Trends in Quantum Mechanics*, eds. H.D. Doebner et al., World Scientific, Singapore (2000), pp.209-226.

DRESSING CHAIN EQUATIONS ASSOCIATED WITH DIFFERENCE SOLITON SYSTEMS

SERGIEI B. LEBLE
Katedra Fizyki Teoretycznej i Metod Matematycznych,
Politechnika Gdańska, ul. Narutowicza 11/12,
80-952 Gdańsk, Poland
E-mail: leble@mifgate.pg.gda.pl

Kaliningrad State University, ul. A. Nevskogo, 14,
236041 Kaliningrad, Russia.

Links of factorization theory, supersymmetry and Darboux transformations understood as isospectral deformations of difference operators are considered in the context of soliton theory. Dressing chain equations for factorizing operators of a spectral problem are derived. The chain equations yield nonlinear systems whose closure generates solutions of the equations themselves and of the nonlinear system if both operators of the corresponding Hirota bilinearization are covariant with respect to Darboux transformation. The latter defines a symmetry of the nonlinear system and the closed chains. Examples of Hirota and Nahm equations are discussed.

1 Introduction

Growing interest in discrete models appeals for a necessity of widening classes of symmetry structures of corresponding nonlinear problems [1]. Theories such as conformal field theory [2] or certain solvable lattice models [3,4] lead to equations which should be studied in this context.

Very recently a good basis for new investigations in the field of differential-difference and difference-difference equations was discovered [5] in the context of classical Darboux transformations (DT). Similarly to the differential operator case it has links to Hirota's bilinearization method [6] and to factorization theory [7] with similar possibilities of applications.

One of promising approaches to construction of solutions is based on the notion of dressing chains [10,11,12,13]. The approach covers soliton, rational, finite-gap and other new solutions within a universal scheme, reducing the problem to solving closed sets of nonlinear ordinary equations with a bi-Hamiltonian structure [13].

In this paper we reformulate the Darboux covariance theorem from [5] introducing a kind of difference Bell polynomials. Such polynomials have natural correspondence with (generalized) differential Bell polynomials in their non-Abelian version [7,8]. Their usage shortens transformation formulas and helps to apply the theory in complicated cases of joint covariance of U-V pairs [9,6,16].

The example of Hirota equation [14], connected with known applications, is studied in [5]. In this context we derive dressing chain equations and study their simplest solutions. In the last section we propose a lattice Lax pair for Nahm equations [15,18,1]. The Lax pair is covariant with respect to combined Darboux-gauge transformations that generate the dressing structure.

2 Darboux Transformations in Associative Ring with Automorphism

In this section we reformulate and analyze the results from [5] in the context of their further use in derivation of chain equations and joint covariance of operator pairs [9]. We begin with general notation. Let $\mathcal{R}$ be an associative ring with an automorphism, implying that there exists a linear innvertible map $T : \mathcal{R} \to \mathcal{R}$ such that for any $\psi(x,t), \varphi(x,t) \in \mathcal{R}$, $x \in \mathbf{R}^n, t \in \mathbf{R}$,

$$T(\psi\varphi) = T(\psi)T(\varphi),\ T(1) = 1. \tag{1}$$

The automorphism with the defining property (1) allows to write down a wide class of functional-differential-difference and difference-difference equations [5] starting from

$$\psi_t(x,t) = \sum_{m=-M}^{N} U_m T^m \psi(x,t). \tag{2}$$

For example, the operators T could be chosen as

$$T\psi(x,t) = \psi(qx+\delta,t), \tag{3}$$

where $q \in GL(n,C)$, $\delta \in \mathbf{R}^n$, or

$$T\psi(x) = W\psi(x)W^{-1}\ , W \in GL(n,C), \tag{4}$$

if $\mathcal{R} = GL(n,C)$ (see also the end of Section 4).

Let us define two Darboux transformations for solutions of (2),

$$D^{\pm}f = f - \sigma^{\pm}T^{\pm 1}f,\ \sigma^{\pm} = \varphi\left(T^{\pm 1}\varphi\right)^{-1}. \tag{5}$$

where φ is a particular solution of the same equation (2). In the case of a differential ring and for $Tf(x,t) = f(x+\delta,t)$, $x, \delta \in \mathbf{R}$ the limit $\partial f = \lim_{\delta\to 0}\frac{1}{\delta}(T-1)f(x,t)$ gives the link to the classical DT.

To derive the DT of potentials U_m it is necessary to evaluate the derivative of the elements $\sigma^{\pm}$ with respect to the variable t (say, time). We shall do it by means of special functions (analogue of differential Bell polynomials) whose introduction is similar to this from [7].

Let us start with the first definition of DT, D^+, expressing $T\varphi = T^{+1}\varphi$ by means of (5), i.e. $T\varphi = (\sigma^+)^{-1}\varphi$. Acting on the result with T, taking into account (1), yields $T^2\varphi = T\left((\sigma^+)^{-1}\right)T\varphi = T\left((\sigma^+)^{-1}\right)(\sigma^+)^{-1}\varphi$. Repeating the action one arrives at

$$T^m\varphi = \prod_{k=0}^{m-1}(T^k(\sigma^+))^{-1}\varphi = B_m^+(\sigma^+)\varphi. \tag{6}$$

Here and below the product is ordered by the index (k) from right to left.

Definition Equation (6) defines the function

$$B_m^+(\sigma) = \prod_{k=0}^{m-1} T^k(\sigma)^{-1})$$

It is convenient to write down the t-derivative of σ by means of functions $B_m^+(\sigma^+)$ connected with generalized Bell polynomials [7],

$$\sigma_t^+ = \sum_{m=-M}^{N}\left[U_m\, B_m^+(\sigma^+)\,\sigma^+ - \sigma^+ T(U_m)\, B_{m+1}^+(\sigma^+)\,\sigma^+\right] \tag{7}$$

The resulting equation (7) is a nonlinear equation associated with (2), reducing to a generalized Miura transformation in the stationary case (see Sec. 3).

The Matveev theorem [5] gives powerful generalizations of the conventional Darboux theorem, proved originally for second order differential equations (for generalizations cf. [17]) and can be formulated by means of the above introduced objects in the following way.

Theorem 1 *Let $\varphi \in \mathcal{R}$ and $\psi \in \mathcal{R}$ satisfy the equation* (2), *then $\psi^+ = D^+\psi$ satisfies*

$$\psi_t^+(x,t) = \sum_{m=-M}^{N} U_m^+ T^m \psi^+(x,t),$$

where the coefficients are evaluated from the following recursive relations

$$U_{-M}^+ = U_{-M} \tag{8}$$

$$U_1^+ - U_0^+\sigma^+ = U_1 - \sigma^+ T U_0 - \sigma_t^+ \tag{9}$$

$$U_m^+ - U_{m-1}^+ T^{m-1}\sigma^+ = U_m - \sigma^+ T U_{m-1} \tag{10}$$

$$U_N^+ = \sigma^+(TU_N)\left(T^N\sigma^+\right)^{-1}. \tag{11}$$

Expressions (8)–(11) *define the DT of the coefficients of the differential equation* (2) *(potentials) by the recurrence. Solving* (10) *by means of* (6) *one finds*

$$U_m^+ = \sum_{l=0}^{m+M} U_{-M+l} - \sigma^+ (TU_{-M+l-1}) B_{-M+l}^+ (\sigma^+) \left(B_m^+ (\sigma^+)\right)^{-1} \quad (12)$$

$$U_N^+ = \sigma^+ (TU_N) \left(T^N \sigma^+\right)^{-1}. \quad (13)$$

The theorem establishes covariance (form-invariance) of the equation (2) with respect to the Darboux transformation.

Proof: For the proof it is necessary to check the additional equality that appears with the term $T^m\psi$, with an essential use of the expression for σ_t^+ from (7). □

The formalism for second DT from (5) may be constructed in a similar way on the basis of the identity

$$T^m\varphi = \prod_{k=0}^{m} T^k (\sigma^-) T^{-1}\varphi = B_m^- (\sigma^-) T^{-1}\varphi. \quad (14)$$

The definition of the second type of lattice Bell polynomials $B_m^- (\sigma^-)$ may be extracted from (14). The evolution equation for σ^- is similar to (7), i.e.

$$\sigma_t^- = \sum_{m=-M}^{N} \left[U_m\, B_m^- (\sigma^-) - \sigma^- T^{-1} (U_m\,) B_{m-1}^- (\sigma^-)\right] \quad (15)$$

and gives a second generalized Miura map for stationary solutions of (2). Explicit formulas for U_m^- are similar to (9), (10), (11), (12), (13).

3 Stationary Equations as Eigenvalue Problems and Chains

A stationary equation that corresponds to (2) appears when solutions of the constraint equations $\psi_t = \lambda\psi$ or $\psi_t = \psi\lambda$ are considered. In the first case

$$\sum_{m=-M}^{N} U_m T^m \psi = \lambda\psi. \quad (16)$$

The derivative

$$\sigma_t = \varphi_t (T\varphi)^{-1} - \varphi (T\varphi)^{-1} (T\varphi)_t (T\varphi)^{-1} = \mu\sigma - \sigma\mu \quad (17)$$

is zero if σ and μ commute. Recall that $\varphi \in \{\psi_{\lambda=\mu}\}$.

Eq. (17) when combined with (7) or (15) leads to an algebraic relation between potentials and σ. If the coefficients in (2) are functions of some unique potential u, the relation allows in principle to express the potential u as a function of σ. This relation has the same form both for the potential u and for the "dressed" one. If one takes the relations for both of them with the corresponding σs and plugs the result into the Darboux transformation, the chain equations follow.

3.1 Example of Generalized Zakharov-Shabat (ZS) Problem

Let us illustrate the algorithm by an example. Let us take the equation (2) in the minimal variant

$$\psi_t(x,t) = \quad (U_0 + U_1 T)\,\psi(x,t). \tag{18}$$

The covariance of this equation with respect to the DT of the first kind (see below) means the invariance of U_0. We rewrite the transform of U_1 by means of (11)

$$U_1^+ = \sigma^+ \left(TU_1\right) \left(T\sigma^+\right)^{-1} \tag{19}$$

For the spectral problem

$$(U_0 + U_1 T)\,\varphi = \mu\varphi \tag{20}$$

the relation between the unique potential U_1 and σ^+ is obtained from (7) and (17), or directly from the equation (20),

$$U_1 = (\mu - U_0)\sigma^+. \tag{21}$$

Introducing the number of iterations n for U_1 and $n+1$ for U_1^+ one finds

$$U_1\,[n+1] = (\mu - U_0)\sigma^+[n+1] = \sigma^+\,[n]\,T\left((\mu - U_0)\sigma^+\ [n]\right)\left(T\sigma^+\,[n]\right)^{-1}. \tag{22}$$

This is the chain equation for the generalized ZS problem. We rewrite the chain equation (22) in a more compact form changing the notation as follows: $U_0 \to J$, $U_1 \to U$, $\sigma^+\,[n] \to \sigma_n$ and supposing that $T(J) = J$. We arrive at

$$\sigma_{n+1} \quad = \quad (\mu - J)^{-1}\sigma_n\,(\mu - J) \tag{23}$$

This dressing, however, is obviously almost trivial. Such a phenomenon is well known in the differential operator case [17]. Alternative and effective transformations appear if in the stationary equation (20) one introduces the element μ that does not commute with σ and changes the order in the RHS between the elements μ and φ (this is the second case mentioned at the very beginning of this section), i.e.

$$(J + UT)\,\varphi = \varphi\mu. \tag{24}$$

The formula for the potential is changed

$$U = \varphi\mu(T\varphi)^{-1} - J\sigma \tag{25}$$

and the role of σ_+ can be played by the function $s = \varphi\mu(T\varphi)^{-1}$. The equation (7) connects U and σ

$$J\sigma + U - \sigma J\sigma + \sigma T(U)(T\sigma)^{-1} = [\mu, \sigma]. \tag{26}$$

The algorithm of the explicit derivation of the chain equation begins with solving the equation (26) with respect to U in an appropriate way. For matrix rings this may be a system of equations for matrix elements, a method which may be effective in low matrix dimensions of (26), as in [12]. Otherwise it opens a special problem.

Let us rewrite (26) and the DT (22) in terms of s, excluding U from (21), denoting the number of iterations by a subscript

$$U[n] = s_n - J\sigma_n.$$

Equation (26) transforms as

$$s - \sigma J\sigma + \sigma T(s)(T\sigma)^{-1} + \sigma J = [\mu, \sigma]. \tag{27}$$

Using of this result one gets for the DT

$$s_{n+1} - s_n = J\sigma_{n+1} + \sigma_n J\sigma_n - [\mu, \sigma_n]. \tag{28}$$

Then, taking the result (28) for two indices (e.g. it could be $(n, n+1)$) one has the chain system. In Section 5 we will give an explicit example for the bilinear Hirota equation.

Let us mention that chain equations for a classical ZS problem and two types of DT were introduced in [12]. Closure of chain equations specifies classes of solutions. Chain equations for a standard Sturm-Liouville problem were derived in connection with quantum mechanical problems in an early Schrödinger paper (cf. the review [10]). In connection with the celebrated scalar KdV equation an important possibility is studied in [11]. A periodic closure of the chains produces integrable bi-Hamiltonian finite-dimensional systems and, in some special cases, the finite-gap potentials [13].

4 Joint Covariance of Equations and Nonlinear Problems

Let us consider a pair of equations of the same type (2) for a function ψ

$$\psi_t(x, y, t) = \sum_{m=-M}^{N} U_m T^m \psi(x, y, t), \tag{29}$$

$$\psi_y(x,y,t) = \sum_{m=-M'}^{N'} V_m T^m \psi(x,y,t). \tag{30}$$

Their compatibility condition is the nonlinear equation

$$U_{sy} - V_{st} = \sum_k V_k T^k (U_{s-k}) - U_{s-k} T^{s-k} (V_k) \tag{31}$$

for $s = -M-M', ..., N+N'$, $k \in \{k' = -M', ..., N'\} \cap \{s-k = -M, ..., N\}$. In the simplest case of ZS operators in both (29) and (30), with the subclass of solutions stationary in y, one has three conditions

$$U_{0t} = V_0 U_0 - U_0 V_0, \tag{32}$$

$$U_{1t} = V_0 U_1 - U_0 V_1 + V_1 T(U_0) - U_1 T(V_0) \tag{33}$$

$$V_1 T(U_1) = U_1 T(U_1) \tag{34}$$

The connection with polynomials of a differential operator and hence with the ordinary theory of Bell polynomials may be found if one changes the definition of potentials. It is clear that if the automorphism T is the shift operator $Tf(x) = f(x+\delta)$, the coefficients of the polynomials in T should be arranged as follows

$$\psi_t = \sum_{m=-M}^{N} u_m \frac{1}{\delta^m} \sum_{r=0}^{m} \binom{m}{m-r} (-1)^{m-r} T^r \psi, \tag{35}$$

where the Newton binomial formula has been used. The recursion equation that defines the usual differential Bell polynomials [8]

$$B_{m+1} = \sum_{r=0}^{m} \binom{m}{r} B_{m-r} y_{r+1}, \tag{36}$$

together with the definition (6) of B_m^+ relates the expressions for these special functions. Let us remark that the transformations for U_m found in Sec. 2 give the transforms for u_m defined by (35). The existence of the inverse transformation depends on independence of functions $(T-1)^n f$ for a given T and the set of functions ψ under consideration. The joint covariance of the system (29), (30) may, therefore, be investigated along the lines of [6,16], where the so-called binary Bell polynomials are used to form a convenient basis.

5 Non-Abelian Hirota System

5.1 *Derivation of the Chain Equation*

Let us consider a pair of ZS type equations,

$$\psi_t(x,y,t) = (V_0 + V_1 T)\,\psi(x,y,t)\,, \tag{37}$$

$$\psi_y(x,y,t) = \left(U_0 + U_{-1} T^{-1}\right)\psi(x,y,t)\,. \tag{38}$$

It differs from the one used in the previous section by the change $T \to T^{-1}$ in the RHS of the equation.

The case of lattice in all the variables is generated by transition to discrete variables: $x, y, t \to n, j, r \in Z$; $f(x,y,t) \to f_n(j,r)$, defined as in [5]. The operator T acts as the shift of n: $Tf_n(j,r) = f_{n+1}(j,r)$. The corresponding equations of the system $(37), (38)$ are

$$f_n(j-1,r) = f_{n+1}(j,r) + v(n,j,r) f_n(j,r) \tag{39}$$

$$f_n(j,r-1) = f_n(j,r) + u(n,j,r) f_{n-1}(j,r) \tag{40}$$

with specified potentials. The compatibility condition of the linear equations $(39), (40)$ has the form

$$u(n,j-1,r) - u(n+1,j,r) = v(n,j,r-1) - v(n,j,r), \tag{41}$$

$$v(n,j,r-1)u(n,j,r) = u(n,j-1,r)v(n-1,j,r). \tag{42}$$

Eq. (42) is automatically valid if

$$u(n,j,r) = \tau_{n+1}(j,r-1)\,\tau_n^{-1}(j,r-1)\,\tau_{n-1}(j,r)\,\tau_n^{-1}(j,r) \tag{43}$$

$$v(n,j,r) = \tau_{n+1}(j-1,r)\,\tau_n^{-1}(j-1,r)\,\tau_n(j,r)\,\tau_{n+1}^{-1}(j,r)\,. \tag{44}$$

We stress that the form of these expressions is valid for the order of entries shown in (42)–(44) (the form differs from that given in [5]). The substitution of (44) to (41) leads to a generalized Hirota bilinear equation [14]. One can also compare this with the generalizations given in [3]

$$\begin{aligned}
&\tau_{n+1}(j-1,r-1)\,\tau_n^{-1}(j-1,r-1)\,\tau_{n-1}(j-1,r)\,\tau_n^{-1}(j-1,r) \\
&\quad - \tau_{n+1}(j-1,r-1))\tau_n^{-1}(j-1,r-1)\,\tau_{n-1}(j,r-1)\,\tau_n^{-1}(j,r-1) \\
&\quad - \tau_{n+2}(j,r-1)\,\tau_{n+1}^{-1}(j,r-1)\,\tau_n(j,r)\,\tau_{n+1}^{-1}(j,r) \\
&\quad + \tau_{n+1}(j-1,r))\tau_n^{-1}(j-1,r)\,\tau_n(j,r)\,\tau_{n+1}^{-1}(j,r) = 0.
\end{aligned} \tag{45}$$

In the scalar case the system reduces to the Hirota bilinear equation [5]

$$\begin{aligned}
&\tau_n(j+1,r)\,\tau_n(j,r+1) - \tau_n(j,r)\,\tau_n(j+1,r+1) \\
&\quad + \tau_{n+1}(j+1,r)\,\tau_{n-1}(j,r+1) = 0.
\end{aligned} \tag{46}$$

The above formulas have applications to quantum transfer matrices for fusion rules [4] and quantum correlation functions [19].

Let us return to the DT theory. The equations (37), (38) are jointly covariant. As a consequence the theory of solving the systems (42) or (45) is based on the symmetry generated by joint covariance of (39), (40) with respect to one of the transformations (5), i.e. to

$$\psi^{-}(j,r) = \psi - \sigma^{-}T^{-1}f,\ \sigma^{-} = \varphi\left(T^{-1}\varphi\right)^{-1}. \tag{47}$$

The form of (39) and (40) is an obvious reduction of equations (37) and (38) with $V_1 = 1$, $V_0 = v$, $U_0 = 1$, $U_{-1} = u$.

We discuss further some details of the proof of the covariance theorem because this shows features important for derivation of the chain equation. Let us start, say, from (40). The conditions of covariance are obtained as the coefficients of ψ, $T^{-1}\psi$, and $T^{-2}\psi$. The first one is valid automatically

$$u^{-} = u - \sigma^{-}(r-1) + \sigma^{-}(r) \tag{48}$$

$$u^{-}T^{-1}\sigma^{-}(r) = \sigma^{-}(r-1)\,u. \tag{49}$$

Eliminating the transformed potential u^{-} one arrives at the equation that links the potential and $\sigma^{-}(r)$

$$u\,T^{-1}\sigma^{-}(r) - \sigma^{-}(r-1)\,u = \left(\sigma^{-}(r-1) - \sigma^{-}(r)\right)T^{-1}\sigma^{-}(r). \tag{50}$$

Note that the expressions in the equalities (48), (49), (50) are still general (i.e. non-Abelian) and may be used in simplest (but definitely rich) closures (e.g. $\sigma^{-}_{n+1} = \sigma^{-}_{n}$). The generic case is outlined at the end of Sec. 3. For some Lie algebraic approaches to non-Abelian chains see also [16].

In the scalar (Abelian) case one can easily solve (50) with respect to the potential

$$u = \frac{\left(\sigma^{-}(r-1) - \sigma^{-}(r)\right)T^{-1}\sigma^{-}(r)}{T^{-1}\sigma^{-}(r) - \sigma^{-}(r-1)}. \tag{51}$$

Equipping the entries of Eq. (51) with the index N (iteration number) and substituting into Eqs. (48), (49) we obtain two equivalent forms of chain equations. For example,

$$u_{N+1} = u_N - \sigma^{-}_{N}(r-1) + \sigma^{-}_{N}(r) \tag{52}$$

yields

$$\frac{\left(\sigma^{-}_{N+1}(r-1) - \sigma^{-}_{N+1}(r)\right)T^{-1}\sigma^{-}_{N+1}(r)}{T^{-1}\sigma^{-}_{N+1}(r) - \sigma^{-}_{N+1}(r-1)} = \frac{\left(\sigma^{-}_{N}(r-1) - \sigma^{-}_{N}(r)\right)\ \sigma^{-}_{N}(r-1)}{T^{-1}\sigma^{-}_{N}(r) - \sigma^{-}_{N}(r-1)} \tag{53}$$

The chain equation (53) generates the chain equation for the specific case of the system (42) by the choice $x \to n, Tf_n\ (j, r\ ) = f_{n-1}\ (j, r\ )$. A solution of the resulting chain equation generates the solution of the system (46) by the use of (51) and the corresponding formula for v. The transition to $\tau_n\,(j, r\ )$ functions is made by (44).

5.2 On Solution of the Chain Equation

If we denote

$$s_N = \frac{\left(\sigma_N^-\,(r-1) - \sigma_N^-\,(r)\right)}{T^{-1}\sigma_N^-\,(r) - \sigma_N^-\,(r-1)}, \tag{54}$$

then the dressing chain equation (53) reads

$$s_{N+1} = s_N \frac{\sigma_N^-\,(r-1)}{T^{-1}\sigma_{N+1}^-\,(r)}. \tag{55}$$

Iterating this recurrence yields

$$s_{N+q} = s_N \frac{\prod_{s=0}^{q-1} \sigma_{N+s}^-\,(r-1)}{\prod_{s=1}^{q} T^{-1}\sigma_{N+s}^-\,(r)}. \tag{56}$$

In analogy with the continuous case let us consider a periodic closure of the chain (53), starting from the simplest case $q=0$, which implies $\sigma_{N+1} = \sigma_N = \sigma$. Since $s_{N+1} = s_N$, one finds

$$T\sigma^-\,(r-1) = \sigma^-\,(r)\,. \tag{57}$$

This means $\sigma^-\,(r+p) = T^p\sigma^-\,(r)$. If a boundary condition at the point $r=0$ is given, then

$$\sigma^-\,(p) = T^p\sigma^-\,(0)\,. \tag{58}$$

The equation for $\varphi(p,x)$ then reads

$$T\varphi(p,x) = T\left(\sigma(p,x)\right)\varphi(p,x). \tag{59}$$

Its solution depends on the choice of T; e.g. if $T\varphi(p,x) = \varphi(p, x+\delta)$, then

$$\varphi(p,x) = \exp\left(Ax\right) \tag{60}$$

with A from

$$A\delta = \ln\left[T^{p+1}\sigma^-\,(0)\right] \tag{61}$$

6 Nahm Equations

Considering the following example we change a little the DT formulas. We show an alternative version, similar to this from [20]. We stress, however, that the formulas from the first section give equivalent results. Some generalization is needed because of reduction constraints and an additional (gauge) transformation denoted by g is employed. This is expressed by the following **Theorem 2** *The equation*

$$\psi_y = uT\psi + v\psi + wT^{-1}\psi \tag{62}$$

is covariant with respect to DT

$$\psi[1] = g(T - \sigma)\psi^{-1}, \tag{63}$$

where $\sigma = (T\phi)\phi^{-1}$ *and* ϕ *is a solution of the same equation* (62). g *is an invertible element of the ring to be defined later. The transforms of the equation's coefficients are*

$$u[1] = gT(u)[T(g)]^{-1}, \tag{64}$$
$$v[1] = gT(v))g^{-1} - g\sigma ug^{-1} + gT(u)T(\sigma)g^{-1} + g_y g^{-1}, \tag{65}$$
$$w[1] = g\sigma w[T^{-1}(g\sigma)]^{-1}. \tag{66}$$

Proof: Substitution of (48) into the transformed equation (62) gives four equations (assuming $T^n\psi$ are independent). Three of them yield transformed potentials (66). The fourth equation is mapped by the transformation into

$$\sigma_y = \sigma F - (TF)\sigma \tag{67}$$

where

$$F = u\sigma + v + w[T^{-1}(\sigma)]^{-1}.$$

One can check the condition by directly substituting the definition of σ and using the equation for ϕ. □

Remark Theorem 2 is valid for the spectral problem

$$\lambda\psi = uT\psi + v\psi + wT^{-1}\psi \tag{68}$$

with only one correction: The last term in the transform $v[1]$ is absent. The equation goes into an analog of the "Riccati equation" for the function σ

$$\mu = u\sigma + v + w[T^{-1}(\sigma)]^{-1}. \tag{69}$$

Note that plugging the element $\sigma = (T\phi)\phi^{-1}$ into (69) transforms it into the spectral problem for ϕ (68) (with the spectral parameter μ).

The Nahm equations can be written in the Lax representation by means of the spectral equation (68) and the evolution equation

$$\psi_y = (q + pT)\psi. \tag{70}$$

Covariance of this equation with respect to the DT (48) may be established similarly to Theorem 1 and taking into account the evolution of the function $\sigma(y)$

$$\sigma_y = T(q)\sigma - \sigma p\sigma + T(p)T(\sigma)\sigma - \sigma q = 0. \tag{71}$$

This implies the following transformation formulas for the coefficients in (70)

$$p[1] = gT(p)[T(g)]^{-1}, \tag{72}$$
$$q[1] = g[T(q) - \sigma p + T(p)T(\sigma)]g^{-1} + g_y g^{-1}. \tag{73}$$

The principle of joint covariance [9] defines the relation of potentials p, q with u, v

$$p = u + \beta I, \qquad q = v/2. \tag{74}$$

It follows that the DT-covariance means integrability of the compatibility condition of (70) and (62)

$$u_y = \frac{1}{2}(uT(v) - vu) + \beta(T(v) - v), \tag{75}$$
$$v_y = uT(w) - wT^{-1}u + \beta(T(w) - w), \tag{76}$$
$$w_y = \frac{1}{2}vw - wT - 1(v). \tag{77}$$

One more possible specification is the use of periodic potentials in the problem (70) with the evolution (62), taking into account the relations (74), resulting in commutators of the RHSs of the equations. Linear transformations and rescalings

$$u = \alpha(-\imath\varphi_1/2 - \varphi_3), \tag{78}$$
$$v = \varphi_3, \tag{79}$$
$$w = \alpha^{-1}(-\imath\varphi_1/2 + \varphi_3), \tag{80}$$

produce the Nahm equations (for periodic functions, $T\varphi_i = \varphi_i$; the latter does not mean periodicity of solutions of the Lax pair ψ, ϕ and the corresponding $\sigma = (T\phi)\phi^{-1}$)

$$\varphi_{k,y} = i\epsilon_{klm}[\varphi_l, \varphi_m] \tag{81}$$

(no summation). The numbers α, and β are free parameters. This system is covariant with respect to the combined DT-gauge transformations if the gauge transformation $g = \exp G$ is chosen as follows

$$G_y = \alpha[(\varphi_3 + \varphi_1/2)T(\sigma) - \sigma(\varphi_3 + \varphi_1/2)]/2, \tag{82}$$

Finally, the following theorem may be formulated.

Theorem 3 *The system* (81) *is invariant with respect to the transformations*

$$\varphi_1[1] = g[(\varphi_1/2 - \imath\varphi_3)T(g)^{-1} + \sigma(\varphi_1/2 + \imath\varphi_3)[T^{-1}(g\sigma)]^{-1}], \tag{83}$$

$$\varphi_2[1] = g[\varphi_2 + \alpha(\imath\sigma\varphi_1/2 - \imath\varphi_1 T(\sigma)/2 + \sigma\varphi_3 - T(\varphi_3\sigma))]g^{-1}, \tag{84}$$

$$\varphi_3[1] = g[(-\imath\varphi_1/2 - \varphi_3)T(g)^{-1} + \sigma(-\imath\varphi_1/2 + \varphi_3)[T^{-1}(g\sigma)]^{-1}] \tag{85}$$

with the function $g = \exp G$, *where* G *is obtained by integrating* (82).

Remark. A similar statement may be formulated for a discrete version (see [21]) of the Nahm system (81) as can be seen from the previous section.

7 Conclusions

Productivity of the method obviously depends on possibilities of solving the chain equations. It is well known that direct application of DT formulas could be effective [17]. Approaches to a direct solution of chain equations are formulated in [13,16] and references therein.

Acknowledgement

The author acknowledges V. Matveev, M. Salle, M. Czachor and N. Ustinov for discussions, the organizers of the SIDE4 conference in Tokyo for an outstanding hospitality and support, and A. Nakamula for valuable information about the Nahm model. The work was supported by the KBN Grant No. 5 P03B 040 20.

References

1. D. Gross, N. Nekrasov, "Solitons in noncommutative gauge theory", Trieste lectures, hep-th/0010090.
2. A. Belavin, A. Fring, "On the fermion quasi-particle interpretation in minimal models of conformal field theory", hep-th/9612049.
3. T. Miwa, *Proc. Japan. Acad.* A **58**, 9-12 (1982); *ibid.* **58**, 339-352 (1982).
4. A. Kuniba, T. Nakanishi, J. Suzuki, *Int. Journ. Mod. Phys.* A **9**, 5215-5312 (1994).

5. V. B. Matveev, "Darboux transformations in associative rings and functional-difference equations", in *The Bispectral Problem*, J. Harnad and A. Kasman, eds. AMS series CRM *Proceedings and Lecture Notes* vol. 14, p. 211-226 (1998).
6. F. Lambert, S. Leble, J. Springael, *Glasgow Math. J* a **43**, 53-63 (2001).
7. A. Zaitsev, S. Leble, math-ph/9903005 (1999); *Rep. Math. Phys.* **46**, 165-174 (2000).
8. R. Schimming, S. Z. Rida *Int. J. Algebra and Computation* **6**, 635–644 (1996).
9. S. Leble, "Darboux Transforms Algebras in 2+1 dimensions", in *Proc. of 7th Workshop on Nonlinear Evolution Equations and Dynamical Systems*, M Boiti et al., eds. p.53-61 (World Scientific, Singapore, 1991). *Computers Math. Applic.*, **35** 73-81, (1998).
10. L. Infeld, T. Hull, *Rev. Mod. Phys.* **23**, 21 (1951).
11. J. Weiss, *J. Math. Phys.* **27**, 2647 (1986).
12. A. Shabat, "Dressing chains anf lattices", in *Nonlinearity, integrability and all that*, M. Boiti et al., eds., 331-342 (World Scientific, Singapore, 2000).
13. A. Veselov, A. Shabat, *Funkt. Analiz Pril* **27**, 1 (1993).
14. R. Hirota, *Journ. Phys. Soc. Japan* **50**, 3875-3791 (1981).
15. W. Nahm, *Phys. Lett.* B **90**, 413 (1980).
16. S. Leble, "Covariance of Lax pairs and integrability of compatibility condition", nlin.SI/0101028; *Theor. Math. Phys.* **128**, 890-905 (2001).
17. V.B. Matveev , M. A. Salle, *Darboux Transformations and Solitons* (Springer, Berlin, 1991).
18. A. Hashimoto, H. Hata, S. Morijama, hep-th/9910196; *J. High Energy Phys.*, 021 (1999).
19. H. Au-Yang, J. H. H Perk, T. T. Wu, *Nucl. Phys.* B **180**, 89-115 (1981).
20. M. Salle, "Darboux autotransformations: Special functions and q-commutations", preprint CSMAE, St. Petersburg (1993).
21. M.K. Murray, M.A. Singer, *Comm. Math. Phys.* **210**, 497-519 (2000)

COVARIANCE APPROACH TO THE FREE PHOTON FIELD

MACIEJ KUNA

Wydział Fizyki Technicznej i Matematyki Stosowanej,
Politechnika Gdańska, ul. Narutowicza 11/12,
80-952 Gdańsk, Poland
E-mail: maciek@mifgate.mif.pg.gda.pl

JAN NAUDTS

Departement Natuurkunde, Universiteit Antwerpen UIA,
Universiteitsplein 1, 2610 Antwerpen, Belgium
E-mail: Jan.Naudts@ua.ac.be

We introduce photon theory following the same principles as for introduction of the quantum theory of a single particle, using a C^*-algebraic approach based on covariance systems. The basic symmetries are additivity of the fields and additivity of test functions. We write down in explicit form a state of this covariance system. It turns out to reproduce the traditional Fock representation of the free photon field, with a Lorentz invariant vacuum. Properties of smeared-out photons are discussed.

1 Introduction

Motivation

This paper is a first attempt to reformulate photon theory. It is motivated by dissatisfaction with expositions in present day textbooks. As Scharf [1] notes, the fact that there are various essentially different methods of quantizing the radiation field shows that there are some difficulties with the subject. Two problems must be recognized.

A first problem arises because of the use of the vector potential $A_\mu(q)$, which is not uniquely determined by the electromagnetic fields. The resulting gauge freedoms can be tackled in many ways. Most often used is the Gupta-Bleuler gauge, which is rather complicated to say the least. Disadvantage of the radiation gauge is the lack of manifest Lorentz covariance. Our treatment of the gauge problem has been influenced by the work of Carey et al [2], which uses the Lorenz gauge. We show that the photon states are invariant under the remaining gauge freedom.

The next problem is that of positivity of the scalar product, in combination with Lorentz invariance. Many textbooks abandon the use of Hilbert spaces for this reason. Here, we give arguments to restrict the set of classical wave functions. A side effect is that the scalar product $\langle\psi|\phi\rangle$ of two classical wave functions ϕ and ψ satisfies the positivity requirement. In addition, we prove explicitly in Appendix D that the vacuum state does satisfy the positivity condition. As a consequence, the quantum probabilistic interpretation is saved in our approach.

Analogy

The simplest example of a classical field is the vibrating string. Its canonical variables are a displacement field $\eta(q)$ and a conjugated momentum field $\pi(q)$. The correct description of the quantized string can be obtained by a limiting procedure starting from a chain of particles interconnected with strings. Alternatively, one can describe the quantized string as a covariance system [3] — see Appendix A — consisting of an abelian algebra $\mathcal{A}$ generated by smeared-out displacement fields, a group G, which is the group of adding fields, and the obvious action of G on $\mathcal{A}$. Such a description is very analogous to the description [4] of standard quantum mechanics as a covariance system, in which case $\mathcal{A}$ is an algebra of functions of position, and G is the group of shifts in position. It is therefore natural to expect that also the quantized electromagnetic field can be described as a covariance system with additivity of fields as the basic symmetry group. The first problem that one encounters when trying such an approach is that the quantized smeared-out field operators $\hat{A}_\mu(\psi)$ do not form an abelian algebra. Indeed, in textbooks [5] one finds the commutation relations

$$\left[\hat{A}_\mu(q), \hat{A}_\nu(q')\right] = -ig_{\mu,\nu}D_0(q-q') \tag{1}$$

with the Pauli-Jordan function $D_0(q)$ defined by

$$D_0(q) = \frac{1}{(2\pi)^3}\int_{\mathbf{R}^3} \mathrm{d}\mathbf{k}\, \exp\left(i\sum_{\alpha=1}^{3}\mathbf{k}_\alpha q_\alpha\right)\frac{1}{|\mathbf{k}|}\sin(q_0|\mathbf{k}|) \tag{2}$$

(we have chosen the sign in (1) in such a way that later on creation and annihilation operators have their usual properties, i.e. the annihilation operator is complex linear in the field). Smearing out (1) with classical wave functions ψ and ϕ gives (see Appendix B)

$$\left[\hat{A}(\psi), \hat{A}(\phi)\right] = 2i\,\mathrm{Im}\int_{\mathbf{R}^3}\mathrm{d}\mathbf{k}\,\frac{1}{2|\mathbf{k}|}\overline{\psi^\mu(\mathbf{k})}\phi_\mu(\mathbf{k}). \tag{3}$$

There are two possible interpretations of these non-trivial commutation relations. It is a tradition in the physics literature to interpret photons as excitations of a harmonic oscillator. In this traditional approach [5], the canonical variables to be quantized are the vector fields $A_\mu(q)$ and their derivatives $\delta_\nu A_\mu(q)$. These correspond with the displacement field $\eta(q)$ and conjugated momentum field $\pi(q)$ of the harmonic string. By integration, one then obtains (3). An alternative interpretation is suggested by recent research on noncommutative spacetime. In the latter context, spacetime positions $\hat{Q}_\mu$ satisfy nontrivial commutation relations

$$\left[\hat{Q}_\mu, \hat{Q}_\nu\right] = i\epsilon_{\mu\nu}. \tag{4}$$

(see e.g. Doplicher et al [6,7] and Naudts and Kuna [8]). The $\hat{Q}_\mu$ are the generators of shifts in momentum space. The commutation relations (4) can [8] be seen as a originating from a projective representation of the group of shifts. By analogy, we can see the operators $\hat{A}(\psi)$ as generators of the group of addition of classical wave functions. In fact, it is clear [4] that the Weyl algebra of the free photon field, as used e.g. in Carey et al [2], can be replaced by a covariance system $(\mathbf{C}, H, \mathbf{I})$ consisting of the algebra $\mathbf{C}$ of complex numbers, the complex vector space of classical wave functions H, and the trivial action $\mathbf{I}$ of the latter on $\mathbf{C}$.

Duality

Because the addition of fields is the basic symmetry, rather than addition of classical wave functions, it is obvious to consider also the generators $\hat{F}(a)$ of the group of adding fields

$$a'(\mathbf{k}) \to a'(\mathbf{k}) + a(\mathbf{k}). \tag{5}$$

Here, the fields are represented by Fourier coefficients $a_\mu(\mathbf{k})$ of the vector potential $A_\mu(q)$ (see next section). In this context a duality between classical wave functions and Fourier transformed vector potentials is of importance. It implies a duality between the operators $\hat{A}(\psi)$ and $\hat{F}(a)$, similar to the duality between position and momentum operators in standard quantum mechanics.

Technically speaking, there is no need to include this duality in the formalism. Indeed, we will find that, in the Fock representation, for each a there exists a ψ such that $\hat{F}(a) = \hat{A}(\psi)$. However, this coincidence is a special property of the photon field and might be absent in more general field theories. For that reason we prefer to clarify the dual rôle of the operators $\hat{A}(\psi)$ and $\hat{F}(a)$.

Notations

We use Greek letters $\mu, \nu, \sigma, \cdots$ for indices that run from 0 to 3, in combination with Einstein's summing convention, i.e., if such an index appears twice then a summation from 0 to 3 is understood. These Greek indices are lowered and raised in the standard way, i.e., by definition is $x^\mu = g^{\mu\nu}x_\nu$, with the metric tensor g equal to the diagonal matrix with eigenvalues $+1, -1, -1, -1$. The Greek index α will be used to label spatial components. Hence it runs from 1 to 3 and no summation convention is used for it. Vectors in $\mathbf{R}^3$ are written in boldface. Quite often, a four-vector q will be written as $(q_0, \mathbf{q})$. The scalar product in $\mathbf{R}^3$ is written as $\mathbf{k} \cdot \mathbf{q} = \sum_{\alpha=1}^{3} \mathbf{k}_\alpha \mathbf{q}_\alpha$. We use the abbreviation $\partial^\mu \equiv \frac{\partial}{\partial q_\mu}$. The (pseudo)-scalar product of two elements ψ and ϕ of a (pseudo)-Hilbert space is denoted $\langle \phi | \psi \rangle$, linear in ψ and anti-linear in ϕ. Throughout the paper, operators carry a hat. The conjugate of $\hat{A}$ is denoted $\hat{A}^*$.

Structure of the Paper

The next section deals with classical electromagnetism. The vector potential $A_\mu(q)$ is represented by Fourier coefficients $a_\mu(\mathbf{k})$. The classical wave functions $\psi_\mu(\mathbf{k})$ are introduced and the duality between classical wave functions $\psi_\mu(\mathbf{k})$ and Fourier coefficients $a_\mu(\mathbf{k})$ is established. Section 3 describes the free photon field as a covariance system. Correlation functions determining the vacuum state are given explicitly. The field operators $\hat{A}(\psi)$ and $\hat{F}(a)$ live in the corresponding G.N.S.-representation. Section 4 discusses the standard Fock representation of the free photon field. Section 5 starts from Poincaré invariance to derive properties of the free photon. In the final section conclusions are drawn.

2 Classical Electromagnetism

The whole section deals with the classical radiation field. The vector potential $A_\mu(q)$ is replaced by Fourier coefficients $a_\mu(\mathbf{k})$. The test functions $f_\mu(q)$, used to smear out the vector potential $A_\mu(q)$, are replaced by the classical wave functions $\psi_\mu(\mathbf{k})$. Finally, a duality between $a_\mu(\mathbf{k})$ and $\psi_\mu(\mathbf{k})$ is established.

Smeared-out Fields

The classical electromagnetic field is described by the vector potential $A(q)$. It has four components $A_\mu(q)$, $\mu = 0, 1, 2, 3$, each of which is a function of position q in $\mathbf{R}^4$. We assume that the Lorenz gauge

$$\partial^\mu A_\mu(q) = 0 \tag{6}$$

is satisfied. Then the Maxwell equations for the free electromagnetic field can be written as a set of four equations

$$\partial^\nu \partial_\nu A_\mu(q) = 0. \tag{7}$$

It is necessary to smear out A using test functions. Given real-valued test functions $f_\mu(q)$, let

$$f(A) = \int_{\mathbf{R}^4} \mathrm{d}q\, f^\mu(q) A_\mu(q). \tag{8}$$

The functions $f(A)$ will become observables of the free photon field.

The electric field $\mathbf{E}$ and the magnetic field $\mathbf{B}$ are related to the vector potential A by

$$\begin{aligned} \mathbf{E}_\alpha(q) &= -\partial^\alpha A_0(q) - \partial^0 A_\alpha, \\ \mathbf{B}_\alpha(q) &= \sum_{\beta,\gamma=1}^{3} \varepsilon_{\alpha\beta\gamma} \partial^\beta A_\gamma(q) \end{aligned} \tag{9}$$

($\varepsilon_{\alpha\beta\gamma}$ is the fundamental antisymmetric tensor).

Note that the smeared-out electromagnetic fields can be obtained from the smeared-out vector potential. Indeed one has

$$\begin{aligned} \int_{\mathbf{R}^4} \mathrm{d}q \sum_{\alpha=1}^{3} f_\alpha(q) \mathbf{E}_\alpha(q) &= -\int_{\mathbf{R}^4} \mathrm{d}q \sum_{\alpha=1}^{3} f_\alpha(q) \partial^\alpha A_0(q) \\ &\quad - \int_{\mathbf{R}^4} \mathrm{d}q \sum_{\alpha=1}^{3} f_\alpha(q) \partial^0 A_\alpha(q) \\ &= \int_{\mathbf{R}^4} \mathrm{d}q\, A_0(q) \sum_{\alpha=1}^{3} \partial^\alpha f_\alpha(q) \\ &\quad + \int_{\mathbf{R}^4} \mathrm{d}q \sum_{\alpha=1}^{3} A_\alpha(q) \partial^0 f_\alpha(q) \\ &= g(A) \end{aligned} \tag{10}$$

with

$$g_0(q) = \sum_{\alpha=1}^{3} \partial^\alpha f_\alpha(q) \quad \text{and} \quad g_\alpha(q) = -\partial^0 f_\alpha(q), \tag{11}$$

and, similarly,

$$\begin{aligned} \int_{\mathbf{R}^4} dq \sum_{\alpha=1}^{3} f_\alpha(q)\mathbf{B}_\alpha(q) &= \int_{\mathbf{R}^4} dq \sum_{\alpha=1}^{3} f_\alpha(q) \sum_{\beta,\gamma=1}^{3} \varepsilon_{\alpha\beta\gamma}\partial^\beta A_\gamma(q) \\ &= -\int_{\mathbf{R}^4} dq \sum_{\alpha,\beta,\gamma=1}^{3} \varepsilon_{\alpha\beta\gamma}\left(\partial^\beta f_\alpha(q)\right) A_\gamma(q) \\ &= h(A) \end{aligned} \tag{12}$$

with

$$h_0(q) = 0 \quad \text{and} \quad h_\gamma(q) = \sum_{\alpha,\beta=1}^{3} \varepsilon_{\alpha\beta\gamma}\partial^\beta f_\alpha(q). \tag{13}$$

Fourier Coefficients

Equation (7) can be solved by Fourier transformation. Let

$$\tilde{A}_\mu(k) = (2\pi)^{-2} \int_{\mathbf{R}^4} dq \exp\left(ik^\nu q_\nu\right) A_\mu(q). \tag{14}$$

Then (7) becomes $k^\nu k_\nu \tilde{A}_\mu(k) = 0$. Hence $\tilde{A}_\mu(k)$ differs from zero only if $k^\nu k_\nu = 0$. Therefore A_μ is of the form

$$\begin{aligned} A_\mu(q) &= (2\pi)^{-2} \int_{\mathbf{R}^4} dk \exp\left(-ik^\nu q_\nu\right) \tilde{A}_\mu(k)\delta(k^\sigma k_\sigma) \\ &= (2\pi)^{-2} \int_{\mathbf{R}^4} dk \exp\left(-ik^\nu q_\nu\right) \tilde{A}_\mu(k)\frac{1}{2|\mathbf{k}|}\left(\delta(k^0 - |\mathbf{k}|) + \delta(k^0 + |\mathbf{k}|)\right) \\ &= (2\pi)^{-2} \int_{\mathbf{R}^3} d\mathbf{k}\, \frac{1}{2|\mathbf{k}|} \exp(i\mathbf{k}\cdot\mathbf{q}) \\ &\quad \times \left(e^{-i|\mathbf{k}|q_0}\tilde{A}_\mu(|\mathbf{k}|,\mathbf{k}) + e^{i|\mathbf{k}|q_0}\tilde{A}_\mu(-|\mathbf{k}|,\mathbf{k})\right). \end{aligned} \tag{15}$$

Here we use the notation $|\mathbf{k}| = \sqrt{\sum_{\alpha=1}^{3} k_\alpha^2}$. We obtain

$$A_\mu(q) = (2\pi)^{-3/2} \int_{\mathbf{R}^3} d\mathbf{k}\, \frac{1}{2|\mathbf{k}|}\, e^{i\mathbf{q}\cdot\mathbf{k}} \left[e^{-iq_0|\mathbf{k}|}a_\mu(\mathbf{k}) + e^{iq_0|\mathbf{k}|}\overline{a_\mu(-\mathbf{k})}\right] \tag{16}$$

with $a_\mu(\mathbf{k}) = (2\pi)^{-1/2}\tilde{A}_\mu(|\mathbf{k}|,\mathbf{k})$. Note that automatically any vector potential of the form (16) satisfies the wave equations (7). Expression (8), in combination with (16), becomes

$$f(A) = \sqrt{2\pi}\int_{\mathbf{R}^3} \mathrm{d}\mathbf{k}\,\frac{1}{2|\mathbf{k}|}\left[a^\mu(\mathbf{k})\overline{\tilde{f}_\mu(|\mathbf{k}|,\mathbf{k})} + \overline{a^\mu(\mathbf{k})}\tilde{f}_\mu(|\mathbf{k}|,\mathbf{k})\right] \tag{17}$$

with

$$\tilde{f}_\mu(k) = (2\pi)^{-2}\int_{\mathbf{R}^4} \mathrm{d}q\,\exp(ik^\nu q_\nu) f_\mu(q). \tag{18}$$

Note that only the values of $\tilde{f}_\mu$ on the light cone $\{k \in \mathbf{R}^4 : k^\mu k_\mu = 0\}$ are of importance.

¿From the Lorenz condition (6) follows

$$|\mathbf{k}|a_0(\mathbf{k}) = \sum_{\alpha=1}^{3} \mathbf{k}_\alpha a_\alpha(\mathbf{k}). \tag{19}$$

This expression can be used to calculate $a_0(\mathbf{k})$ in function of $a_\alpha(\mathbf{k})$.

Classical Wave Functions

The first gauge problem that arises is that two different sets of test functions $f_\mu(q)$ and $g_\mu(q)$ may define functions $f(A)$ and $g(A)$ which coincide on all vector potentials A that satisfy (6) and (7). To avoid this non-uniqueness we make use of the so-called classical wave functions of the photon. Given test functions f_μ the classical wave functions ψ_μ are defined by

$$\begin{aligned}\psi_\mu(\mathbf{k}) &= \sqrt{2\pi}\tilde{f}_\mu(|\mathbf{k}|,\mathbf{k})\\ &= (2\pi)^{-3/2}\int_{\mathbf{R}^4} \mathrm{d}q\,\exp\left(iq_0|\mathbf{k}| - i\mathbf{q}\cdot\mathbf{k}\right) f_\mu(q).\end{aligned} \tag{20}$$

Note that these are complex functions over $\mathbf{R}^3$. Two sets of test functions f_μ and g_μ can give rise to the same classical wave functions ψ_μ. In fact, this will be the case if and only if $f(A) = g(A)$ for all A satisfying (6) and (7). Indeed, (17) can be written as

$$f(A) = \int_{\mathbf{R}^3} \mathrm{d}\mathbf{k}\,\frac{1}{2|\mathbf{k}|}\left[a^\mu(\mathbf{k})\overline{\psi_\mu(\mathbf{k})} + \overline{a^\mu(\mathbf{k})}\psi_\mu(\mathbf{k})\right] = -2\,\mathrm{Re}\,\langle a|\psi\rangle \tag{21}$$

where the bilinear form $\langle\cdot|\cdot\rangle$ is given by

$$\langle a|\psi\rangle = -\int_{\mathbf{R}^3} \mathrm{d}\mathbf{k}\,\frac{1}{2|\mathbf{k}|}\,\overline{a^\mu(\mathbf{k})}\psi_\mu(\mathbf{k}). \tag{22}$$

This shows that $f(A)$ depends only on the classical wave functions ψ and on the Fourier coefficients a.

Duality

The Lorenz gauge (6) does not suffice to fix uniquely the vector potential A corresponding with a given electromagnetic field. The gauge transformations $A_\mu \to A'_\mu$ with

$$A'_\mu = A_\mu + \partial_\mu \chi \tag{23}$$

with $\chi(q)$ an arbitrary solution of $\partial^\mu \partial_\mu \chi = 0$ leave the electric and magnetic fields invariant. We use this gauge freedom to derive a condition on the classical wave functions. If two fields A and A' differ only by a gauge transformation as given by (23), then no classical wave function ψ should be able to distinguish A from A', i.e. $\mathrm{Re}\,\langle a|\psi\rangle = \mathrm{Re}\,\langle a'|\psi\rangle$ should hold. The result of this condition, deduced below, is

$$|\mathbf{k}|\psi_0(\mathbf{k}) = \sum_{\alpha=1}^{3} \mathbf{k}_\alpha \psi_\alpha(\mathbf{k}). \tag{24}$$

This condition implies a duality between Fourier coefficients $a(\mathbf{k})$ determining the vector potential A and classical wave functions $\psi(\mathbf{k})$ determining test functions f. Indeed, both are sets of four complex functions satisfying similar conditions (19) respectively (24).

Because χ is a solution of $\partial^\mu \partial_\mu \chi = 0$ it can be written as (see (16))

$$\chi(q) = \int_{\mathbf{R}^3} \mathrm{d}\mathbf{k}\, \frac{1}{2|\mathbf{k}|}\, e^{i\mathbf{q}\cdot\mathbf{k}} \left[e^{-iq_0|\mathbf{k}|} c(\mathbf{k}) + e^{iq_0|\mathbf{k}|}\overline{c(-\mathbf{k})} \right], \tag{25}$$

with $c(\mathbf{k})$ an arbitrary complex function of $\mathbf{k} \in \mathbf{R}^3$. From (23) and (16) follows then that

$$a'_\alpha(\mathbf{k}) = a_\alpha(\mathbf{k}) + ic(\mathbf{k})\mathbf{k}_\alpha, \qquad \alpha = 1,2,3. \tag{26}$$

Using (22) and (19) one obtains

$$\begin{aligned}\langle a|\psi\rangle &= -\int_{\mathbf{R}^3} \mathrm{d}\mathbf{k}\, \frac{1}{2|\mathbf{k}|} \left(\overline{a_0(\mathbf{k})}\psi_0(\mathbf{k}) - \sum_{\alpha=1}^{3} \overline{a_\alpha(\mathbf{k})}\psi_\alpha(\mathbf{k}) \right) \\ &= -\int_{\mathbf{R}^3} \mathrm{d}\mathbf{k}\, \frac{1}{2|\mathbf{k}|^2} \sum_{\alpha=1}^{3} \overline{a_\alpha(\mathbf{k})} \left(\psi_0(\mathbf{k})\mathbf{k}_\alpha - |\mathbf{k}|\psi_\alpha(\mathbf{k}) \right)\end{aligned} \tag{27}$$

and a similar expression for $\langle a'|\psi\rangle$. Hence the condition $\mathrm{Re}\,\langle a'|\psi\rangle = \mathrm{Re}\,\langle a|\psi\rangle$ yields

$$0 = \mathrm{Re}\int_{\mathbf{R}^3} \mathrm{d}\mathbf{k}\,\frac{1}{2|\mathbf{k}|^2}\sum_{\alpha=1}^{3} i\mathbf{k}_\alpha\overline{c(\mathbf{k})}\,(\psi_0(\mathbf{k})\mathbf{k}_\alpha - |\mathbf{k}|\psi_\alpha(\mathbf{k}))\,. \tag{28}$$

Because the latter should hold for all choices of $c(\mathbf{k})$ one concludes that (24) holds.

Radiation Gauge

Note that the scalar product $\langle a|\psi\rangle$ is degenerate. Indeed, condition (24), to be satisfied by classical wave functions, was derived precisely by requiring that, if a gauge transformation maps a onto b then $\langle a|\psi\rangle = \langle b|\psi\rangle$ holds for all ψ. By duality, we say that ψ and ϕ are equivalent if $\langle a|\psi\rangle = \langle a|\phi\rangle$ holds for all a. As noted by Carey et al [2], in each set of equivalent ψ one can select a unique representative ψ satisfying

$$\psi_0(\mathbf{k}) = 0 \qquad \text{and} \qquad \sum_{\alpha=1}^{3}\mathbf{k}_\alpha\psi_\alpha(\mathbf{k}) = 0. \tag{29}$$

See Appendix C. These conditions are called the *radiation gauge*. However, by selecting such a representative one breaks the property of manifest Lorentz invariance. Therefore, we will use this gauge only to discuss the physical content of certain formulas.

3 Quantum Description

This section gives a description of the electromagnetic field as a quantum system. We start from explicit correlation functions and use the generalized GNS-theorem to construct a representation in Hilbert space.

Covariance Approach

In the previous section the smeared-out vector potential $f(A)$ could be written as $-2\,\mathrm{Re}\,\langle a|\psi\rangle$ (see (21)), where $a(\mathbf{k})$ are Fourier coefficients representing the vector potential $A_\mu(q)$ and $\psi(\mathbf{k})$ is a classical wave function representing the test functions $f_\mu(q)$. In the quantum theory both $a(\mathbf{k})$ and $\psi(\mathbf{k})$ become operators in Hilbert space. They will be denoted $\hat{F}(a)$ and $\hat{A}(\psi)$, respectively.

We want to derive the quantum description of the electromagnetic field in a way similar to the quantum description of a single particle. The quantity corresponding with a function $f(q)$ of the position q of the particle is the function $f(A)$ considered as a function of the vector potential A_μ. Hence, in the obvious quantum description quantum mechanical wave functions would be complex square integrable functions of A_μ (replacing q-dependent functions) and the $f(A)$ is mapped onto an operator $f(\hat{A})$ (replacing $f(\hat{q})$, with $\hat{q}\psi(q) = q\psi(q)$). A more common notation, replacing $f(\hat{A})$, is $\hat{A}(\psi)$, with ψ the classical wave function corresponding with f. As discussed in the introduction, the problem with this approach is that the operators $\hat{A}(\psi)$ are expected not to be mutually commuting, so that they cannot be simple multiplication operators, as in the case of quantum mechanics of a single particle. The solution adopted here is to see the operators $\hat{A}(\psi)$ as generators of the group of adding test functions. An additional advantage of this point of view is that it is then natural to consider also the group of adding fields. The generators of the latter group are the operators $\hat{F}(a)$. Another advantage is that the resulting formalism is very close to the C^*-algebraic approach using Weyl algebras [2,9].

Correlation Functions

The classical wave functions ψ form a linear space, denoted H. The Fourier coefficients a belong to the dual space H^*. In what follows we will consider H^* as a real linear space, and not a complex one, because only multiplication of a_μ with a real constant corresponds with multiplication of the vector potential A_μ with the same constant. As a consequence, also H will be considered as a real linear space.

Consider $H^* \times H$ as an additive group. A state of the covariance system $(\mathbf{C}, H^* \times H, \mathbf{I})$ is determined by correlation functions $\mathcal{F}(a, \psi; b, \phi)$. We make the following choice:

$$\begin{aligned}\mathcal{F}(a,\psi;b,\phi) &= \exp\left(-\frac{i}{2\eta}\operatorname{Im}\langle b+i\eta\phi|a+i\eta\psi\rangle\right)\\ &\quad\times\exp\left(-\frac{1}{4\eta}\langle b-a+i\eta(\phi-\psi)|b-a+i\eta(\phi-\psi)\rangle\right)\end{aligned} \tag{30}$$

with η a positive number. The proof that these correlation functions have the necessary properties to define a state of $(\mathbf{C}, H^* \times H, \mathbf{I})$ is given in Appendix D. The generalized GNS-theorem [3] implies that there exists a projective representation $\hat{W}(a, \psi)$ of $H^* \times H$ in a Hilbert space $\mathcal{H}$, and a normalized wave function Ω in $\mathcal{H}$, for which

$$\mathcal{F}(a,\psi;b,\phi) = \langle \hat{W}(b,\phi)^*\Omega|\hat{W}(a,\psi)^*\Omega\rangle \tag{31}$$

holds.

The cocycle

From the *ansatz*

$$\hat{W}(a,\psi)\hat{W}(b,\phi) = \exp\left(\frac{i}{2}\, s(a,\psi;b,\phi)\right)\hat{W}(a+b,\psi+\phi) \tag{32}$$

follows, using that $\hat{W}(a,\psi)^* = \hat{W}(-a,-\psi)$,

$$\begin{aligned}\mathcal{F}(a,\psi;b,\phi) &= \langle\Omega|\hat{W}(b,\phi)\hat{W}(-a,-\psi)\Omega\rangle \\ &= \exp\left(-\frac{i}{2}\, s(b,\phi;a,\psi)\right)\langle\Omega|\hat{W}(b-a,\phi-\psi)\Omega\rangle \\ &= \exp\left(-\frac{i}{2}\, s(b,\phi;a,\psi)\right)\mathcal{F}(a-b,\psi-\phi;0,0).\end{aligned} \tag{33}$$

Using the definition of $\mathcal{F}$ one obtains

$$s(a,\psi;b,\phi) = \eta^{-1}\,\mathrm{Im}\,\langle a+i\eta\psi|b+i\eta\phi\rangle. \tag{34}$$

This function $s(a,\psi;b,\phi)$ is a symplectic form, as it should be. It is anti-symmetric under exchange of (a,ψ) and (b,ϕ). It is real linear in its arguments. Note that it is degenerate because $\langle\cdot|\cdot\rangle$ is degenerate. The function $\exp\left((i/2)s(a,\psi;b,\phi)\right)$, appearing in (32), is a cocycle of the additive group $H^* \times H$.

Field Operators

The operators $\hat{A}(\psi)$ and $\hat{F}(a)$ are introduced as self-adjoint operators satisfying

$$\hat{W}(0,\lambda\psi) = \exp\left(i\lambda\hat{A}(\psi)\right) \quad \text{and} \quad \hat{W}(\lambda a,0) = \exp\left(i\lambda\hat{F}(a)\right) \tag{35}$$

for all real λ. The commutation relations for these operators can be obtained from

$$\hat{W}(a,\psi)\hat{W}(b,\phi) = \exp\left(is(a,\psi;b,\phi)\right)\hat{W}(b,\phi)\hat{W}(a,\psi) \tag{36}$$

by inserting real numbers, as in (35), and taking derivatives. One obtains

$$\begin{aligned}\left[\hat{F}(a),\hat{F}(b)\right] &= -i\eta^{-1}\,\mathrm{Im}\,\langle a|b\rangle \\ \left[\hat{F}(a),\hat{A}(\phi)\right] &= -i\,\mathrm{Re}\,\langle a|\phi\rangle \\ \left[\hat{A}(\psi),\hat{A}(\phi)\right] &= -i\eta\,\mathrm{Im}\,\langle\psi|\phi\rangle.\end{aligned} \tag{37}$$

Comparison with the traditional result (3) gives $\eta = 2$.

The unitary operator $\hat{W}(a)$ implements adding (or subtracting) a field. Indeed, one calculates

$$\begin{aligned}\hat{W}(a,0)\hat{A}(\psi)\hat{W}(-a,0) &= -i\frac{\mathrm{d}}{\mathrm{d}\lambda}\bigg|_{\lambda=0} \hat{W}(a,0)\hat{W}(0,\lambda\psi)\hat{W}(-a,0) \\ &= -i\frac{\mathrm{d}}{\mathrm{d}\lambda}\bigg|_{\lambda=0} \exp\left(i\lambda s(a,0;0,\psi)\right)\hat{W}(0,\lambda\psi) \\ &= \hat{A}(\psi) + \mathrm{Re}\,\langle a|\psi\rangle \\ &= \hat{A}(\psi) - f(A)\end{aligned} \tag{38}$$

with $f_\mu(q)$ the test functions corresponding with ψ and $A_\mu(q)$ the vector potential corresponding with a. Formally, this can be rewritten as

$$\hat{W}(a,0)\hat{A}_\mu(q)\hat{W}(-a,0) = \hat{A}_\mu(q) - A_\mu(q). \tag{39}$$

Similarly, the unitary operator $\hat{W}(0,\psi)$ implements adding wave functions. Indeed, one finds

$$\begin{aligned}\hat{W}(0,\psi)\hat{F}(a)\hat{W}(0,-\psi) &= \hat{F}(a) - \mathrm{Re}\,\langle\psi|a\rangle \\ &= \hat{F}(a) + f(A).\end{aligned} \tag{40}$$

Real Additivity and Identification

Let us calculate

$$\begin{aligned}&\langle\hat{W}(b,\phi)^*\Omega|\,\hat{A}(\chi)\hat{W}(a,\psi)^*\Omega\rangle \\ &= i\frac{\partial}{\partial\lambda}\bigg|_{\lambda=0} \langle\hat{W}(b,\phi)^*\Omega|\,\hat{W}(0,\lambda\chi)^*\hat{W}(a,\psi)^*\Omega\rangle \\ &= i\frac{\partial}{\partial\lambda}\bigg|_{\lambda=0} \exp\left(-(i\lambda/2)\,\mathrm{Re}\,\langle\chi|\,a+i\eta\psi\rangle\right)\mathcal{F}(a,\psi+\lambda\chi;b,\phi) \\ &= \frac{1}{2}\left[\langle a+i\eta\psi|\,\chi\rangle + \langle\chi|\,b+i\eta\phi\rangle\right]\langle\hat{W}(b,\phi)^*\Omega|\,\hat{W}(a,\psi)^*\Omega\rangle\end{aligned} \tag{41}$$

and, similarly,

$$\begin{aligned}&\langle\hat{W}(b,\phi)^*\Omega|\,\hat{F}(d)\hat{W}(a,\psi)^*\Omega\rangle \\ &= \frac{i}{2\eta}\left[\langle d|\,b+i\eta\phi\rangle - \langle a+i\eta\psi|\,d\rangle\right]\langle\hat{W}(b,\phi)^*\Omega|\,\hat{W}(a,\psi)^*\Omega\rangle.\end{aligned} \tag{42}$$

These expressions show that the operators $\hat{A}(\chi)$ and $\hat{F}(d)$ are real linear functions. Moreover, comparison of the two expressions yields $\eta\hat{F}(i\chi) = \hat{A}(\chi)$ for all χ. One concludes that in the Hilbert space representation determined by the correlation functions (30) the generators of adding fields, respectively of adding wave functions, coincide.

4 Fock Representation

In this section the Hilbert space representation determined by the correlation functions (30) is identified with the Fock space in which photon states are created by repeated application of creation operators onto the vacuum state.

Creation and Annihilation Operators

From (41) follows that

$$\langle \hat{W}(b,\phi)^*\Omega|\hat{A}(\psi)\Omega\rangle = \frac{1}{2}\langle \hat{W}(b,\phi)^*\Omega|\Omega\rangle\langle\psi|b+i\eta\phi\rangle. \tag{43}$$

Hence one has

$$\langle \hat{W}(b,\phi)^*\Omega|(\hat{A}(\psi)-i\hat{A}(i\psi))\Omega\rangle = 0. \tag{44}$$

Since b and ϕ are arbitrary this implies that

$$(\hat{A}(\psi)-i\hat{A}(i\psi))\Omega = 0. \tag{45}$$

It is therefore obvious to define *annihilation operators* $\hat{A}_-(\psi)$ by

$$\hat{A}_-(\psi) = \frac{1}{2}\,\hat{A}(\psi) - \frac{i}{2}\,\hat{A}(i\psi), \tag{46}$$

which means that also

$$\hat{A}_-(\psi) = \frac{1}{2}\,\hat{A}(\psi) + \frac{i\eta}{2}\,\hat{F}(\psi). \tag{47}$$

The latter expression resembles the definition $Q+iP$ of the annihilation operator by means of a pair of position and momentum operators, in the context of the harmonic oscillator. Note that $\hat{A}_-(\psi)$ is a complex linear function of ψ. As shown above, the annihilation operators satisfy

$$\hat{A}_-(\psi)\Omega = 0. \tag{48}$$

One verifies that these operators are commuting

$$\left[\hat{A}_-(\psi),\hat{A}_-(\phi)\right] = 0. \tag{49}$$

The conjugate operator $\hat{A}_+(\psi) = \hat{A}_-(\psi)^*$ is the *creation operator*. From the definition follows immediately that

$$\hat{A}(\psi) = \hat{A}_+(\psi) + \hat{A}_-(\psi). \tag{50}$$

The commutation relations between creation and annihilation operators are found to be

$$\begin{aligned}\left[\hat{A}_+(\psi), \hat{A}_-(\phi)\right] &= \frac{1}{4}\left[\hat{A}(\psi) + i\hat{A}(i\psi), \hat{A}(\phi) - i\hat{A}(i\phi)\right] \\ &= -\frac{\eta}{2}\langle\psi|\phi\rangle. \end{aligned} \tag{51}$$

With $\eta = 2$ this relation gives to the operator $\hat{A}_+(\psi)\hat{A}_-(\phi)$ the usual interpretation of number operator.

One-Photon States

A one-photon state is determined by an element of the Hilbert space of the form $\hat{A}_+(\xi)\Omega$, where ξ is a classical wave function, not equivalent to zero. It is straightforward to verify that two equivalent classical wave functions ψ and ϕ determine the same one-photon state. Indeed, by definition they satisfy $\langle a|\psi\rangle = \langle a|\phi\rangle$ for all a. From (43) then follows that $\hat{A}(\psi)\Omega = \hat{A}(\phi)\Omega$ so that $\hat{A}_+(\psi)\Omega = \hat{A}_+(\phi)\Omega$. Hence ϕ and ψ determine the same wave function in the Hilbert space, and hence, the same physical state.

A short calculation gives

$$\begin{aligned}||\hat{A}_+(\psi)\Omega||^2 &= \langle\Omega|\hat{A}_-(\psi)\hat{A}_+(\psi)\Omega\rangle \\ &= -\langle\Omega|\left[\hat{A}_+(\psi), \hat{A}_-(\psi)\right]\Omega\rangle \\ &= \frac{\eta}{2}\langle\psi|\psi\rangle \\ &= \frac{\eta}{2}\int_{\mathbf{R}^3} \mathrm{d}\mathbf{k}\,\frac{1}{2|\mathbf{k}|^3}\sum_{\alpha,\beta=1}^{3}\overline{\psi_\alpha(\mathbf{k})}\left(|\mathbf{k}|^2\delta_{\alpha\beta} - \mathbf{k}_\alpha\mathbf{k}_\beta\right)\psi_\beta(\mathbf{k}) \\ &\geq 0. \end{aligned} \tag{52}$$

The last steps of this calculation use results of Appendix D. In particular, $||\hat{A}_+(\psi)\Omega|| = 0$ holds if $\psi_\mu(\mathbf{k})$ is of the form

$$\psi_\alpha(\mathbf{k}) = \frac{\mathbf{k}_\alpha}{|\mathbf{k}|}\psi_0(\mathbf{k}), \qquad \alpha = 1, 2, 3. \tag{53}$$

The usual interpretation of this result is that there do not exist photon states for which the electric and magnetic fields are not perpendicular to the wave vector $\mathbf{k}$.

5 Poincaré Invariance

In this section we study the action of the proper Poincaré group in Fock space. The generators of this group determine physical quantities like energy, momentum, mass, and spin of the photon.

Shifts in Spacetime

A shift with vector x in spacetime maps the vector potentials $A_\mu(q)$ onto vector potentials $A^x_\mu(q)$ given by

$$A^x_\mu(q) = A_\mu(q - x). \tag{54}$$

Using (16), one finds that the Fourier coefficients a_μ transform into a^x_μ given by

$$a^x_\mu(\mathbf{k}) = \exp\left(i|\mathbf{k}|x_0 - i\mathbf{k}\cdot\mathbf{x}\right) a_\mu(\mathbf{k}). \tag{55}$$

The corresponding transformation of the classical wave functions $\psi_\mu(\mathbf{k})$ is

$$\psi^x_\mu(\mathbf{k}) = \exp\left(i|\mathbf{k}|x_0 - i\mathbf{k}\cdot\mathbf{x}\right) \psi_\mu(\mathbf{k}). \tag{56}$$

With this choice of action the correlation functions $\mathcal{F}(a,\psi;b,\phi)$ are invariant under shifts. This is what we want because the corresponding state of the system is the vacuum state. Note that the smeared-out fields $f(A)$ are invariant under shifts.

A unitary representation of the group of shifts $\mathbf{R}^4,+$ is defined by

$$\hat{U}(x)\hat{W}(a,\psi)\Omega = \hat{W}(a^x,\psi^x)\Omega \tag{57}$$

(see Appendix E). The generators of this representation are denoted $\hat{K}_\mu$ and are defined by

$$\hat{U}(x) = \exp(-ix^\mu \hat{K}_\mu). \tag{58}$$

By convention, the momentum operators P_μ equal $\hbar\hat{K}_\mu$. The energy operator is $cP_0 = c\hbar\hat{K}_0$. The vector Ω corresponds with the vacuum state and is invariant under shifts. In particular, $\hat{K}_\mu\Omega = 0$ holds. Hence the energy and momentum of the vacuum are zero, contrary to what is claimed in textbooks, based on the harmonic oscillator picture of the photon.

Energy and Momentum of a One-Photon State

Let us calculate

$$\begin{aligned}&\langle \hat{W}(b,\phi)^*\Omega|\hat{K}_0\hat{W}(a,\psi)^*\Omega\rangle\\ &= i\frac{\partial}{\partial x_0}\bigg|_{x=0} \langle \hat{W}(b,\phi)^*\Omega|\hat{U}(x)\hat{W}(a,\psi)^*\Omega\rangle\end{aligned}$$

$$
\begin{aligned}
&= i\frac{\partial}{\partial x_0}\bigg|_{x=0} \mathcal{F}(a^x, \psi^x; b, \phi) \\
&= -\mathcal{F}(a, \psi; b, \phi) i\frac{\partial}{\partial x_0}\bigg|_{x=0} \bigg(\frac{i}{2\eta}\,\mathrm{Im}\,\langle b + i\phi | a^x + i\psi^x\rangle \\
&\quad + \frac{1}{4\eta}\langle b - a^x + i\eta(\phi - \psi^x) | b - a^x + i\eta(\phi - \psi^x)\rangle\bigg) \\
&= \frac{1}{2\eta}\mathcal{F}(a, \psi; b, \phi)\langle a + i\eta\psi|\,|\mathbf{k}|\,(b + i\eta\phi)\rangle. \qquad (59)
\end{aligned}
$$

Similarly, one shows that, with $\alpha = 1, 2, 3$,

$$
\langle \hat{W}(b,\phi)^*\Omega | \hat{K}_\alpha \hat{W}(a,\psi)^*\Omega\rangle = \frac{1}{2\eta}\mathcal{F}(a,\psi;b,\phi)\langle a + i\eta\psi | \mathbf{k}_\alpha (b + i\eta\phi)\rangle. \qquad (60)
$$

Consider now a one-photon state. From (59) and (60) it follows that

$$
K_0 \hat{A}_+(\psi)\Omega = \hat{A}_+(|\mathbf{k}|\psi)\Omega \quad \text{and} \quad K_\alpha \hat{A}_+(\psi)\Omega = \hat{A}_+(\mathbf{k}_\alpha \psi)\Omega. \qquad (61)
$$

To see this, use that

$$
\langle W(b,\phi)^* |\, A_+(\psi)\Omega\rangle = \frac{1}{2}\langle \psi | b + i\eta\phi\rangle \mathcal{F}(0,0;b,\phi). \qquad (62)
$$

Here we work with photon states smeared out with classical wave functions. It is tradition to associate the notion of photon with states that are not smeared out. These are idealized states which are not represented by wave functions in Hilbert space. In order to approach such a photon state we have to select classical wave functions that converge to a Dirac measure concentrated at a single wave vector $\mathbf{k}$. The energy of such a photon is then equal to $\hbar c|\mathbf{k}|$, the momentum is equal to $\hbar\mathbf{k}$. In particular, this implies that the mass of such an idealized photon is exactly equal to zero.

Lorentz Transformations

The discussion of Lorentz transformations is not very easy because both Fourier coefficients $a_\mu(\mathbf{k})$ and classical wave functions $\psi_\mu(\mathbf{k})$ depend on a wave vector $\mathbf{k}$ in $\mathbf{R}^3$, instead of covariant vectors in $\mathbf{R}^4$, and do not obey easy transformation rules. However, both the vector potential $A_\mu(q)$ and the test functions $f_\mu(q)$ transform as vectors so that the smeared-out vector potential $f(A)$ is invariant under Lorentz transformations. Since $f(A) = -\mathrm{Re}\,\langle a|\psi\rangle$ holds, this shows that the pseudo-scalar product is invariant under Lorentz transformations. Hence the correlation functions (30) are invariant under Lorentz transformations.

Let Λ denote a proper Lorentz transformation. Under its action the vector potential $A_\mu(q)$ transforms into $A'_\mu(q)$ given by

$$A'_\mu(q) = \Lambda_\mu^{\ \nu} A_\nu(\Lambda^{-1}q). \tag{63}$$

Assume first that Λ is a spatial rotation described by the 3-by-3 matrix R. Then the Fourier coefficients $a_\mu(\mathbf{k})$ transform into $a'_\mu(\mathbf{k})$ given by

$$a'_\mu(\mathbf{k}) = \Lambda_\mu^{\ \nu} a_\nu(R\mathbf{k}). \tag{64}$$

Next consider a boost in direction 3. The non-zero matrix elements are $\Lambda_{00} = \Lambda_{33} = \cosh(\chi)$, $\Lambda_{03} = \Lambda_{30} = \sinh(\chi)$, $\Lambda_{11} = \Lambda_{22} = 1$. Then one obtains

$$\begin{aligned}
a'_0(\mathbf{k}) &= \cosh(\chi)a_0(\mathbf{k}') - \sinh(\chi)a_3(\mathbf{k}')\\
a'_1(\mathbf{k}) &= a_1(\mathbf{k}')\\
a'_2(\mathbf{k}) &= a_2(\mathbf{k}')\\
a'_3(\mathbf{k}) &= -\sinh(\chi)a_0(\mathbf{k}') + \cosh(\chi)a_3(\mathbf{k}')
\end{aligned} \tag{65}$$

with

$$\mathbf{k}' = (\mathbf{k}_1, \mathbf{k}_2, \cosh(\chi)\mathbf{k}_3 + \sinh(\chi)|\mathbf{k}|). \tag{66}$$

Together, the spatial rotations and the boosts in direction 3 generate the proper Lorentz group. Hence, the above formulas represent the action of the proper Lorentz group on the Fourier coefficients $a(\mathbf{k})$. The classical wave functions $\psi(\mathbf{k})$ transform in a similar way. A unitary operator $\hat{V}(\Lambda)$ is now defined by

$$\hat{V}(\Lambda)\hat{W}(a,\psi)\Omega = \hat{W}(a',\psi')\Omega. \tag{67}$$

These unitary operators form a representation of the proper Lorentz group. To show this one uses the same arguments as in case of the group of shifts.

The Generators of Spatial Rotations

The six generators of the proper Lorentz group are denoted $\hat{M}_{\mu\nu} = -\hat{M}_{\nu\mu}$. Three of them correspond with spatial rotations, the other three with boosts.

Consider now a rotation by an angle χ around the third coordinate axis. The Fourier coefficients transform like

$$\begin{aligned}
a'_0(\mathbf{k}) &= a_0(\mathbf{k}')\\
a'_1(\mathbf{k}) &= \cos(\chi)a_1(\mathbf{k}') - \sin(\chi)a_2(\mathbf{k}')\\
a'_2(\mathbf{k}) &= \sin(\chi)a_1(\mathbf{k}') + \cos(\chi)a_2(\mathbf{k}')\\
a'_3(\mathbf{k}) &= a_3(\mathbf{k}')
\end{aligned} \tag{68}$$

with

$$\mathbf{k}' = (\cos(\chi)\mathbf{k}_1 + \sin(\chi)\mathbf{k}_2, -\sin(\chi)\mathbf{k}_1 + \cos(\chi)\mathbf{k}_2, \mathbf{k}_3). \tag{69}$$

We calculate

$$\begin{aligned}
&\langle \hat{W}(b,\phi)^*\Omega | \hat{M}_{12}\hat{W}(a,\psi)^*\Omega\rangle \\
&= i\frac{\partial}{\partial\chi}\bigg|_{\chi=0} \langle \hat{W}(b,\phi)^*\Omega | V(\Lambda)\hat{W}(a,\psi)^*\Omega\rangle \\
&= i\frac{\partial}{\partial\chi}\bigg|_{\chi=0} \mathcal{F}(a',\psi';b,\phi) \\
&= -\frac{i}{2\eta}\mathcal{F}(a,\psi;b,\phi)\frac{\partial}{\partial\chi}\bigg|_{\chi=0} \\
&\quad \left(i\,\mathrm{Im}\,\langle b+i\eta\phi|a'+i\eta\psi'\rangle + \frac{1}{2}\langle b-a'+i\eta(\phi-\psi')|b-a'+i\eta(\phi-\psi')\rangle \right) \\
&= \frac{1}{2\eta}\mathcal{F}(a,\psi;b,\phi)\langle a+i\eta\psi|(S_{12}+L_{12})(b+i\eta\phi)\rangle
\end{aligned} \tag{70}$$

with $S_{\mu\nu}$ the 4-by-4-matrix with i at position μ,ν, $-i$ at position ν,μ, and zeroes everywhere else, and with

$$L_{12} = i\left(\mathbf{k}_1\frac{\partial}{\partial\mathbf{k}_2} - \mathbf{k}_2\frac{\partial}{\partial\mathbf{k}_1}\right). \tag{71}$$

The Generators of Boosts

Now let Λ be a boost in direction 3, as given by (65). The corresponding generator M_{03} is calculated as follows.

$$\begin{aligned}
&\langle \hat{W}(b,\phi)^*\Omega | \hat{M}_{03}\hat{W}(a,\psi)^*\Omega\rangle \\
&= -i\frac{\partial}{\partial\chi}\bigg|_{\chi=0} \langle \hat{W}(b,\phi)^*\Omega | V(\Lambda)\hat{W}(a,\psi)^*\Omega\rangle \\
&= -i\frac{\partial}{\partial\chi}\bigg|_{\chi=0} \mathcal{F}(a',\psi';b,\phi) \\
&= -\frac{i}{2\eta}\mathcal{F}(a,\psi;b,\phi)\frac{\partial}{\partial\chi}\bigg|_{\chi=0} \\
&\quad \left(-i\,\mathrm{Im}\,\langle b+i\eta\phi|a'+i\eta\psi'\rangle - \frac{1}{2}\langle b-a'+i\eta(\phi-\psi')|b-a'+i\eta(\phi-\psi')\rangle \right) \\
&= -\frac{1}{2\eta}\mathcal{F}(a,\psi;b,\phi)\langle a+i\eta\psi|(S_{03}-L_{03})(b+i\eta\phi)\rangle.
\end{aligned} \tag{72}$$

with

$$L_{0\alpha} = i|\mathbf{k}|\frac{\partial}{\partial \mathbf{k}_\alpha}. \tag{73}$$

Spin of the Photon

From expressions (70, 72) it is clear that there are two different types of contributions to the generators $M_{\mu\nu}$ of the Lorentz group. These are called the spin part $\hat{S}$, respectively the orbital part $\hat{L}$. The spin contribution originates from the vector character of the electromagnetic vector potential $A_\mu(q)$, the orbital part follows from the transformation of Minkowski space. The operators $\hat{L}_{23}$, $\hat{L}_{31}$ and $\hat{L}_{12}$ are the components of angular momentum, the operators $\hat{S}_{23}$, $\hat{S}_{31}$ and $\hat{S}_{12}$ are the components of the spin of the photon. Note that these notions are not covariant. Worse is that the splitting of $\hat{M}$ into $\hat{S}$ and $\hat{L}$ is not gauge invariant. This implies that the components of $\hat{S}$ and $\hat{L}$ are not physically observable. Hence one could say that the mechanical spin of the photon is not observable. See Jauch and Rohrlich [5] for a discussion of these points.

However, it is common to say that the photon is a spin-1 particle. In order to understand this statement let us assume that a one-photon state $\hat{A}_+(\psi)\Omega$ is an eigenstate of the operator $\hat{S}_{12}$

$$\hat{S}_{12}\hat{A}_+(\psi)\Omega = \lambda\hat{A}_+(\psi)\Omega. \tag{74}$$

From

$$\langle\hat{W}(b,\phi)^*\Omega|\hat{S}_{12}\hat{W}(a,\psi)^*\Omega\rangle = \frac{1}{2\eta}\mathcal{F}(a,\psi;b,\phi)\langle a+i\eta\psi|S_{12}(b+i\eta\phi)\rangle \tag{75}$$

follows

$$\langle\hat{W}(b,\phi)^*\Omega|\hat{S}_{12}A_+(\psi)\Omega\rangle = \frac{1}{2}\langle\hat{W}(b,\phi)^*\Omega|\Omega\rangle\langle\psi|S_{12}(b+i\eta\phi)\rangle. \tag{76}$$

In combination with the assumption that the one-photon state is an eigenstate of $\hat{S}_{12}$ with eigenvalue λ there follows

$$\lambda\langle\hat{W}(b,\phi)^*\Omega|A_+(\psi)\Omega\rangle = \frac{1}{2}\langle\hat{W}(b,\phi)^*\Omega|\Omega\rangle\langle\psi|S_{12}(b+i\eta\phi)\rangle. \tag{77}$$

On the other hand is

$$\langle\hat{W}(b,\phi)^*\Omega|A_+(\psi)\Omega\rangle = \frac{1}{2}\langle\hat{W}(b,\phi)^*\Omega|\Omega\rangle\langle\psi|b+i\eta\phi\rangle. \tag{78}$$

Comparison of both expressions gives the condition

$$S_{12}\psi(\mathbf{k}) = \lambda\psi(\mathbf{k}). \tag{79}$$

Now, the eigenvalues of the matrix S_{12} are +1, -1, and 0 (two-fold degenerated). Hence, the space of classical wave functions H can be split into three real-linear subspaces H_+, H_0, and H_- with the properties that $S_{12}\psi = \pm\psi$ if ψ is in H_+, respectively in H_-, and $S_{12}\psi = 0$ if ψ is in H_0. As a consequence, also the one-photon subspace of Fock space can be written as a direct sum of three subspaces which consist of eigenvectors of $\hat{S}_{12}$ corresponding to the eigenvalues $+1, 0, -1$. Spin-operators with a spectrum $+1, 0, -1$ are associated with spin-1 particles.

Polarization of the Photon

As stated earlier in section 2, each class of equivalent classical wave functions contains a representative ψ satisfying $\psi_0 = 0$. None of these representatives belongs to H_0. Indeed, if $S_{12}\psi(\mathbf{k}) = 0$ holds for all $\mathbf{k}$ then $\psi_1 = \psi_2 = 0$ and $|\mathbf{k}|\psi_0(\mathbf{k}) = \mathbf{k}_3\psi_3(\mathbf{k})$. But because of $\psi_0 = 0$ also $\psi_3 = 0$ follows. Hence, if a representative ψ belongs to H_0 then all of its components are zero. Classical wavefunctions in $H_\pm$ satisfy

$$0 = (\mathbf{k}_1 \mp i\mathbf{k}_2)\psi_1(\mathbf{k}) = \mathbf{k}_3\psi_3(\mathbf{k}) \quad \text{and } \psi_2(\mathbf{k}) = \mp i\psi_1(\mathbf{k}) \tag{80}$$

with $\psi_1(\mathbf{k})$ and $\psi_2(\mathbf{k})$ not identically zero. The only solutions of these conditions are idealized photons with wave vector parallel to the third direction and with $\psi_3(\mathbf{k}) = 0$. This implies that the electric and magnetic fields lie in the plane orthogonal to the wave vector $\mathbf{k}$. Two independent solutions are allowed. They correspond with the two independent polarizations of the electromagnetic field. A more detailed analysis can be found in Jauch and Rohrlich [5].

6 Conclusions

We have shown in this paper that the standard theory of the free photon field can be derived within the covariance approach to quantum mechanics. Typical for this approach is that it starts from the action of a group in a C^*-algebra and from correlation functions describing a state of the covariance system. In the case of the free radiation field the C^*- algebra is the algebra of complex numbers, the group is the group of adding fields times the group of adding test functions. The Lorenz gauge is used to eliminate part of the redundancy. The action is trivial. The correlation functions describe a Lorentz invariant vacuum state. The state vectors of the induced Fock representation are invariant under the remaining gauge freedoms.

The present approach has several advantages. In the first place the formalism is mathematically rigorous. The development of photon theory is crystal clear and there is no need for hand waving arguments. Both the gauge problem and the problem of positivity of the scalar product are solved in a satisfactory manner. The approach is generic. It is obvious how to apply it to other fields than the electromagnetic one.

We have stressed that there exists a duality between classical wave functions ψ and Fourier coefficients a. As a consequence of this duality there exist, besides the usual field operators $\hat{A}(\psi)$, also operators $\hat{F}(a)$, labeled with Fourier transformed vector potentials. However, in the representation of the electromagnetic vacuum state the field operators $\hat{F}(a)$ and $\hat{A}(\psi)$ coincide. Hence, this duality has no practical consequences for the description of the vacuum. We do not know if this degeneracy continues to exist in other representations of the electromagnetic radiation field.

The formalism considers photons smeared out with classical wave functions. These differ from the idealized photons discussed in most text books. The reason for smearing-out is of course that the strictly localized objects $\hat{A}(q)$ cannot be defined as operators in Fock space, while the smeared-out equivalents $\hat{A}(\psi)$ are nicely defined self-adjoint operators. Finally, the generators of the Lorentz group can be decomposed into a sum of an orbital part and a spin part. The space of classical wave functions can be split into three parts corresponding with spin 1, 0 , and -1 respectively. Because of gauge freedom only two independent polarizations of the idealized photons occur.

Up to now, we did not consider electromagnetic fields in presence of external charges and currents. The first question that arises in this context is whether all states of the covariance system of the free radiation field (i.e. the one used in the present paper) describe radiation fields, or whether states can be found which describe fields produced by charges and currents. If the latter is not the case, then the covariance system has to be modified. Another topic for further investigation is the description of massive photons in terms of the present formalism (see e.g. section 6-5 of Jauch and Rohrlich [5]). Our ultimate goal is of course a combination of electron and photon fields within the same covariance approach.

Acknowledgement

We thank Marek Czachor for his interest in the present work.

Appendix A: Covariance Systems

A covariance system $(\mathcal{A}, X, \sigma)$ consists of a C^*-algebra $\mathcal{A}$, a locally compact group X, and an action σ of this group as automorphisms of $\mathcal{A}$. For each $a \in \mathcal{A}$ the map $x \in X \rightarrow \sigma_x a$ should be continuous. In the present paper, the C^*-algebra $\mathcal{A}$ is the algebra $\mathbf{C}$ of complex numbers. In this case the only possible action of X is the trivial one, leaving the complex numbers invariant. The resulting covariance system is denoted $(\mathbf{C}, X, \mathbf{I})$ and is rather trivial. Still, the notions of state and of representation of a covariance system apply, and are nontrivial.

A state [3] of a covariance system $(\mathcal{A}, X, \sigma)$ is determined by correlation functions $\mathcal{F}(a, x, y)$ depending on $a \in \mathcal{A}$ and $x, y \in X$. They satisfy conditions of positivity, normalization, covariance, and continuity. In the present context, where the C^*-algebra is the algebra of complex numbers, the dependence on elements of $\mathcal{A}$ can be omitted, and the conditions reduce to

- (positivity) For all $n > 0$ and for all possible choices of $\lambda_1, \ldots, \lambda_n$ in $\mathbf{C}$, of $x_1, \ldots, x_n$ in X, is

$$\sum_{j,k=1}^{n} \lambda_j \overline{\lambda_k} \mathcal{F}(x_j, x_k) \geq 0. \tag{81}$$

- (normalization) $\mathcal{F}(e, e) = 1$ (e is the neutral element of X).

- (covariance) $\mathcal{F}(xz, yz) = \mathcal{F}(x, y)\xi(x, z)\overline{\xi(y, z)}$ for all x, y, z in X, where $\xi(x, z)$ is a cocycle of X.

- (continuity) the map $x, y \rightarrow \mathcal{F}(x, y)$ is continuous in a neighborhood of the neutral element of X.

A representation of the covariance system $(\mathbf{C}, X, \mathbf{I})$ is nothing but a projective representation U of the group X as unitary operators of a Hilbert space $\mathcal{H}$, with the property that the map $x \rightarrow U(x)$ is strongly continuous for x in a neighborhood of the neutral element of X. In this context the generalized G.N.S.-theorem states that for each state of $(\mathbf{C}, X, \mathbf{I})$, described by the correlation functions $\mathcal{F}(x, y)$, there exists a representation U of $(\mathbf{C}, X, \mathbf{I})$ in a Hilbert space $\mathcal{H}$ and an element Ω of $\mathcal{H}$ with the property that

$$\mathcal{F}(x, y) = \langle U(y)^*\Omega | U(x)^*\Omega \rangle \tag{82}$$

holds for all x and y in X.

Appendix B: Smeared-out Field Operators

Here we discuss the relation between expressions (1) and (3). The obvious relation between $\hat{A}(\psi)$ and $\hat{A}_\mu(q)$ is

$$\hat{A}(\psi) = \int_{\mathbf{R}^4} \mathrm{d}q\, f^\mu(q)\hat{A}_\mu(q). \tag{83}$$

where the test functions f_μ correspond with classical wavefunctions ψ_μ by (20). Similarly, let ϕ_μ correspond with test functions g_μ. Then one obtains, using the commutation relations (1) and definition (2) of $D_0(q)$,

$$\begin{aligned}
\left[\hat{A}(\psi), \hat{A}(\phi)\right] &= \int_{\mathbf{R}^4} \mathrm{d}q \int_{\mathbf{R}^4} \mathrm{d}q'\, f^\mu(q) g^\nu(q') \left[\hat{A}_\mu(q), \hat{A}_\nu(q')\right] \\
&= -i \int_{\mathbf{R}^4} \mathrm{d}q \int_{\mathbf{R}^4} \mathrm{d}q'\, f^\mu(q) g_\mu(q') D_0(q-q') \\
&= -i \frac{1}{(2\pi)^3} \int_{\mathbf{R}^4} \mathrm{d}q \int_{\mathbf{R}^4} \mathrm{d}q'\, f^\mu(q) g_\mu(q') \\
&\quad \times \int_{\mathbf{R}^3} \mathrm{d}\mathbf{k}\, \exp\left(i\mathbf{k}\cdot(\mathbf{q}-\mathbf{q}')\right) \frac{1}{|\mathbf{k}|} \sin((q_0 - q_0')|\mathbf{k}|) \\
&= -\frac{1}{(2\pi)^3} \int_{\mathbf{R}^3} \mathrm{d}\mathbf{k}\, \frac{1}{2|\mathbf{k}|} \int_{\mathbf{R}^4} \mathrm{d}q\, f^\mu(q) \exp\left(i\mathbf{k}\cdot\mathbf{q}\right) \\
&\quad \times \int_{\mathbf{R}^4} \mathrm{d}q'\, g_\mu(q') \exp\left(-i\mathbf{k}\cdot\mathbf{q}'\right) \\
&\quad \times \left(\exp\left(i(q_0 - q_0')|\mathbf{k}|\right) - \exp\left(-i(q_0 - q_0')|\mathbf{k}|\right)\right).
\end{aligned} \tag{84}$$

Using the definition (20) of classical wave functions one obtains

$$\left[\hat{A}(\psi), \hat{A}(\phi)\right] = -\int_{\mathbf{R}^3} \mathrm{d}\mathbf{k}\, \frac{1}{2|\mathbf{k}|} \left(\psi^\mu(\mathbf{k})\overline{\phi_\mu(\mathbf{k})} - \overline{\psi^\mu(\mathbf{k})}\phi_\mu(\mathbf{k})\right). \tag{85}$$

The latter implies (3).

Appendix C: Representative Classical Wave Functions

Here we show that each class of equivalent classical wave functions contains a representative satisfying the radiation gauge. Given the classical wave functions $\psi_\mu(\mathbf{k})$, let

$$\lambda(\mathbf{k}) = \frac{1}{|\mathbf{k}|^2} \sum_{\alpha=1}^{3} \mathbf{k}_\alpha \psi_\alpha(\mathbf{k}) \tag{86}$$

and

$$\phi_0(\mathbf{k}) = 0, \qquad \phi_\alpha(\mathbf{k}) = \psi_\alpha(\mathbf{k}) - \lambda(\mathbf{k})\mathbf{k}_\alpha, \qquad \alpha = 1,2,3. \tag{87}$$

Then one calculates

$$\sum_{\alpha=1}^{3} \mathbf{k}_\alpha \phi_\alpha(\mathbf{k}) = \sum_{\alpha=1}^{3} \mathbf{k}_\alpha \psi_\alpha(\mathbf{k}) - \lambda(\mathbf{k})|\mathbf{k}|^2 = 0. \tag{88}$$

Hence $\phi_\mu(\mathbf{k})$ are classical wave functions satisfying the radiation gauge.

Consider now an arbitrary set of Fourier coefficients $a_\mu(\mathbf{k})$ satisfying (19). Then one finds

$$\begin{aligned}\langle a|\phi\rangle &= \int_{\mathbf{R}^3} d\mathbf{k}\,\frac{1}{2|\mathbf{k}|}\sum_{\alpha=1}^{3}\overline{a_\alpha(\mathbf{k})}\phi_\alpha(\mathbf{k})\\ &= \int_{\mathbf{R}^3} d\mathbf{k}\,\frac{1}{2|\mathbf{k}|}\sum_{\alpha=1}^{3}\overline{a_\alpha(\mathbf{k})}\psi_\alpha(\mathbf{k})\\ &\quad-\int_{\mathbf{R}^3} d\mathbf{k}\,\frac{1}{2|\mathbf{k}|}\lambda(\mathbf{k})\sum_{\alpha=1}^{3}\overline{a_\alpha(\mathbf{k})}\,\mathbf{k}_\alpha.\end{aligned} \tag{89}$$

Using (19) this becomes

$$\begin{aligned}\langle a|\phi\rangle &= \int_{\mathbf{R}^3} d\mathbf{k}\,\frac{1}{2|\mathbf{k}|}\sum_{\alpha=1}^{3}\overline{a_\alpha(\mathbf{k})}\psi_\alpha(\mathbf{k})\\ &\quad-\frac{1}{2}\int_{\mathbf{R}^3} d\mathbf{k}\,\lambda(\mathbf{k})\overline{a_0(\mathbf{k})}\\ &= \int_{\mathbf{R}^3} d\mathbf{k}\,\frac{1}{2|\mathbf{k}|}\sum_{\alpha=1}^{3}\overline{a_\alpha(\mathbf{k})}\psi_\alpha(\mathbf{k})\\ &\quad-\int_{\mathbf{R}^3} d\mathbf{k}\,\frac{1}{2|\mathbf{k}|^2}\overline{a_0(\mathbf{k})}\sum_{\alpha=1}^{3}\mathbf{k}_\alpha\psi_\alpha(\mathbf{k})\\ &= \langle a|\psi\rangle.\end{aligned} \tag{90}$$

To obtain the latter, (24) has been used. This shows that ϕ and ψ are equivalent.

Appendix D: Existence of the Vacuum State

The difficult point to verify is positivity of the state. Note that the symbols ψ_j and a_j below have each 4 components $(\psi_j)_\mu$ respectively $(a_j)_\mu$. One verifies, using the notation $b_j = a_j + i\psi_j$,

$$\sum_{jj'} \lambda_j \overline{\lambda_{j'}} \mathcal{F}(a_j, \psi_j; a_{j'}\psi_{j'}) = \sum_{jj'} \lambda_j \overline{\lambda_{j'}} \exp\left(-\frac{i}{2\eta}\,\mathrm{Im}\,\langle b_{j'}|b_j\rangle\right)$$

$$\times \exp\left(-\frac{1}{4\eta}\left\langle(b_{j'}-b_j|b_{j'}-b_j\right\rangle\right)$$

$$= \sum_{jj'} \mu_j \overline{\mu_{j'}} \exp\left(\frac{1}{2\eta}\langle b_j|b_{j'}\rangle\right) \tag{91}$$

with

$$\mu_j = \lambda_j \exp\left(-(1/4\eta)\langle b_j|b_j\rangle\right). \tag{92}$$

Positivity of (91) follows by means of Schur's lemma, provided we can show that the matrix with elements $\langle b_j|b_{j'}\rangle$ is positive-definite. But the latter is clear because of the positivity of the scalar product, for which we now give a proof. Using the Lorenz condition (19, 24), one writes

$$\begin{aligned}\langle\phi|\phi\rangle &= \int_{\mathbf{R}^3} d\mathbf{k}\,\frac{1}{2|\mathbf{k}|}\left(-\overline{\phi_0(\mathbf{k})}\phi_0(\mathbf{k}) + \sum_{\alpha=1}^{3}\overline{\phi_\alpha(\mathbf{k})}\phi_\alpha(\mathbf{k})\right)\\ &= \int_{\mathbf{R}^3} d\mathbf{k}\,\frac{1}{2|\mathbf{k}|^3}\sum_{\alpha,\beta=1}^{3}\overline{\phi_\alpha(\mathbf{k})}\left[|\mathbf{k}|^2\delta_{\alpha,\beta} - \mathbf{k}_\alpha\mathbf{k}_\beta\right]\phi_\beta(\mathbf{k})\\ &\geq 0.\end{aligned} \tag{93}$$

The latter follows because the matrix $|\mathbf{k}|^2\delta_{\alpha,\beta} - \mathbf{k}_\alpha\mathbf{k}_\beta$ is positive-definite.

Appendix E: Unitary Representation of the Group of Shifts

We show here that a unitary representation of the group of shifts $\mathbf{R}^4, +$ is determined by (57).

Let us show that $U(x)$ is well-defined. Assume that

$$\hat{W}(a^x, \psi^x)\Omega = \hat{W}(b^x, \phi^x)\Omega. \tag{94}$$

By taking the inner product with $\hat{W}(c,\chi)^*\Omega$ one obtains

$$\mathcal{F}(a^x, \psi^x; c, \chi) = \mathcal{F}(b^x, \phi^x; c, \chi). \tag{95}$$

Note now that $\langle a^x|\psi^x\rangle = \langle a|\psi\rangle$ so that

$$\mathcal{F}(a^x, \psi^x; c, \chi) = \mathcal{F}(a, \psi; c^{-x}, \chi^{-x}). \tag{96}$$

Hence (95) becomes

$$\mathcal{F}(a, \psi; c^{-x}, \chi^{-x}) = \mathcal{F}(b, \phi; c^{-x}, \chi^{-x}). \tag{97}$$

This implies

$$\langle \hat{W}(c^{-x}, \chi^{-x})^*\Omega|(\hat{W}(a,\psi)^* - \hat{W}(b,\phi)^*)\Omega\rangle = 0. \tag{98}$$

Since c and χ are arbitrary, it follows that $\hat{W}(a,\psi)^*\Omega = \hat{W}(b,\phi)^*\Omega$. This shows that $U(x)$ is well-defined.

It is now straightforward to show that $U(x)U(y) = U(x+y)$ and that $U(x)$ is isometric. Therefore, $U(x)$ is a unitary representation of $\mathbf{R}^4, +$.

References

1. G. Scharf, *Finite quantum electrodynamics*, Springer-Verlag (1989).
2. A.L. Carey, J.M. Gaffney, and C.A.Hurst, "A C^*-algebra formulation of the quantization of the electromagnetic field", *J. Math. Phys.*, **18**, 629-640 (1977).
3. J. Naudts, M. Kuna, "Covariance systems", math-ph/0009031, *J. Phys. A: Math. Gen.*, **34**, 9265-9280 (2001).
4. J. Naudts, "Covariance approach to quantum theory", *Proceedings of the conference 'Quantum Theory and Symmetries'*, Krakow, July 2001.
5. J.M. Jauch and F. Rohrlich, *The theory of photons and electrons, 2nd ed.*, Springer-Verlag (1980).
6. S. Doplicher, K. Fredenhagen, J.E. Roberts, "Spacetime quantization induced by classical gravity", *Phys. Lett.*, **B331**, 39-44 (1994).
7. S. Doplicher, K. Fredenhagen, J.E. Roberts, "The quantum structure of spacetime at the Planck scale and quantum fields", *Commun. Math. Phys.*, **172**, 187-220 (1995).
8. J. Naudts, M. Kuna, "Model of a particle in spacetime", *J. Phys. A: Math. Gen.*, **34**, 4227-4239 (2001).
9. D. Petz, *An Invitation to the Algebra of Canonical Commutation Relations*, Leuven University Press, Leuven (1990).